CONFRONTING THE CLIMATE CRISIS

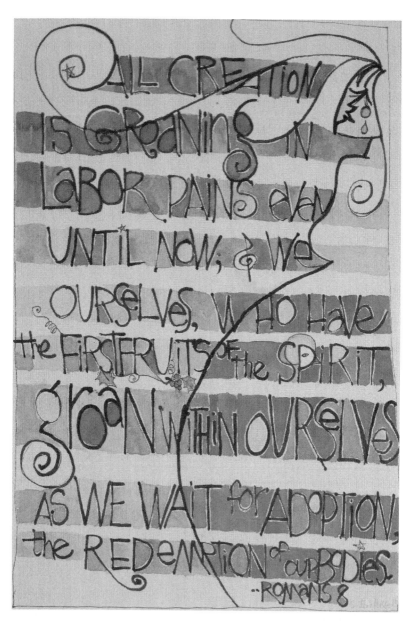

Jane Elyse Russell, O.S.F., 2008. Used with permission.

CONFRONTING THE CLIMATE CRISIS

CATHOLIC THEOLOGICAL PERSPECTIVES

Jame Schaefer

EDITOR

MARQUETTE
UNIVERSITY
PRESS

MARQUETTE STUDIES IN THEOLOGY

NO. 75

ANDREW TALLON, SERIES EDITOR

LIBRARY OF CONGRESS CATALOGING-IN-PUBLICATION DATA

Confronting the climate crisis : Catholic theological perspectives / Jame Schaefer, editor.

p. cm. — (Marquette studies in theology ; no. 75)

Includes bibliographical references and index.

ISBN-13: 978-0-87462-791-6 (pbk. : alk. paper)

ISBN-10: 0-87462-791-5 (pbk. : alk. paper)

1. Human ecology—Religious aspects—Catholic Church. 2. Global warming—Religious aspects—Christianity. 3. Ecotheology. I. Schaefer, Jame.

BX1795.H82C67 2011

261.8'8—dc23

2011031390

Cover illustration "Illumination" © Mary Southard, C.S.J.

Association of American University Presses

MARQUETTE UNIVERSITY PRESS
MILWAUKEE

The Association of Jesuit University Presses

CONTENTS

Dedicated to Catholic theologians in the present and future who strive to address ecological and biospheric problems in meaningful, relevant, and helpful ways.

INTRODUCTION

Jame Schaefer

Confronting the Climate Crisis: Catholic Theological Perspectives is the culmination of a three-year study by participants in the Catholic Theology and Global Warming Interest Group of the Catholic Theological Society of America (CTSA). This interest group was established in 2008 when several members of the CTSA shared a desire to respond theologically to the complex and vexing problem of global climate change. Some had already been addressing environmental issues from the perspectives of the various sub-disciplines of theology in which we have been trained (e.g., biblical, historical, systematics, ethics) because inherent in our vocation as theologians is a calling by God to be responsive to the moral issues and signs of our times. Furthermore, we value our role in relation to the teaching authority of the Church and wish to be responsive to the bishops who stress living in harmony with God's creation as "a moral responsibility" (John Paul II, 1989)[1] and

---

1   Pope John Paul II's 1990 World Day of Peace Message was the first papal statement dedicated to human responsibility for living compatibly with the natural environment, a teaching he continued in his subsequent encyclicals, messages, statements, and actions to make Vatican City ecologically "friendly." Pope Benedict XVI, who has been dubbed "the green pope" because of his many teachings on the human relation to God's creation and continuing his predecessors' initiatives in Vatican City, urged world leaders to "[s]ave the planet before it's too late" when speaking at the 2007 world youth rally. Dressed in a green vestment when celebrating Mass in Loreto, Italy, he underscored the responsibility that humans have toward the natural environment: "New generations will be entrusted with the future of the planet...which bears clear signs of a type of development that has not always protected nature's delicate equilibrium." He continued: "A decisive 'yes' is needed in decisions to safeguard creation as well as a strong commitment to reverse tendencies that risk leading to irreversible situations of degradation," and he urged "a strong alliance between man and earth" (Pullella 2007b).

who specifically urged biblical experts, theologians, and ethicists "to help explore, deepen, and advance the insights of our Catholic tradition and its relation to the environment," and especially "the relationship between this tradition's emphasis upon the dignity of the human person and our responsibility to care for all of God's creation" (U.S. Conference of Catholic Bishops 1991, 13).[2] Our efforts to address the climate crisis in this anthology is an example of a way in which theologians can respond to the Church magisterium.

The first step of our newly-initiated interest group was an examination of the scientific evidence through experts available to each of us on our campuses, other institutions, and reports by climate experts. Following this individual endeavor was a group consultation with David Quesada, Ph.D., Associate Professor of Physics at the University of St. Thomas in Miami who specializes in atmospheric and mathematical physics and has been gathering and interpreting data pertaining to climate change. Our cumulative conclusions were and continue to be indisputable based on the plethora of data collected and verified by scientists throughout the world: (1) changes in the global climate are accelerating; (2) these changes cannot be attributed exclusively to natural causes; (3) in-depth investigations of possible unnatural causes point to emissions of carbon dioxide, methane, nitrous oxide, water vapor, and other greenhouse gases from human activities; (4) adverse effects from these emissions are ongoing and affecting others who have little part in causing these climate disruptions; and, (5) future generations will be affected directly and also indirectly through effects on other species, ecological systems, and the biosphere. The fact that effective measures are not underway to mitigate these problems, the need to adapt to what cannot be mitigated, and to avoid suffering especially of the poor and vulnerable prompted us to probe sources in the Catholic theological tradition from our various perspectives with the hope of providing some relevant, meaningful, and helpful ways of addressing the climate crisis.

---

2   Other bishops individually and regionally have issued similar calls and statements pertaining to the need for humans to accept their responsibilities in relation to other species and the natural environment. Statements by Catholic bishops are available from www.inee.mu.edu.

COMPELLING SCIENTIFIC FINDINGS

What is this evidence that prompted our study of the climate crisis and compelled our production of this anthology? Climatologists and other scientists recognize that some extraterrestrial and terrestrial factors are primarily responsible for most of the past episodes of changes in Earth's climate. Extraterrestrial factors include solar output, Earth-Sun geometry, and interstellar dust, while terrestrial factors from oceans, the atmosphere and land include volcanic emissions, mountain-building, continental drift, atmospheric chemistry, atmospheric reflectivity, land reflectivity, and atmosphere/ocean heat exchange (Physical Geography.net 2010). However, these factors cannot fully account for the climate changes that are occurring today, accelerating, and predicted for the future with various levels of certainty issued by the Intergovernmental Panel on Climate Change (IPCC), an international association established by the United Nations Environmental Programme and the World Meteorological Organization in 1988 to assess peer-reviewed scientific, technical, and socio-economic reports by scientists throughout the world. On the basis of these data, the IPCC is charged with the arduous task of providing "a clear scientific view on the current state of knowledge" about changes in the global climate and its "potential environmental and socio-economic impacts" (IPCC 2010a).

The key scientific fact with which the IPCC has had to grapple is a sharp increase in Earth's average surface temperature by approximately 1.4 °F (about 0.75 °C) since the mid-nineteenth century (IPCC 2007a, 22-23).[3] The key culprits responsible for this increase are human activities through which carbon dioxide, methane, nitrous oxide, and other "greenhouse" gases are emitted into the atmosphere, intensify the natural greenhouse effect that warms Earth, and increase the global temperature (Mastrandrea and Schneider 2010, 12-14; IPCC 2007d). According to climate scientists who gather their data from various sources including ice cores that span thousands of years, tree rings, the extent of mountain glaciers, changes in coral reefs, and pollen in lake beds, atmospheric concentrations of carbon dioxide, methane, and nitrous oxide have increased markedly since 1750 as a result of human activities and far exceed pre-industrial levels (IPCC 2007d).

---

3 See also National Oceanic and Atmospheric Administration (NOAA) 2010 and National Aeronautics and Space Administration (NASA) 2010.

Scientific reports analyzed by the IPCC indicate that the largest growth in these gaseous emissions between 1970 and 2004 came from energy production, other industries, and transportation, while gases from residential and commercial buildings, forestry (including deforestation), and agriculture sectors grew at a lower rate (IPCC 2007e, 36). Thus, our activities are "forcing" changes on the global climate that exceed the effects caused by natural factors.

Among these changes are increases in the average global air, land, and ocean temperatures in the northern hemisphere that are causing decreases in snow cover and thickness of sea ice; widespread melting of snow and ice that is causing a rise in the sea level; increases in precipitation in eastern parts of North and South America, northern Europe and northern and central Asia; decreases in precipitation in the semi-arid land south of the Sahara Desert, the Mediterranean area, southern Africa and parts of southern Asia; and, an increase in the intensity of weather events including more extreme tropical cyclone activity in North America (IPCC 2007e, 36-38).

Physical systems are also adversely affected. For example, glacial lakes have enlarged and increased in numbers as glaciers and ice melts. Ground instability has increased in permafrost regions while rock avalanches have increased in mountain regions. Increased runoff is occurring in many glacier and snow-fed rivers. And, lakes and rivers have warmed in many regions causing changes in the thermal structure of bodies of water and degraded water quality (IPCC 2007e, 30-33).

Effects on ecological systems are also significant. Within land systems, spring events (e.g., leaf unfolding, bird migration, and egg laying) are occurring earlier, and ranges in plant and animal species are shifting upward. Within marine and freshwater systems, rising water temperatures are causing higher levels of salinity, shifts in ranges of algal, plankton and fish abundance, earlier fish migrations in rivers, and stresses on coral reefs. The loss of coastal wetlands and mangroves are attributed to a combination of climate changes and human adaptations of these ecological systems (IPCC 2007e, 32-33).

Adverse effects on human health from human-forced climate change have also been identified. Among these are increased heat-related mortality in Europe, changes in infectious disease vectors in parts of Europe, and earlier onset of and increases in seasonal production of allergenic pollen in high and mid-latitudes of the Northern Hemisphere (IPCC 2007e, 33).

A conclusion from a plethora of climate studies over the past decade is important to recognize: the sharp increase in the global temperature has occurred rapidly with uneven effects across geographic, economic, and social divisions. As John Holdren, Director of Harvard University's Science, Technology, and Public Policy Program and Co-Chair of the U.S. President's Council of Advisors on Science and Technology explained recently, the rise in the global temperature is "rapid compared with ordinary historic rates of climate change" and "rapid compared with the adjustment times of ecosystems and human society" (2010, 1). Furthermore, the effects of global climate change "on human well-being are and undoubtedly will remain far more negative than positive" (ibid.). Our choices are to mitigate the effects of human-driven climate changes, adapt to them, and suffer (ibid. 3).

Predictions of Future Effects

Because the major greenhouse gases from human activities remain in the atmosphere for periods ranging from decades to centuries and more will inevitably be emitted, the IPCC, the U.S. Environmental Protection Agency, and other major research institutions throughout the world anticipate that the average surface temperature of Earth is likely to increase between 1.1-6.4 °C/2 to 11.5 °F (USEPA 2010; IPCC 2007a, chap 10). This range is the outcome of computer models into which known data are entered and a range of likelihoods obtained (see Mastrandrea and Schneider 2010, 23-26). How much and how quickly Earth's temperature will actually increase remains unknown given the uncertainty of the amount of future greenhouse gases, the planet's response to changing conditions, and natural influences (e.g., changes in the sun and volcanic activity) (USEPA 2010). Putting these variables into computer models has yielded a "best estimate" range of a 3.2 to 7.2 °F (1.8-4.0 °C) increase in Earth's temperature by the end of the 21st century (USEPA 2010; IPCC 2007a, chap. 10). This increase will not be evenly distributed: land areas will warm more than oceans; high latitudes will warm more than low latitudes; warming is expected to occur during the winter months in northern North America and northcentral Asia; and, summer warming over continents may be accompanied by drier soils in many regions (NOAA 2007).[4] The USEPA issued this foreboding conclusion based on research collected by the IPCC: "The average rate of warming over each

4   See also USEPA 2010 and Meehle et al. 2007.

inhabited continent is very likely to be at least twice as large as the rate of warming experienced during the 20th century" (2010).

Even if emissions of greenhouse gases by human activities are kept constant at levels that were recorded at the beginning of this century, the effects will increase because, as already noted, the major gases already emitted remain in the atmosphere for decades to centuries. More greenhouse gas emissions at or above levels during the year 2000 would cause further warming and induce many changes in the global climate system during the 21st century that would "very likely" be larger than those observed during the 20th century (IPCC 2007e, 45). Among these changes are shrinking sea ice in the Arctic and Antarctic, hot extremes in temperature, heat waves and heavy precipitation in high latitudes, more intense tropical typhoons and hurricanes, and increased flooding of coastal regions (ibid.).[5]

Predictions beyond the 21st century are more dire when considering the effect on sea levels, even if greenhouse gas concentrations are stabilized, because adding more gases into the atmosphere already containing them in significant qualtities would cause the sea level to rise for centuries due to the time scales associated with climate processes and feedbacks (IPCC 2007e, 46; Mastrandrea and Schneider 2010, 21-23)). Prompted by the melting of glaciers and ice sheets on polar lands, the rising sea level would flood coastlines and inundate low-lying areas, the greatest effects of which will be in river deltas and low-lying islands (Pachauri 2007; Gleick 2010). Increasing atmospheric concentrations of greenhouse gases, especially carbon dioxide, will also further acidify the oceans (IPCC 2007e, 52; Turley 2010)).

The survival of some species will also be threatened. According to IPCC Chairperson Pachauri, approximately "20-30% of the species assessed in 2007 would be at increased risk of extinction if the global average temperature increase exceeds 1.5 to 2.5 °C," while an increase that exceeds about 3.5 °C suggests "significant extinctions" (40-70% of species assessed) around the planet (Pachauri 2008). Scientists anticipate accelerated adaptations by wild plants and species as they shift their ranges to inhabit areas with more tolerable temperatures, begin spring activities earlier, invade habitats of other species, and change genetically (Leemans 2010).

Closely related to species endangerment and extinction are risks to ecological systems (Leemans 2010; Turley 2010). The adaptive ability

5   See also NOAA 2007 and USEPA 2010.

of forests, grasslands, coastal wetlands, and other ecosystems to adapt to flooding, drought, wildfires, insect infestations, ocean acidification, and other disturbances caused by climate changes will be thwarted. Adaptation is particularly difficult when these disturbances are combined with other drivers of ecological degradation, including land use changes, pollution, and overexploitation of natural sources. Major changes in the structure and function of ecosystems are anticipated with predominantly negative consequences for biological diversity and for water, food supply, clean air, and other ecosystem goods (IPCC 2007e).

Effects on People, Especially Poor & Vulnerable

Adverse effects of global climate change on ecological systems also puts our species at risk. According to predictive models developed by scientists, human-forced climate change will have "serious effects on the sustainability of several ecosystems and the services they provide to human society (Pachauri 2007). One of the major effects anticipated is a decrease in the availability of potable water in the mid-latitudes and semi-arid areas of the planet. While an increase in water availability is anticipated in the moist tropics and at high latitudes, people in the Hindu-Kush, Himilayan, and Andes mountain ranges where more than one-sixth of the world's population currently lives will experience difficulty in obtaining the water they need as will people in the Mediterranean Basin, western United States, southern Africa, northeastern Brazil and other semi-arid areas. In already poverty-stricken Africa, between 75 and 250 million people will be exposed to increased water stresses by the year 2020, including limited access to potable water (ibid.; Gleick 2010).

Food scarcity will also be exacerbated in many areas by the year 2020 due to a decline in crop productivity caused by changes in the global climate. Effects on crop productivity is particularly problematic in seasonally dry and tropical regions where it is projected to decrease with small local temperature increases of 1-2 °C). Yields from rain-fed agriculture in some African countries could be reduced by up to 50 percent. While initial slight increases in crop productivity are projected at mid to high latitudes where the local average temperature rises from 1 to 3 °C, productivity is expected to decrease subsequently (IPCC 2007e). The same pattern of an initial increase in crop productivity followed by a decrease is anticipated globally (ibid.).

Millions of people are projected to experience yearly floods due to sea level rise by the year 2080 than people experience today. Particularly vulnerable are megadeltas of Asia whose major cities, including Shanghi, Dhaka, and Kolkata, cities on the coast, and cities in river flood plains whose industries and businesses are closely linked with climate-sensitive resources (Pachauri 2008; IPCC 2007e). Small islands off the coasts of Asia and Africa are especially vulnerable. Within these at-risk areas, poor communities will be most affected (IPCC 2007e; 2007b) because their sensitivity to the adverse effects of climate change is exacerbated by their poverty, food scarcity, malnutrition, and inaccessibility to other necessities of life. Adding to their impoverished circumstances, the poor are especially vulnerable to debilitating trends in economic globalization, regional conflicts, and diseases, including HIV/AIDS (Pachauri 2007).

The health of future people is also at risk as the global climate changes. Malnutrition, deaths, diarrhoeal and other diseases, injuries due to extreme weather events, and increased frequency of cardio-respiratory diseases due to higher concentrations of ground-level ozone in urban areas are included in these projections (IPCC 2007e).[6] While residents in temperate areas may benefit initially by fewer deaths from cold exposure and combinations of other factors (IPCC 2007e), any benefits will be outweighed by the negative health effects of rising temperatures, especially in developing countries (ibid.; 2007b). Poor people are most vulnerable to adverse health effects caused by forced climate change, and education, health care programs, and public health initiatives are crucial for shaping healthy populations to prevent more adverse effects (IPCC 2007e; 2007b).

When accepting the 2007 Nobel Peace Prize on behalf of the Intergovernmental Panel on Climate Change, Chairman Rajendra Pachauri underscored the equity implications of climate change and how they will affect some of the poorest and most vulnerable communities of the world:

> One of the most significant aspects of the impacts of climate change, which has unfortunately not received adequate attention from scholars in the social sciences, relates to the equity implications of changes that are occurring and are likely to occur in the future. In general, the impacts of climate change on some of the poorest and the most vulnerable communities in the world could

6   See IPCC 2007a and 2007b for in-depth discussions of these risks.

prove extremely unsettling. And, given the inadequacy of capacity, economic strength, and institutional capabilities characterizing some of these communities, they would remain extremely vulnerable to the impacts of climate change and may, therefore, actually see a decline in their economic condition, with a loss of livelihoods and opportunities to maintain even subsistence levels of existence (Pachauri 2007).

He also called attention to the potential conflicts resulting from human-forced climate change and lamented the IPCC's failure to provide directions for mitigating them. He expressed concern about conflicts that may arise when access to clean water, food availability, stable health conditions, ecosystem resources, and secure settlements are disrupted by changes in the climate. A major potential source of conflict that he identified is the migration and movement of people from one area to another. Though usually temporary and often from rural to urban areas in response to floods, famine and warfare, the migration and movement of people from climate change impacts may become highly problematic for them, for the people of the regions to which they are relocating, and for efforts to establish a peaceful global society. To address these potential threats, Pachauri urged conducting an "in-depth analysis of risks to security among the most vulnerable sectors and communities impacted by climate change across the globe" (ibid.).

EFFORTS, HOPES, AND ERRORS

Though climate scientists have been collecting and measuring data for several decades that point to an increase in the global temperature and, more recently, to the ongoing effects on people, other species, ecosystems, and the biosphere, and though they have developed increasingly sophisticated models for predicting the likelihood of future effects that are ominous even at a low range of certainty, little has been accomplished at national and international levels of decision-making to mitigate human-forced climate change. The United States government has been especially remiss by failing to pass comprehensive legislation aimed at addressing the climate crisis, though anticipation was high during the first two years of the Obama Administration when the House of Representatives and the U.S. Senate were controlled by Democrats who campaigned on platforms committed to mitigating this crisis. Little hope lingers for effective climate legislation during the 2011-2013 congressional year. Meanwhile, the U.S. has retained the dubious distinction of emitting the largest amount of greenhouse gases per capita (Netherlands Environmental Assessment Agency 2007; Knickerbocker 2007). This is not surprising given our nation's highly industrialized economy that thrives on energy generated primarily with non-renewable sources and uses energy less efficiently than possible.

Internationally, efforts to address climate change have been officially ongoing since 1988 when, as already noted, the United Nations collaborated with the National Oceanic and Atmospheric Administration (NOAA) to establish the IPCC that they charged with collecting, assessing, and drawing conclusions from peer-reviewed reports to serve as basis for informed decision-making. Another milestone occurred in 1992 when the UN established the Framework Convention on Climate Change (UNFCCC) through which environmental ministers of UN countries that signified their desire to participate as Parties of the Framework Convention[7] would meet to discuss and take joint action on climate-related issues. At their third meeting (Conference of the Parties–COP 3) held in Kyoto, Japan in 1997, the ministers developed and signed on behalf of their nations a protocol that committed them to reducing greenhouse gas emissions to 1990 levels by 2012 through methods each nation determined (UNFCCC 1997).[8] The United States was among the first signatories of the Kyoto Protocol, but neither President Bill Clinton nor President George Bush submitted it to the U.S. Senate for ratification, ostensibly because elected officials feared jeopardizing the economic well-being of the U.S if attempts were made to meet Kyoto emission targets. Detailed rules for implementing the Protocol were adopted at COP 7 in Marrakesh (UNFCCC 2001), nearly two hundred nations have ratified and/or accepted Kyoto Protocol goals, but the extent to which they will be met has yet to be ascertained.

---

7   The signatory Parties of the Convention number 194–193 nations and one regional economic integration organization; see list of parties, dates signed, ratified, and effective at http://unfccc.int/essential_background/convention/status_of_ratification/items/2631.php.

8   The Kyoto Protocol placed a heavier burden on industrially developed nations under the international law principle of "common but differentiated responsibilities" and specified three market-based mechanisms through which to meet their emission targets in cost-effective ways: emissions trading, clean development, and joint implementation.

Despite high hopes for COP 15 held in Copenhagen in 2009, involvement of many non-government agencies and youth groups (in which a former outstanding environmental ethics student of mine participated), a stimulating pre-conference featuring the latest climate science, and eager anticipation that the top leaders of the world's nations would meet on the last day to bind their nations in a rigorous plan to address the climate crisis, little progress was made over the two weeks of the summit. The international media, UNFCCC web site, and my reliable student reported extensively on friction between developing and developed nations, and the social communication network was alive with minute-to-minute accounts of the proceedings, encounters, and disappointments. Representatives of developing nations balked at highly industrialized nations' pressuring them to curtail their emissions of greenhouse gases without offering adequate financial and technical assistance to facilitate their opting for cleaner coal technology, energy efficient strategies, and ability to tap renewable energy sources. Developing and developed nations also strongly disagreed over means of monitoring, verifying, and reporting reductions in emissions of greenhouse gases. Shortly before COP 15 was scheduled to end in failure, the heads of state arrived and endeavored to broker a positive outcome from the bleak scenario, and an "accord" was drafted by the United States, China, India, Brazil, and South Africa. The accord was vigorously debated by opponents and proponents, recognized (not adopted), and eventually signed by representatives of nations who committed to reporting by January 31 their individual plans for reducing greenhouse gas emissions. The twelve provisions of the Copenhagen Accord began with the following commitment:

> We underline that climate change is one of the greatest challenges of our time. We emphasise our strong political will to urgently combat climate change in accordance with the principle of common but differentiated responsibilities and respective capabilities. To achieve the ultimate objective of the Convention to stabilize greenhouse gas concentration in the atmosphere at a level that would prevent dangerous anthropogenic interference with the climate system, we shall, recognizing the scientific view that the increase in global temperature should be below 2 degrees Celsius, on the basis of equity and in the context of sustainable development, enhance our long-term cooperative action to combat climate change. We recognize the critical impacts of climate change and the potential impacts of

response measures on countries particularly vulnerable to its adverse effects and stress the need to establish a comprehensive adaptation programme including international support (UNFCCC 2009, #1).

Though reported in the media as a failure,[9] the nations signing the Copenhagen Accord recognized for the first time the need to stay below a 2 °C increase in the global temperature, the possibility that this increase may be too high, and the need for a reassessment in 2015 to consider "strengthening the long-term goal" from 2 °C to 1.5 °C (UNFCCC 2009, #12). Among other articles in this accord was a commitment by the developed nations to provide "adequate, predictable and sustainable" financial and technological support to developing nations so they can reduce their emissions. The signatories also committed to establish a "Copenhagen Green Climate Fund" through which the developed countries will be able to secure funding from public and private sources to finance "meaningful mitigation actions" by "the most vulnerable developing countries, such as the least developed countries, small island developing States and Africa" (ibid. #8-10). By January 31, 138 nations had signed the Copenhagen Accord, some plans to reduce emissions had been submitted and appended to it, and others were at various stages of completion. Because signing this accord was voluntary instead of binding as some COP 15 ministers and many watchers had hoped, the first chairperson of the UNFCCC described the accord outcome as a "shock" that he hope will stimulate serious discussion about what nations are willing to do to mitigate human-forced climate change (Cutajar 2010).

As I write this introduction, COP 16 has just ended in Cancun, Mexico where 192 nations were represented. Apparently the environmental ministers had not been sufficiently shocked by COP 15. An overarching and binding agreement to slash emissions of greenhouse gases was not struck, and disagreements between developing and developed nations that stymied the ministers' progress in Copenhagen

---

9   After emerging from 13 hours of talks with other world leaders and reaching agreement on the principles of the accord, President Obama was quoted widely in the media as stating that the progress made was "meaningful" but "not enough." British Prime Minister Gordon Brown viewed the accord as a "vital first step we are taking towards a green and low carbon future for the world." He also stated: "I know what we really need is a legally binding treaty as quickly as possible" (Vidal et al. 2009).

also stymied the 2010 proceedings—the amount of emissions that need to be cut and how much the developed nations would contribute to developing nations' mitigation and adaptation efforts (e.g., Rothbard and Rucker 2010). Especially contentious was a proposal to extend the Kyoto Protocol commitment period beyond 2012 by which most developed nations committed to cut their greenhouse gas emissions or to inaugurate a new commitment period that might attract nations like the United States that had not ratified the Kyoto Protocol (Buckley and Wynn 2010). After extending the conference another day so the delegates could try to produce a positive outcome, the ministers approved[10] "urgent action" to cap the global temperature rise to no more than 2 °C/3.6 °F over pre-industrial levels while asking for a study on strengthening the commitment to a lower degree rise, creating the "Green Climate Fund" for which they would aim to raise $100 billion by 2020 for aiding developing countries' efforts to mitigate and adapt to climate change, and initiating several measures aimed at protecting tropical forests and sharing new clean energy technologies (UNFCCC 2010c). However, they put off until COP 17 a decision on extending the Kyoto Protocol period or inaugurating another new legally-binding protocol (Vidal and Goldenberg 2010; DeSousa 2010).

Plaguing deliberations at Copenhagen and Cancun were questions about the believability of climate scientists' research that surfaced before COP 15 officially began and carried over into COP 16. The tenor at Copenhagen was marred considerably by the disclosure of e-mails from a climate research laboratory at the University of East Anglia suggesting that some scientists were attempting to suppress findings by other climatologists that did not agree with their own. Following this disclosure was subsequent evidence of improprieties reflected in the 2007 IPCC reports pertaining to the rate at which the Himilayan glaciers are melting and direct connections made between catastrophic events and human-forced climate change (e.g., Leake 2010). After investigating the Himilaya glacier issue reported in a paragraph of the 978-page Working Group II report and a companion reference in the Synthesis Report, the IPCC Chairman, vice-chairs, and co-chairs of the report expressed their "regrets" for the "poorly substantiated" conclusions that resulted from improperly following "well-established standards of evidence" and "poor application of IPCC procedures" when preparing those sections of the report (IPCC 2010b). The

10   COP 16 *approval* is significant when compared with COP 15 *recognition*.

IPCC proceeded to establish guidelines for using scientific literature in its reports, addressing errors in previous assessment reports, and treating uncertainties consistently (IPPC 2010c).

Collection and assessment of peer-reviewed research are underway for the next IPCC report, and we can only hope that lessons learned from 2007 are reflected in it so the bases for decision-making by the nations are well grounded in facts, methods for assessing them and identifying ranges of certitude about present and future effects are instituted and carefully used, and options for addressing these effects are well substantiated. Every climate scientist and every climate research institution should be able to benefit in some way from errors that have been committed and from allegations levied by climate change skeptics (e.g., MacRae 2010; Holland 2007)[11] who may help purge the process from errors and improve methodology in assessing research findings. Theologians who feel compelled to address the climate crisis will also benefit through verifiable scientific findings and interpretations thereof upon which to reflect in meaningful ways. We rely upon scientists to provide the facts and realize that we need a heavy dose of the virtue of fortitude to face and act on them.

Nevertheless, despite the few but significant improprieties discovered in the IPCC's 2007 reports and the points made by skeptics who argue against attributing any aspect of the warming planet to human causes, the consensus of the scientific community persists: the temperature of Earth is increasing; human activities that spew greenhouse gases into the atmosphere are forcing the temperature rise; and, adverse effects are experienced now and projected for the future at various levels of certainty. Scientists have banded together to issue statements about this consensus (e.g., "Open Letter" 2010; American Institute of Biological Sciences 2009) so decision-makers accept the fact that human activities are forcing the temperature of our planet to rise, adverse effects are occurring now and projected to continue in the future, and changes in how we are functioning must be made now to mitigate these adversities.

In the meantime, industrialized countries continue to spew excessive greenhouse gases into the atmosphere while China, Malaysia, and other countries that are developing industrial economies are emitting

---

11   Other skeptics and points they are making are accessible from the Australian Climate Science Coalition site at http://www.climatescienceinternational.org/.

increasing amounts. China is leading the way as the largest emitter overall, while the U.S. remains the largest emitter per capita (Netherlands Environmental Assessment Agency 2007; Knickerbocker 2007). As the latest IPCC report indicates, all countries with portfolios of greenhouse gas emissions have options for reducing them. These options include energy conservation and energy efficiency improvement, the use of more renewable energy sources, recovery and abatement of methane and nitrous oxide emissions of various sources, sequestration of carbon in forests and soils, and carbon capture and storage (2007c). However, developing nations need help in tapping their options, and the Green Climate Fund outlined in the Copenhagen Accord and established by the Cancun Accord is surfacing as a vital vehicle to helping them help themselves. Hopefully, their efforts will have the effect of slowing the rate at which Earth's temperature increases.

CATHOLIC ACTION

As indicated in texts by the U.S. Catholic bishops and popes John Paul II and Benedict XVI, the Church magisterium is cognizant of climate science and concerned about human activities that are forcing the temperature of Earth to increase. Their teachings are covered at length in two essays of this anthology. One by Connie Lasher and Msgr. Charles Murphy explores the Christian humanism of popes John Paul II and Benedict XVI and its significance for addressing human-forced climate change. In the second pertinent essay, I explore the Catholic social teachings of popes from Leo XIII and successor popes on the principles of subsidiarity, solidarity, and preferential option for the poor. References are made to papal texts and pastoral statements by bishops in other essays of this anthology, thereby demonstrating our engagement with magisterial teachings.

Action is not limited to these teachings, however. During the pontificates of John Paul II and Benedict XVI, Vatican City has become progressively "green" through conservation efforts that include recycling, using photovoltaic cells on buildings to produce electricity, and hosting a scientific conference to discuss global warming (Pullella 2007a). At a conference organized by the Vatican's Council for Justice and Peace, Bishop Bernd Uhl of Freiburg told the scientists, environment ministers, and leaders of various religions from twenty countries who had gathered in Vatican City: "Climate change is one of the signs of the times affecting the Catholic Church as a global organization.

The Catholic Church must take a stand on this present-day and urgent question" (Pullella 2007a).

Efforts have been initiated at national and diocesan levels to study the climate change issue, to advocate activities aimed at mitigating human-forced change, and to lobby for appropriate legislation with special attention to minimizing the suffering of poor and vulnerable people who will be most severely affected by climate change. Having issued *Global Climate Change: A Plea for Dialogue, Prudence, and the Common Good* in 2001, the text of which appears below, the U.S. Catholic Conference of Bishops (USCCB) through its Environmental Justice Program has been assisting these efforts. A major national initiative by the USCCB in partnership with Catholic Relief Services and other organizations is the Catholic Coalition on Climate Change which encourages "a more thoughtful dialogue about ways the Catholic community can respond to climate change" and "invites Catholics to participate in a new initiative offering a distinctively Catholic perspective on global climate change" (USCCB and CRS 2010). The impressive array of Catholic organizations involved with the U.S. Catholic bishops and Catholic Relief Services in the Coalition[12] (CCCC 2010) should inspire hope for action aimed at mitigating the effects of human-forced climate change. Among the thoughtful activities of the Coalition are the Catholic Climate Covenant's encouraging the faithful to take the St. Francis Pledge to pray and reflect on the duty to care for God's Creation and protect the poor and vulnerable; to learn about and educate others on the causes and moral dimensions of climate change; to assess how we as individuals and in our families, parishes, and other social and economic affiliations contribute to climate change by our energy use, consumption, and wastefulness; to change our choices and behaviors by reducing the ways in which we contribute to climate change; and, to advocate for Catholic principles and priorities in climate change discussions and decisions, especially as they impact those who are poor and vulnerable (CCCC 2010). Through its web site, the Coalition shares inspiring stories about actions other people are taking and gives practical, timely tips for minimizing each person's

---

12   The National Council of Catholic Women, the Catholic Health Association of the U.S., Catholic Charities USA, the National Catholic Rural Life Conference, the Association of Catholic Colleges and Universities, the Franciscan Action Network, Carmelite NGO, the Leadership Conference of Women Religious, and the Conference of Major Superiors of Men.

impact on the climate in daily activities. One of the Coalition's most recent projects is training Catholic Climate Ambassadors—leaders from across the nation who become immersed in climate science and Catholic teaching on caring for God's creation and the poor and who offer presentations to parishes, schools, diocesan ministry professionals, and others on the Church's teachings.

As a participating organization in the Catholic Climate Change Coalition, Catholic Relief Services (CRS) is helping "to strengthen the ability of the most vulnerable communities in the developing world to respond to and prepare for the effects of climate change" by providing education and training to people in poor communities on ways to reduce their vulnerability to the effects of climate change in the areas in which they live (USCCB and CRS 2010). CRS recognizes that people who live in poverty "contribute least to climate change but they are likely to suffer its worst consequences" (e.g., increasingly limited access to water, reduced crop yields, and more widespread disease) with few resources with which to adapt and respond, thereby making their lives more difficult and increasing their suffering (ibid.).

Thus, the teaching authority of the Church and Catholic organizations are busily engaged in efforts to address climate change at many levels of endeavor. Theologians can help, as the U.S. Catholic bishops urged in 1991, by using our skills to identify, explain, and demonstrate various theological perspectives from which to think about the human person in relation to one another and to other species, ecological systems, and the biosphere of our planet that are imperiled now and will be imperiled into the future by our climate disruptive activities.

RESPONDING THEOLOGIANS

In this anthology, we share what we have found and reflected upon that we think is helpful toward addressing the climate crisis. Our first three essays explore promising themes in the Bible. *Suzanne Franck* focuses on Wisdom literature that conveys belief in God's dynamic presence as wisdom that permeates the world, establishes an interrelationship among humans and other creatures that constitute the world, heightens that relationship by becoming incarnate as Jesus Christ, and requires that we intrinsically value the world as a reflection of God. Challenging us to live wisely in the world today in ways that avoid forcing an increase in the global temperature, she suggests several approaches including reflection on the sacramental character of the world from the perspective of the Incarnation to motivate us to live wisely, accepting the overwhelming evidence that our climate-disruptive activities are jeopardizing the opportunity for people in the future to experience God's presence in the world, and engaging conscientiously in the ongoing story of creation as co-creators with God who assure that future generations will be able to live in a world that is charged with the grandeur of God. *Ann Marie Kidder* demonstrates a new hermeneutical approach used by biblical scholars to read Romans 8 and pertinent parts of other letters written by St. Paul and his followers for their theological significance today. Finding in these texts an intricate interplay and shared destiny of creation and covenant in the scheme of salvation through Christ, she points to Paul's understanding that God's glory is manifested first in humans and second in God's creation. Humans are distinguished by knowing our connection with other living and non-living creatures and knowing that they suffer when we sin and rejoice when we are redeemed. Applying this motif to the climate crisis, Kidder explains that we can experience the groaning of Earth in tandem with the Spirit of Christ working within us to have compassion for other species, the land, the air, and the waters whose role as "cheerleaders" of our salvation is jeopardized by our climate-forcing actions. *Vincent Pizzuto* explores the Christ hymn in Colossians 1 from an ecological-hermeneutical lens to understand the significance of the Pauline author's depiction of humans in relation to other creatures, and he finds a soteriology in which humans are created with the world and saved with the world through the presence and power of Christ who permeates and heads the cosmic body. While all creatures are interconnected through Christ in this scheme of creation and redemption, Pizzuto explains, and all are members of Christ's cosmic body who together reflect God's presence in the world, human creatures are most reflective of Christ's headship through our self-reflective consciousness and moral agency that requires us to act justly in relation to others with whom we constitute the cosmos. Harming Earth by forcing changes in the global climate constitutes harming the body to which we belong, and we are morally responsible for attending to the wounds we are inflicting.

The next four essays draw primarily upon the work of two revered theologians in the Catholic tradition—St. Bonaventure of Bagnoregio (1217-1274) and St. Thomas Aquinas (1224/5-1274). Focusing on Bonaventure's trinitarian theology with special attention to his Christology, *Dawn Nothwehr* emphasizes his hope-filled vision of the intimately related and interconnected reality of God, the divine-human Christ, and the world in which humans are called to serve as its guardians and co-creators with God. Bonaventure knew long ago that the faithful need to recognize this reality within which we live, acknowledge our creaturely status as loved and sustained by God, and act in ways that show our respect for the interconnections and inter-dependence of all creatures. Answering the question "How shall we live?" from the perspective of Bonaventure's Christology, Nothwehr urges us to develop the virtue of humility so we can realize that God desires a relationship not only with us but also with the entire world, to develop the virtue of poverty by releasing ourselves from our ar-rogance and consumerism, and to develop the virtue of obedience by living lovingly within the world that our loving God has held in dynamic existence for nearly 14 billion years. *Daniel Scheid* finds in Aquinas' theology of creation a "robust" understanding of the com-mon good that unites God, the universe, and humanity in a holistic paradigm that is helpful for addressing the climate crisis. From Aqui-nas's tripartite understanding that the whole universe surpasses in excellence any individual creature, the most valuable feature of the universe is the orderly functioning of its constituents, and the entire universe best glorifies God, Scheid urges respect for the goodness of Earth, cooperation with the other constituents of Earth for our mu-tual sustainability, humility about our place in the world in relation to the contributions that other species, ecological systems, and the bio-sphere make to the planetary common good, and upholding the life and dignity of human persons who have the unique capacity to love one another. *William French* draws upon Aristotle's ethic of the "fitting response" and Thomas Aquinas' understanding of self-mastering hu-man acts, vincible and invincible ignorance, and the virtue of prudence to stress the need for knowing the full range of internal and external costs incurred when human actions disrupt the global climate. Failing to include external costs of production and use imposed on others or future generations thwart an individual's or a society's ability to know the full impact of their actions and to make decisions about mitigat-ing their effects. Arguing that these "hidden" costs need to be internal-ized through green taxation schemes and tax shifting policies, French points to the educational and consciousness-raising benefits they can provide by helping us recognize the nature and ramifications of our actions. *William George* brings his continuing interest in international law to the question of how to respond theologically to the climate crisis in cogent and constructive ways. Stressing the need for theologians to be as informed as possible about international law principles and institutions that deal with climate issues, he draws upon Aquinas and one of his intellectual heirs in the 20th century, Bernard Lonergan, to explore two interrelated points of convergence between Catholic theology and international law that can open theologians to reflection—a method they share and virtues they affirm (e.g., justice, charity, and prudence). George explores in depth Aquinas' systematic treatment of the moral virtue of prudence and applies each component to pertinent aspects of international law for addressing the climate crisis. Though foresight provides the most direct link between prudence and intergenerational justice in theology and international law, other aspects of prudence work in tandem with foresight, including memory, understanding, docility, shrewdness, reason, circumspection, and caution. He recognizes with Aquinas how little humans can accomplish in addressing the climate crisis apart from God's grace.

In the next group of essays, the works of the scientist-mystic Teilhard de Chardin (1881-1955), theologians Karl Rahner (1904-1984), Bernard Lonergan (1904-1984), and Hans Urs von Balthasar (1905-1988), and the self-styled "geologian" Thomas Berry (1914-2009) are explored. Having immersed himself in Teilhard's writings for many years, *Robert Faricy* reflects on the value-supportive world view of the paleontologist and mystic who integrated evolutionary biology and Christian theology to produce an understanding that God is continuously creating the world of many, suffusing it with the divine presence heightened by the Incarnation, and influencing its convergence from a multiplicity to its fulfillment and unification in Christ. Humans have emerged out of this process with the capacity to be responsible participants who should feel motivated to mitigate human-forced climate change. *Richard Kropf* has devoted decades to studying Teilhard and focuses his essay on the confluence of the Club of Rome/MIT study on limits to growth and Teilhard's analysis of human evolutionary development and the final transformation of the material energies of the universe into "Ultra-humanity." Particularly problematic to reaching this culmination, Kropf notes, is the human population increase that is stretching planetary resources and forcing us to make some major choices today about the fate of our species. Most poignant in Kroft's analysis is his pointing to Teilhard's thinking late in his life that God might have a greater purpose for the universe than one planet or one species. Aiming to clarify the kind of dialogue needed for theologians to address the climate crisis, *Denis Edwards* draws from Karl Rahner's theology five interpretive principles grounded in the belief that the world is the place where we encounter God and five interpretive principles from a scientifically informed ecological consciousness grounded in the reality that our actions are forcing changes in the global climate. Both "sides" need to be heard for a fruitful dialogue, Edwards insists, challenges they make to one another recognized, and critical questions they raise for one another answered to yield helpful outcomes. *Richard Liddy* presents three levels of Bernard Lonergan's heuristic framework for analyzing questions that arise when addressing the climate crisis: the basic world view of emergent probability that surfaces from an analysis of contemporary scientific findings; a fundamental analysis of human biases against recognizing how we "ought" to act from what "is"; and, an illustration for answering questions about God as the absolute intelligence and basis for human intelligence when faced with lack of human intelligence and absurdities that are driving an increase in Earth's temperature. Lonergan's analysis of the dynamics of redemption and conversion through God's grace is also helpful, Liddy explains, for bringing about change of our minds and in our hearts about how we are functioning within our planet. *Anthony Sciglitano* finds in "deep ecology" a suitable dialogue partner for Balthasar who also critiques viewing other species exclusively as instruments for human use, recommends contemplation of the environment, and believes it should inspire responses of wonder, reverence, and joy. Though finding these convergences, Sciglitano points to challenges Balthasar makes to deep ecology's antipathy for divine transcendence and low esteem for human moral responsibility. Balthasar contributes a theological perspective to a serious difficulty within secular ecological thought generally and deep ecology in particular, Sciglitano argues, and he names and explains Balthasar's "doxological ecology" which supports a communal practice rooted in contemplation that respects Earth in all its diversity as having an intrinsic value and needing human care. Finally, *Peter Ellard* outlines key aspects of the groundwork Thomas Berry provides for "a dark green Catholic theology" that stresses the history of the unfolding of the universe and the reality of marvels that we encounter through our senses, fail to recognize, and destroy. With Berry, Ellard endorses thinking about the universe as "a community of subjects, not a collection of objects," the need to reinvent the human within the context of this community, and engagement in "the great work" of developing an intimate relationship with Earth and its constituents as subjects instead of merely objects for human use. Ellard also endorses Berry's suggestion that we need to put problematic aspects of the Bible, doctrines, and ideas within the Catholic theological tradition "on the shelf" for a while until we are knowledgeable about the universe story from which our species emerged, transform our destructive and instrumental attitude toward our planetary home, and acknowledge the human place within her.

Our next two essays emerge variously from feminist theology. *Colleen Mary Carpenter* finds highly problematic our failure to recognize the world as revelatory of God's goodness and God's presence. Concerned that this sacramental sense of God in and through the world is threatened by the damage we are inflicting on Earth through our climate-forcing actions, she draws upon two different options that have surfaced in the recently revived field of pneumatology to imagine how we can think about God's presence in the midst of climate change. One option proposed by Mark Wallace is to understand the Holy Spirit as wounded, traumatized, and in agony over ecological degradation, while the other proffered by Elizabeth Johnson understands the Holy Spirit as a vivifying presence that gives us hope in the midst of the destruction and death we have unleashed on the world. Carpenter examines these options critically, opts for aspects of both, and urges embracing them imaginatively in liturgies with the hope of moving beyond the climate-disruptive and planet-destructive behaviors that have become "normal" in our lives. A pioneer in the development and understanding of feminist theology, *Anne Clifford* turns to another kind of experience in a particularly vulnerable part of the world—sub-Sahara Africa—where biologist and Nobel laureate Wangari Maathai has led disempowered and subordinated women to become proficient in growing and planting thousands of trees that are indigenous to the region but were almost decimated by European colonizers. Clifford explains Matthai's concern about and efforts to mitigate the negative effects the colonial patriarchal order had on the natural environment and women, identifies convergences in Maathai's eco-feminist perspective and Catholic social teaching on the climate crisis, and points to the hope the Green Belt Movement has given to women for their improved status in Kenyan society and a more life-sustaining climate.

In the final group are essays and documents dedicated to Catholic social teaching. *Connie Lasher* and *Msgr. Charles Murphy* analyze the Christian humanism that surfaces from the theology-environment dialogue in which popes John Paul II was engaged and Benedict XVI is currently engaged, finding features that are both noteworthy and promising for confronting the climate crisis. Among these features is an understanding of the world as God's "gift" to all humans and our species' distinct capacity among creatures to accept this gift by contemplating the world through which the mystery of God as the source of all reality is revealed, by respecting the integrity of the world as a common heritage of humans now and in the future, and by acting together with a sense of shared responsibly for the natural environment. My essay provides an overview of three Catholic social teaching principles: solidarity as an understanding that all people should be engaged in seeking their common good; subsidiarity as an method for making decisions beginning at the most local level; and opting to give preference to the poor in decision-making as demonstrated in the life and teachings of Jesus Christ. Though these principles can help guide our efforts to address the climate crisis in the interests of humans, I propose expanding solidarity to include all species and systems that constitute our planet—Earth solidarity, reconfiguring subsidiarity to conform to biological regions in which humans function with other species and abiota—bioregional subsidiarity, and including future people, species and ecological systems among the poor for whom we must show preference when making decisions about mitigating and adapting to climate change. The next entry is *Global Climate Change: A Plea for Dialogue, Prudence, and the Common Good* that was issued by the U.S. Catholic bishops in 2001 and is introduced here by *Jane Russell* who underscores the attention the bishops give to the virtue of prudence for confronting the climate crisis. Closing this anthology is Pope Benedict XVI's 2010 World Day of Peace Message, *If You Want to Cultivate Peace, Protect Creation,* a document he issued twenty years after the first papal statement dedicated to the human-Earth relationship. We are grateful for permission to include these magisterial documents in *Confronting the Climate Crisis: Catholic Theological Perspectives.*

SOURCES

American Institute of Biological Scientists (AIBS). 2009. "Scientists Issue Statement on Climate Change: Scientific Organizations Reaffirm Consensus on Climate Change." AIBS Public Policy, October 21. Accessed from http://www.aibs.org/position-statements/20091021_scientists_issu. html on December 10.

Benedict XVI, Pope. 2009.

Brown, Donald A. 2010. Climate Progress. Center for American Progress Action. Accessed from http://climateprogress.org/2010/02/09/climate-change-policy-ethics/?utm_sour on December 1.

Buckley, Chris, and Gerard Wynn. 2010. "UN Climate Text Hints at Kyoto Compromise." Reuters Africa, December 10. Accessed from http://af.reuters.com/article/energyOilNews/idAFN1011198820101210   on December 10.

Catholic Coalition on Climate Change. 2010.

Cutajar, Michael Zammit. 2010. Interview on Climate Change.TV. June 10. Accessed from http://www.climate-change.tv/michael-zammit-cutajar-june-2010-2 on December 5.

De Souza, Mike. 2010. "World Leaders Approve New Global Climate Package, Fund." Accessed from http://www.canada.com/business/World+leaders+approve+global+climate+package+fund/3960935/story.html#ixzz17vYcx2FU on December 11.

Gleick, Peter H. 2010. "Water." In *Climate Change Science and Policy*, eds. Stephen H. Schneider, Armin Rosencranz, Michael E. Mastrandrea, and Kristin Kuntz-Duriseti, 74-81. Washington, DC: Island Press.

Hanley, Charles J. 2010. "Cancun Climate Talks Hunt for Compromises." ABC News, December 9. Accessed from http://abcnews.go.com/Technology/wireStory?id=12354064 on December 10.

Holdren, John. 2010. Introduction. In *Climate Change Science and Policy*, eds. Stephen H. Schneider, Armin Rosencranz, Michael E. Mastrandrea, and Kristin Kuntz-Duriseti, 1-7. Washington, DC: Island Press.

Holland, David. 2007. "Bias and Concealment in the Ipcc Process: the "Hockey-stick" Affair and its Implications." *Energy and Environment* 18 (7/8): 951-83.

Intergovernmental Panel on Climate Change (IPCC). 2007a. *Climate Change 2007: The Physical Science Basis. Contribution of Working Group I to the Fourth Assessment Report of the Intergovernjmental Panel on Climate Change.* Edited by S. Solomon, D. Qin, M. Manning, Z. Chen, M. Marquis, K.B. Averyt, M. Tignor and H.L. Miller. Cambridge and New York: Cambridge

University Press. Also available at http://www.ipcc.ch/publications_and_data/ar4/wg1/en/contents.html.

———. 2007b. *Climate Change 2007: Impacts, Adaptation and Vulnerability: Contribution of Working Group II to the Fourth Assessment Report of the Intergovernmental Panel on Climate Change.* Edited by M. L. Parry, O.F. Canziani, J.P. Palutikof, P.J. van der Linden and C.E. Hanson. Cambridge and New York: Cambridge University Press. Also available at http://www. ipcc.ch/publications_and_data/ar4/wg2/en/contents.html.

———. 2007c. *Climate Change 2007: Contribution of Working Group III to the Fourth Assessment Report of the Intergovernmental Panel on Climate Change.* Edited by B. Metz, O.R. Davidson, P.R. Bosch, R. Dave, and L.A. Meyer. Cambridge and New York: Cambridge University Press. Also available at http://www.ipcc.ch/publications_and_data/ar4/wg3/en/contents.html.

———. 2007d. Summary for Policymakers. In: *Climate Change 2007: The Physical Science Basis. Contribution of Working Group I to the Fourth Assessment Report of the Intergovernmental Panel on Climate Change.* Edited by S. D. Solomon, M. Qin, M. Manning, Z. Chen, M. Marquis, K.B. Averyt, M.Tignor and H.L. Miller. Cambridge: Cambridge University Press. Also available at http://www.ipcc.ch/pdf/assessment-report/ar4/wg1/ar4-wg1-spm.pdf.

———. 2007e. *Climate Change 2007: Synthesis Report, Contribution of Working Groups I, II and III to the Fourth Assessment Report.* Edited by Rajendra K. Pachauri and Andy Reisinger. Geneva: Intergovernmental Panel on Climate Change. Also available at http://www.ipcc.ch/publications_and_data/publications_ipcc_fourth_assessment_report_synthesis_report.htm.

———. 2010a. Organization. Accessed from http://www.ipcc.ch/organization/organization.shtml on December 2.

———. 2010b. IPCC Statement on The Melting of Himalayan Glaciers. Geneva, Switzerland, January 20. Accessed from http://www.ipcc.ch/pdf/presentations/himalaya-statement-20january2010.pdf on December 10.

———. 2010c. Decisions Taken by the Panel at its 32nd Session With regards to the Recommendations Resulting from the Review of the IPCC Processes and Procedures by the InterAcademy Council (IAC). Busan, Republic of Korea, October 11-14. Accessed from http://www.ipcc.ch/meetings/session32/ipcc_IACreview_decisions.pdf on November 27.

John Paul II, Pope. 1989. *Peace with God the Creator, Peace with All of Creation.* Message for the 1990 World Day of Peace, December 8. Accessed from http://www.vatican.va/holy_father/john_paul_ii/messages/peace/

documents/hf_jp-ii_mes_19891208_xxiii-world-day-for-peace_en.html on September 5, 2010.

Kintisch, Ed. 2009. "Projections of Climate Change Go From Bad to Worse, Scientists Report." *Science* 323, 20 March, 1546-47. Accessed from http://www.sciencemag.org/content/323/5921/1546.full?sid=18f367ef-70ae-4a74-b7b7-95a176895bce on December 10.

Knickerbocker, Brad. 2007. "China now world's biggest greenhouse gas emitter." *The Christian Science Monitor*, 28 June. Accessed from http://www.csmonitor.com/2007/0628/p12s01-wogi.html on December 5, 2010.

Leemans, Rik. 2010. "Ecosystems." In *Climate Change Science and Policy*, ed. Stephen H. Schneider et al., 56-65. Washington, DC: Island Press.

MacRae, Paul. 2010. *False Alarm: Global Warming—Facts Versus Fears.* : Spring Bay Press.

Mastrandrea, Michael D., and Stephen H. Schneider. 2010. Climate Change Science Overview. In *Climate Change Science and Policy*, eds. Stephen H. Schneider et al., 11-27. Washington, DC: Island Press.

Meehl, G.A., T.F. Stocker, W.D. Collins, P. Friedlingstein, A.T. Gaye, J.M. Gregory, A. Kitoh, R. Knutti, J.M. Murphy, A. Noda, S.C.B. Raper, I.G. Watterson, A.J. Weaver and Z.-C. Zhao. 2007. Global Climate Projections. In *Climate Change 2007: The Physical Science Basis. Contribution of Working Group I to the Fourth Assessment Report of the Intergovernmental Panel on Climate Change*, ed. S. Solomon, D. Qin, M. Manning, Z. Chen, M. Marquis, K.B. Averyt, M. Tignor and H.L. Miller. Cambridge and New York: Cambridge University Press. Also accessible from http://www.ipcc.ch/publications_and_data/ar4/wg1/en/contents.html.

National Oceanic Atmosphere Administration. 2007. "Patterns of Greenhouse Warming." *GFDL Climate Modeling Research Highlights* 1.6 (January): 1. Accessed at http://www.gfdl.noaa.gov/cms-filesystem-action/user_files/kd/pdf/gfdlhighlight_vol1n6.pdf?_rewrite_sticky=research/climate/highlights/PDF/GFDLhighlight_Vol1N6.pdf on December 4, 2010.

Netherlands Environmental Assessment Agency. 2007. "Chinese CO2 Emissions in Perspective." June 22. Accessed from http://www.pbl.nl/en/news/pressreleases/2007/20070622ChineseCO2emissionsinperspective.html on December 7, 2010.

Open Letter: Climate Change and the Integrity of Science." 2010. *Manchester Guardian*, May 6. Accessed from http://www.guardian.co.uk/environment/2010/may/06/climate-science-open-letter/print on November 5.

Oreskes, Naomi. 2004. "The Scientific Consensus on Climate Change." *Science* 306, 3 December, 1686. Accessed from http://www.sciencemag.org/content/306/5702/1686.full on December 5, 2010.

Pachauri, R. K. 2007. Acceptance Speech for the Nobel Peace Prize Awarded to the Intergovernmental Panel on Climate Change (IPCC), Oslo, Norway, December 10. Accessible from http://www.ipcc.ch/graphics/speeches/nobel-peace-prize-oslo-10-december-2007.pdf on December 5, 2010.

———. 2008. Opening Session of the World Economic Forum, Davos, Switzerland, January 23. Accessed from http://www.ipcc.ch/graphics/speeches/pachauri-davos-January-2008.pdf on December 5, 2010.

Partridge, Earnest. 2009. "A Convenient Delusion." Monday 28 December 2009. Accessed from http://www.truth-out.org/1229093? print on December 10.

Physical Geography.net. 2010. Factors that Influence the Earth. s Climate. Accessed from http://www.physicalgeography.net/fundamentals/7y.html on December 2.

Pullella Philip. 2007a. "Religion Must Help Protect Planet: Conference." *Reuters*, April 27. Accessed from http://www.reuters.com/article/idUS-PAR75418020070430 on December 5, 2010.

———. 2007b. "Pope Leads Eco-friendly Youth Rally," *Reuters*, September 2. Accessed from http://www.reuters.com/article/idUSL0111597220070901 on September 3. Also see video from http://www.reuters.com/news/video?videoId=65392.

Root, Terry L., and Elizabeth S. Goldsmith. 2010. "Wild Species and Extinction." In *Climate Change Science and Policy*, ed. Stephen H. Schneider et al., 44-55. Washington, DC: Island Press.

Rothbard, David, and Craig Rucker. 2010. "The Cancun, Mexico, climate summit had barely begun when the topic du jour became how much climate: Climate realism for the developing world." *The Washington Times*. December 7. Accessed from http://www.washingtontimes.com/news/2010/dec/7/the-cancun-mexico-climate-summit-had-barely-begun-/ on December 7.

Schneider, Stephen H., Armin Rosencranz, Michael E. Mastrandrea, and Kristin Kuntz-Duriseti, eds. 2010. *Climate Change Science and Policy*. Washington, DC: Island Press.

"Scientists Agree Human-Induced Global Warming Is Real, Survey Says." 2009. *ScienceDaily*, January 21. Accessed from http://www.sciencedaily.com/releases/2009/01/090119210532.htm on December 10, 2010.

Spotts, Peter N. 2009. "Copenhagen Accord: Essential Beginning To Some, Shaky Foundation to Others." *The Christian Science Monitor*, December 19. Accessed from http://www.csmonitor.com/World/2009/1219/Copenhagen-accord-essential-beginning-to-some-shaky-foundation-to- others on December 3, 2010.

Turley, Carol. 2010. "Marine Ecosystems." In *Climate Change Science and Policy*, ed. Stephen H. Schneider et al., 66-73. Washington, DC: Island Press.

United Nations Framework Convention on Climate Change (UNFCCC). 1997. Kyoto Protocol. COP 3, December 11. Accessed from http://unfccc.int/resource/docs/convkp/kpeng.html on December 5, 2010.

———. 2001. Marrakesh Accords and Declaration. COP 7, November 10. Accessed from http://unfccc.int/cop7/documents/accords_draft.pdf on December 5, 2010.

———. 2009. Copenhagen Accord. Convention of the Parties 15, December 18. Accessed from http://unfccc.int/resource/docs/2009/cop15/eng/l07.pdf on December 5, 2010.

———. 2010a. Status of Ratification of the Kyoto Protocol. Accessed from http://unfccc.int/kyoto_protocol/status_of_ratification/items/2613.php on December 5.

———. 2010b. Official Website of the United Nations Climate Change Conference Cancun - COP 16 / CMP 6. Accessed from http://unfccc.int/2860.php on December 11.

———. 2010c. Press Release: UN Climate Change Conference in Cancún Delivers Balanced Package of Decisions, Restores Faith in Multilateral Process. Accessed from http://unfccc.int/files/press/news_room/press_releases_and_advisories/application/pdf/pr_20101211_cop16_closing.pdf on December 12.

United States Conference of Catholic Bishops (USCCB) and Catholic Relief Services (CRS). 2010. "Catholics Confront Global Poverty: Global Climate Change and Our Catholic Response." Accessed from http://www.usccb.org/sdwp/globalpoverty/pdfs/Climate_Change.pdf on December 5.

U.S. Environmental Protection Agency. 2010. Climate Science: Future Temperature Changes. Accessed from http://www.epa.gov/climatechange/science/futuretc.html on December 4.

Vidal, John, Allegra Stratton, and Suzanne Goldenberg. 2009. "Low Targets, Goals Dropped: Copenhagen Ends in Failure." *The Guardian*, December 19. Accessed from http://www.guardian.co.uk/environment/2009/dec/18/copenhagen-deal on December 10, 2010.

Vidal, John, and Susan Goldenberg. 2010. "Deal is Reached at Cancún Summit." *The Guardian*, December 11, 2010. Accessed from http://www.guardian.co.uk/environment/2010/dec/11/mexico-cancun-environment-climate-summit on December 11.

BIBLICALLY INSPIRED PERSPECTIVES

I

SOPHIA WISDOM & CLIMATE CHANGE

Suzanne Franck, C.S.J.

T he Christian churches awoke to the environmental crises in the 1970s. In 1990, Pope John Paul II issued his 1990 World Day of Peace Message, *The Ecological Crisis: A Common Responsibility*, stressing the need for the faithful to transform their indifferent attitudes toward Earth. Catholic bishops around the world responded with pastoral letters acknowledging ecological degradation as a moral problem and identifying personal over-consumption of Earth's natural goods and industry's adverse effects on the planet as contrary to Church teachings. A new understanding of the relationship between theology and ecology was sought, environmental education program were initiated, and the faithful were encouraged to examine their individual and communal lifestyles. Bishops in the United States and Australia explored the connection between concern for the human person and caring for Earth through biblical texts, drawing upon Genesis 1 and 2, Wisdom, Colossians, Philippians, Ephesians, and the Gospel of John (U.S. Bishops 1996, 223-43; Australian Bishops 1996, 245-57). From this effort, *Sophia* wisdom surfaced as a promising source for thinking about this connection.

In this essay, I draw upon theologian Denis Edwards' exemplary examination of *Sophia* wisdom as a basis for ecological theology today. Emphasis is placed on the parallels Edwards' identified between *Sophia* wisdom in the Old Testament and Jesus Christ in the New Testament, yielding a sense of God's presence in the world that becomes incarnate in Jesus. I proceed to apply Edwards' insights about *Sophia* wisdom to the climate crisis by drawing upon the poetry of Gerard Manley Hopkins who juxtaposes the natural world as filled with God's presence and as suffering from human degradation. By submitting to *Sophia* wisdom in the world, we are able to change our destructive actions and live in God's presence with hope for a life-sustaining climate.

SOPHIA WISDOM AND ECOLOGICAL THEOLOGY

In his exploration of the biblical texts about wisdom in the Old Testament and Jesus in the New Testament, Edwards identifies Jesus as personified wisdom to show a biblical foundation for a new ecological "*Wisdom* Christology." For Edwards, "*Sophia's* role in creating, sustaining and redeeming all things" and the early "Christian identification of *Sophia* and her cosmic role with Jesus Christ" suggest a basis for a new ecological theology (1995, 43). Radically associated with all creation, *Sophia* wisdom in Jesus of Nazareth pitches a tent among all creation. The prologue of John's Gospel attributes the characteristics of creative wisdom to the Word, and the entirety of this gospel represents Christ as the wisdom of God. From the second century, the apologist St. Justin the Martyr, devoted his life to the task of harmonizing the wisdom of the world with those greater riches revealed in true faith.

Edwards argues that wisdom Christology can begin to show the interrelation between the expanding, interconnected, and self-organizing universe with all its creatures and the saving work of Jesus Christ. "Because *Wisdom* Christology is a creation theology, a cosmic theology, and a relational theology, it can be understood as an ecological theology" (Edwards 1995, 19). Creation theology is the act whereby God brings the entire substance of an entity into existence from a state of nothing (*ex nihilo*). God cared enough to bring about creation. A cosmic theology is as organic as humanity itself, continuously evolving, growing, and changing. Relationship is at the heart of the Christian faith, and humans are intended to be in relationships with one another. A relational theology speaks to our need for a personal relationship with Jesus Christ and all creation. Out of love God created the universe and invites us to be co-creators in the dynamic development through which we are united with God and all creation. We are made of the same dust created in the furnaces of stars billions of years ago, thus relating and interconnecting all creation in the web of life.

Wisdom was with God in the beginning, John the Evangelist tell us in his Gospel, and God created all things in her. Wisdom is thereby present in all creation, and Jesus is the embodied wisdom of God. Placing Jesus at the center of the evolutionary story, Edwards' ecological theology develops from a focus on God revealed in Jesus and the interdependence of God, humankind, and all other creation. Jesus reveals God's presence in the world, and his death resulted from a life lived in fidelity to the demands of love, the cross being the final expression of the unconditional character of Jesus' commitment to creation. Edwards interprets the risen Jesus as the expression of God's continuing presence. Crucified and risen, "Jesus is not only the universe reaching out to God but also God reaching out to embrace the universe" (Edwards 1995, 380).

In Proverbs, wisdom both transcends creation and yet is immanently within it as the instrument of its production (Prov 8:22-31). Wisdom participates in the activity of creation. Born before all else, *Sophia* wisdom was by God's side, "a master craftsperson, delighting him day after day, ever at play in his presence, at play everywhere on his Earth, delighting to be with the children of men[sic]" (Prov 8:30-31). Wisdom rejoices in the newly fashioned world and in the humans who live on it. Wisdom is the mother of all good things (Wis 7:12). Within her a spirit intelligent, holy, unique, manifold, subtle, mobile, incisive, unsullied, lucid, invulnerable, benevolent, shrewd, irresistible, beneficent, friendly to human beings, steadfast, dependable, unperturbed, almighty, all surveying, penetrating, pure and most subtle (Wis 7:22). Wisdom identifies with the divine spirit (Wis 1: 6-7; 9:17), who is utterly pure, intelligent, unique, loving the good and providing all things. Wisdom ascribes unlimited power to *Sophia*: "Although she is alone, she can do everything, and while remaining in herself, she renews the world; and generation after generation passing into holy souls, she makes them into God's friends and prophets (Wis 7:27).

The unity between *Sophia* and God assures continuous creation. She is "sharing God's life, for the Maker of All has always loved her. Indeed she shares the secrets of God's knowledge, and she chooses what he will do" (Wis 8:3-4). As her tradition unfolds, wisdom becomes more and more a female personification of God. Supporting Edwards, Elizabeth Johnson adds that "*Sophia's* activity is none other than the activity of God.... *Sophia* [with God] personifies divine reality...[and] is an expression of the most intense divine presence in the world" (Johnson 1992, 91-92). Ordering creation and pervading its every development, wisdom dominates the activity of creation; she is the giver of life (Prov 4:13) and works to establish justice in the world (Prov 8:32-36).

Edwards contends that the unity between *Sophia* wisdom and God heightens in chapter 9 of the Book of Wisdom wherein humans are saved by *Sophia* (saved by practicing wisdom) (v. 18) and in chapter 10 which retells Israel's salvation history and attributes to wisdom saving acts usually attributed to God. She is both Savior and Creator. To relate to her is, therefore, to relate to God. Acknowledging the unity of wisdom and God in their pervasive presence in all creation unifies God and all creation.

Drawing on the work of Raymond Brown, Edwards identifies three parallels in the New Testament between *Sophia* wisdom and Jesus Christ (Brown 1966, cxxii-cxxviii). One parallel is that *Sophia* and *Logos* are with God in the beginning. Since the Book of Sirach (written in Hebrew in 180 B.C. and translated into Greek in 132 B.C.) and the Wisdom of Solomon (written in Greek around 50 B.C.) were written shortly before the time of Jesus, first generation Christians viewed Jesus as the figure of wisdom. As recorded in Colossians 1, followers of the Way believed Jesus to be God incarnate and the firstborn of creation. In Edwards' view, the early Christian community saw Jesus as personified wisdom, the mediator, sustainer, and reconciler. The liturgical texts of these early communities used wisdom themes that now remain in Christian scripture. These early hymns reflect the community's faith in Jesus Christ risen from the dead and attribute to him a cosmic role in both creation and redemption. The doctrine of redemption presents Jesus as the representative of the whole universe. Among the early hymns, Col 1:15-20 contributes to an ecological theology by offering insight into the relationship between the resurrection of Jesus and the material universe:

> He is the image of the invisible God, the firstborn of all creation; for in him all things in heaven and on earth were created, things visible and invisible, whether thrones or dominions or rulers or powers- all things have been created through him and for him. He himself is before all things and in him all things hold together. He is the head of the body, the church; he is the beginning, the firstborn from the dead, so that he might come to have first place in everything. For in him all the fullness of God was pleased to dwell, and through him God was pleased to reconcile to himself all things, whether on earth or in heaven, by making peace through the blood of his cross (Col 1:15-20; see Prov 8:27-30).

This hymn reveals the belief that all of creation from the beginning is in Christ and that the risen Christ has the role of cosmic reconciliation. The Colossians hymn shows that the theology of redemption is not just for human sin and forgiveness but rather that the whole universe is caught up in the Christ event. Christ is the power at work in the whole cosmos. Edwards shows that this hymn moves from God's work of creation to God's redemption of all things in Christ (Edwards 1995, 81). Anne Clifford, Elisabeth Schüssler Fiorenza, and Raymond Brown illuminate our understanding when comparing the Colossians hymn with the Prologue of John's Gospel (John 1:1-5, 10-14). When professing faith in the divinity of Jesus, John uses the masculine *Logos* rather than the feminine *Sophia*. Edwards explains from Raymond Brown's commentary on the Prologue of John's Gospel that by drawing a "portrait of Jesus as divine revelation descended from above and become incarnate, the evangelist has capitalized on the identification of Jesus with personified *Wisdom*" (Edwards 1995, 42). Brown claims "it was thus through God's *word* that things came into being. To this idea of God's creative word, we may join the concept of divine *Wisdom*. Divine yet almost distinct from God, wisdom had a role in creation; wisdom was sent forth from the mouth of God and helps to save all people. In the Word of the prologue, we have a union of wisdom and God's word, a divine person uncreated and existing with the Father" (1966, 15-16). Scripture scholars set statements in Colossians about the Pre-existent One directly alongside those about the human Jesus and his death on the cross. Just as Wisdom is the reflection, glory, and mirror of God, so also is Christ (Edwards 1995, 80).

A second parallel between *Sophia* and *Logos* is their activity. Like the *Logos* in the Prologue of the Gospel of John, *Sophia* was portrayed as an active agent in creation. This identification of Jesus and *Sophia*, says Edwards, is the "bridge whereby the community which believed that God had raised up Jesus of Nazareth came to see this Jesus as the pre-existent one" (Edwards 1995, 33). "*Wisdom* Christology was the bridge to the theology of incarnation" (ibid.). *Sophia/Logos* is the One who was with God in the beginning, was an agent in creation, became truly human, died on the cross, rose from the dead, is exalted above all powers and is the source of universal reconciliation. In Proverbs, *Sophia* asserts that she is present with God and cooperates with God in the creation of all creatures: "I was beside God like a master worker" (Prov 8:27; Wis 7:22; 9:9). Through *Sophia*, God not only creates the universe but also sustains all things and relates to all creation. Identifying Jesus with *Sophia* includes *Sophia's* care for the *oikos*, the household of living creatures.

A third parallel between Jesus and wisdom comes from John's Gospel and the Book of Sirach. In his prologue, John attributes the characteristics of creative wisdom to the Word and throughout represents Christ as the wisdom of God. In the prologue the Logos (Word) became flesh and lived among us. Setting up a dwelling place among humans, wisdom received a command to dwell in Jacob and Israel where she sets up her tent as a resting place in the beloved city (Sir 24:8-11). Wisdom, God's agent in creation, eventually becomes flesh in Jesus Christ.

Much of the later tradition, including John's prologue, replaced *Sophia* with *Logos*. At the time of John's composition (100 A.D.), Greek androcentrism played a significant role in the dominance of the Greek *Logos* language and its philosophy. However, Wis 9:1-2 equates *logos* and *sophia* in the act of creation. Because Wisdom of Solomon was written by a Hellenistic Jew (50 B.C.), the Gospel *logos* makes an easy link with Hellenistic rather than Jewish philosophy. *Logos* was the more concrete concept to provide a bridge of communication between wisdom's Jewish monotheism and Greek androcentrism. Prior to John's gospel, Christian reflection easily associated both the historical Jesus and the risen Christ with *Sophia*. The concept of a personified wisdom as a literary device in Prov 14:1 was further developed in the post-exilic period when polytheism was no longer a threat to Judaism (*The Jerusalem Bible* 1966, 943g). In Prov 1:20-23, 3:16-19, and Wisdom 8-9, wisdom is represented as a person. Here wisdom herself reveals her origin (created before all other creatures (22-26), the active part she plays in the creation and the function she discharges among all in leading them to God (31, 35-36). Paul's first letter to the Corinthians, for example, claimed an identity between Jesus and divine wisdom: "For Jews demand signs and Greeks desire wisdom, but we proclaim Christ crucified, a stumbling block to Jews and foolishness to Gentiles, but to those who are called, both Jews and Greeks, Christ the power of God and the *wisdom of God*. For God's foolishness is wiser than human wisdom, and God's weakness is stronger than human strength… He is the source of your life in Christ Jesus, who became for us *wisdom from God*, and righteousness and sanctification and redemption, in order that, as it is written, Let the one who boasts, boast in the Lord" (1 Cor 1:22-24, 30-31). Wis 7:22-8:11 gives the impression that wisdom, an outpouring of God's glory, shares in the divine nature. The use of wisdom categories in Christian scripture allowed Christian communities to attribute cosmic significance to the crucified Christ. Therefore, it is necessary to retrieve *Sophia* wisdom, present for the Jews with God and an active agent in creation, to reestablish an understanding of her relationship with God and creation. Edwards states that "it was only through identifying Jesus as divine *Wisdom* that the theological insight could be reached that the *Wisdom* who is God's agent in creation is made flesh in Jesus of Nazareth" (Edwards 1995, 85). As Edwards contends, early Christians may have seen Jesus as the herald of wisdom and the wisdom prophet. Echoing *Sophia* wisdom's message, Jesus speaks as divine *Sophia* herself: "Come to me, all you that are weary and are carrying heavy burdens, and I will give you rest. Take my yoke upon you, and learn from me; for I am gentle and humble in heart, and you will find rest for your souls. For my yoke is easy, and my burden is light (Matt 11:28-30).[1] Edwards' understanding of the scripture reveals that *Sophia* wisdom is God's presence in the universe. *Sophia* wisdom is concerned with the interrelationship between humanity and the rest of creation. Wisdom was with God in the beginning and is dynamically present in all creation. Creation, then, is a reflection of the divine, and is intrinsically valuable because of its relationship with God. Divine wisdom becomes incarnate in Jesus of Nazareth.

TOWARD A NEW ECOLOGICAL THEOLOGY IN RESPONSE TO THE CLIMATE CRISIS

Through images, words, and experiences we come to a new awareness of the relationship between *Sophia* wisdom and the present climate crisis. How can we apply the insights of *Sophia* wisdom in a practical way to the problem of human-induced climate change? The poetry of Gerard Manley Hopkins offers a modern challenge to our understanding of the impact of climate change and environmental devastation. His *God's Grandeur* describes the natural world filled with God's presence, but also suffering at the hands of contemporary human life,

---

1   See also Wis 3:1-3; Sir 24:19, 51:27, 53.

transforming the landscape through industry and prioritization of the economic over the spiritual.

> The world is charged with the grandeur of God,
>   It will flame out, like shining from shook foil;
>   It gathers to a greatness, like the ooze of oil
> Crushed. Why do men then now not reck his rod?
> Generations have trod, have trod, have trod;
>   And all is seared with trade; bleared, smeared, with toil;
>   And wears man's smudge and shares man's smell:
>   the soil
> Is bare now, nor can foot feel, being shod.
> And for all this, nature is never spent;
>   There lives the dearest freshness deep down things;
> And though the last lights off the black west went
>   Oh, morning, at the brown brink eastward, springs-
> Because the Holy Ghost over the bent
>   World broods with warm breast and with ah!
>   bright wings (Hopkins 1877).

For Hopkins, however, there is always hope. Permeating nature is a deep "freshness" that testifies to the continual renewing of God's creation. Hopkins bemoans nature's fate at the hands of the Western industrialism that ignores the world as the gift of Gods' wisdom and finds troubling both the separation of humans from nature ("why do men then now not reck his rod?...the soil is bare now, nor can foot feel, being shod") and the desecration of God's nature by human technology ("Generations have trod, have trod, have trod; and all is seared with trade; bleared, smeared with toil and wears man's smudge and share's man's smell"). What ought a world "charged" with God's wisdom be? How can we see *Sophia* in every aspect of creation? Has our disregard marred the world with destruction and devastation? Our hope for the "bent world" does not lie in nature's own redemptive powers; rather, it rests on the Holy Spirit's "warm breast and ah! bright wings." Hopkins reminds us that "no matter how bad things get, there is hope, not because of human beings or even nature itself, but because the power of life and love that was at the beginning of creation is within us still as our source and our savior" (quoted in McFague 2008, 160). "Nature is never spent" and "there lives the dearest freshness deep down things" because of the sustaining power and love of God's wisdom.

Yet how do we put the hope claimed by Hopkins into action? How do we live *Sophia* wisdom in our daily lives? I suggest several approaches, each requiring an acceptance of the reality of human- induced climate change and a commitment to action, both personal and social. First, a more intense focus on the Christian doctrine of the Incarnation may help us to understand the material world in a more "sacramental" way than we currently do. Creation is not separate from God. Because *Sophia* wisdom is both transcendent to and immanent within creation, we may say that the divine is united with the material and that all therefore is holy, demanding our respect and love. A focus on the Incarnation might also challenge us to test our presuppositions about the nature of God in a more critical way and with more radical questions. Who is this God who our faith tells us offers God's self to us through the mediation of the most material, the most corporeal, often the least humane aspects of our human experience? Is the very crisis itself God's asking us profoundly to rethink the very nature of our relationship with God? We need only think of Noah, Moses and the burning bush speaking for God, and so few years ago, the persecution of the Jews.

Sophia wisdom was with God from the beginning of creation and in Jesus as the embodied wisdom of God. The divine unites with the material. We live in a sacramental universe, not just the seven moments named by the Catholic Church. In this understanding human personal spirituality as well as communal spirituality should recognize the interrelatedness of all creation. Human responsibility to preserve our Earth becomes apparent. The glory of God is revealed in each aspect of creation.

Second, unless we engage in a consciousness shift as well as a willingness to step out of our comfort zone, minimal changes to the present devastating direction of climate change will succeed. As I traveled across the country five years ago, particularly in the mid and northwest, I saw the devastation from global warming in the dried up waterfalls of Montana and the rapidly melting glaciers of Glacier Bay. The increased rate of calving in Alaska provides a visual of the pervasive climate crisis. There is a growing concern among climate scientists who believe that it is already too late to avoid significant damage to the atmosphere. However, we still have opportunities to save human lives, the lives of diverse species, and ecosystems that sustain our beautiful Earth. Motivation may come from the realization that we are inflicting harm on ourselves as well as our Earth, but motivation for change should also come from the teaching of *Sophia*: God is present in all creation.

Third, we are co-creators in the evolving, developing universe. As co-creators, we have an obligation to engage in deliberate, positive creativity. The Climate Crisis Coalition facilitates, participates in, and promotes local actions that raise awareness to the effects of climate change that is forced by human activities. Supporting the ecological efforts advocated by various organizations such as Green Peace, World Watch Institute, Friends of the Earth, and the Climate Crisis Coalition is essential to the restoration of God's creation. We must accept ownership of the problems we are forcing on the global climate. The actions we undertake today provide for a new direction to be set, away from the bleak picture of forced migration, extinction of species, destruction of the natural landscape, and increased health and food issues that are negatively affecting us. Embracing the wisdom of God revealed in all creation, churches and parishes, as spirituality centers among us, serve as a locus for education and support for action. We must educate and challenge our neighbors to make a commitment to sustainability. To fail Earth, we also fail humanity.

Interconnected physically with all creation, we share a mutual relationship with all creation. We are part of the ordered harmony of our planet. We need to learn *from* nature as well as *about* nature. We need to recognize the sacred in our midst. We must reinvent our view of the world from human-centered to *Sophia*-centered, yet our lack of imagination limits the efforts to arrest and reverse the climate crisis. Living as one with all creation must become a way of life for humans. We are caretakers of Earth that was gifted by the Creator.

We need to face up to the reality of human-forced climate change as one of the most serious moral crises of the twenty-first century. Humans must acknowledge that we must change our self-destructive behavior. We must undergo a moral conversion from egocentrism to theocentrism. Humans must relearn that we and all creation are reflections of God's wisdom. "We must desire to be what we truly are- made in the image of God, and thus able to live justly and sustainable on earth with all other creatures" (McFague 2008, 161). Grace helps us acknowledge our total dependence on God who is life and the giver of life. We are totally dependent on Earth and all its interlocking systems. Because we are part of the web of life, we must accept responsibility for Earth's well- being so also to be the reflection of God. Accepting this responsibility is an awesome task. We are supposed to be co-creators with God, and we must acknowledge that we are not functioning as co-creators when we are forcing changes in the global climate. We must become informed about the climate crisis and our actions that are forcing a rise in the temperature of Earth. As McFague urges, "[w]e must take care of the earth that is taking care of us" (McFague 2008, 167).

Fourth, if we need to rethink how we think about God, we also must revise how we think about culture. In *Ethics and Politics*, Aristotle makes clear that a culture's primary concerns are human fundamental needs for food, clothing, and shelter. To satisfy these needs, cultures build up a network of human activity in which the skills of individuals are employed in cooperation with nature. When these cultures lose their simplicity and operate only in complexity, these networks disappear and cooperation becomes abstract. Human community disappears. Distinct from culture, civilization is concerned primarily with the continual extension of material luxury, often at the cost of the health and happiness of those who pursue it. Civilizations and the capitalism that encourages and supports it have manufactured the artificial needs whose satisfaction can only be met by exploiting nature, not by cooperating with nature. How do we make our own choices? To what extent do we understand God's creation well enough to resist our civilization's demands? Do we allow *Sophia* wisdom to lead us to her Spirit's warm breast or do we simply want to get a more expensive pair of shoes?

A fifth action to enlighten human consciousness and to combat human-induced climate change is to attempt to transcend the dualistic categories that have dominated Western civilization almost since its beginning. Transcending dualisms that separate humans from other species and ecosystems might serve as a foundation on which the application of *Sophia* wisdom may rest. Even though Christian thinking seems to have ignored St. Paul, he was suggesting that the dualistic categories that dominated pre-Christian culture do not express reality after we become one in Christ: "there are no more distinctions between Jew and Greek, slave and free, male and female, but all of you are one" (Gal 3:28). Another way must be found to describe this wholeness, a wholeness that both preserves singularity and unites it. The major symbol of that wholeness is of course the Trinity itself in whose relational image and likeness we have been made. Among the Catholic theologians who have found a new way are Bede Griffiths and Ramon Pannikar, both of whom find Western dualisms useless, but borrow the discourse of the East to explore the relational nature of creation more closely. In *Universal Wisdom: The Scriptures of Hinduism, Buddhism, Taoism, Sikhism, Islam, Judaism* and *Christianity* (1994), Griffiths shows that only by going beyond the differences that dualistic thinking demands can we grasp not only our relational life with God, with our neighbors, and with our enemy, but also, in the case of the climate crisis, our relational life with what remains of God's creation. If we too easily pamper ourselves, we treat God's creation destructively as a foreign object. As the Triune God shares their lives, we too share our existence with creation, a creation which we must preserve since we are destroying ourselves along with it. Only our devotion to *Sophia* wisdom will save us from these destructive dualisms and help us heal the world. The divine is united with the material and all is made holy. God has embraced all creation. God the divine has come among humans and all creation in the being of Jesus Christ. God affirms the goodness of *all* creation and through Jesus' death and resurrection draws *all* creation back to God's self.

Many writers including Sallie McFague, Elizabeth Johnson, James Nash, and Larry Rasmussen, recognize that our lack of wisdom has put our planet in peril. They argue that humans must reestablish a new moral relationship with nature and must see ourselves as part of the web within the cosmos and life, so that we and our planet continue to flourish together. Earlier writers include Pierre Teilhard deChardin who wrote about the unfolding of the material cosmos from primordial particles to the development of life, the human species, and the noosphere and then to the Omega Point of supreme consciousness. Teilhard wrote "The body of Christ…constitutes a world that is natural and new, an organism that is animate and in motion, one in which we are all united, physically and biologically. The exclusive task of the world is the physical incorporation of the faithful in Christ who is of God" (1961, 81).

Karl E. Peters insists that the inter-relatedness of humanity and the rest of nature imply that we are all part of a single cosmic journey. He proposes a religious conceptualization of the human-Earth-God relationship in an age of science: "the earth is a child of God and we humans are the developing mind of the earth, with the power to shape its destiny, which includes our own destiny" provides another insightful approach to our ecological dilemma (2002, 137). Peters points to the use of birthing image in Scripture as a vivid metaphor of the activity of God's ongoing creation. Earth as the child of an evolving wise universe suggests "the evolution of the universe as a coherent single event with many subevents along the way" (138). The birth event undergoes random interactions, unpredictable events that influence the outcome of the birth. Peters' metaphor suggests the intrinsic value of all creation and the inter-relatedness of humans and the rest of nature. The child is Earth in its entirety with humans as its self-consciousness, its reflective consciousness. Earth in its entirety reflects God's Wisdom and has intrinsic value, just as an entire human child has intrinsic value. Although the wise Earth does much on its own, humans are able to use this child of God in a way that leads either to its deterioration or to its flourishing. Until recently, the self-consciousness of humans acted in a childish manner expecting all our needs to be met by the Earth-child. With the rise of global environmental problems, we, as the reflective part of Earth, realize that we are not wise. We act in a way that contributes to the destruction of the body (Earth). Humans now need to find ways to act in concert with creation's own wisdom. As an "Earth adult," we must reach out into the universe, to other children of God (142). As "[h]eirs of the sacred creative dance that has been going on for fourteen billion years, giving birth to the Earth child," we are part of the original promise of the future (143). If we accept ourselves as part of the consciousness of the planet, we can make moral choices that result in a flourishing for all. As Peters says, "In loving the Earth, we love the web of life in us and of which we are a part" (144).

The unfolding of the universe continues in the human person and in human community. It is also manifest in human interaction with the rest of creation. However, God's creative action is not something extrinsic to the cosmic process. As Edwards explains, "God has chosen to create in such a way that all possibilities are already contained within the original fireball" (1995, 38). The movement of self-transcendence at the core of cosmic processes only reaches its fulfillment in the "embrace between creator and creatures that is called grace" (29). The experience of grace—God's offering of self to humans—opens us to the mystery at the heart of human existence: *Sophia* wisdom dwells among us and we are forever interconnected.

CONCLUSION

Solomon recognized his need for wisdom, prayed for it, and received wisdom, valuing it above any other gift (Wis 7:7-8). God continues to give us the gift of creation that is ongoing. As personified wisdom, Jesus challenges us to live in harmony with a constantly changing world (Wis 16: 24-25). *Sophia* wisdom has made her home within us, and we must welcome her into our dwelling. Only in dialogue with science can Christian theology understand the nature of the relationship between humans and Earth as a unified sharing in God's creative action. By consciously engaging in the ongoing story of creation, we can know and experience God's presence in every aspect of the universe. As co-creators with God in this story, we must guarantee that future generations will have the opportunity to live in a world that is charged with the grandeur of God. We need to realize and accept the overwhelming evidence that our climate-disruptive actions are jeopardizing their opportunity because we lack the wisdom to respect our Christian and biological identities. Only by submitting to this wisdom can we radically change our destructive actions and live in God's presence with hope for a life-sustaining climate in the future.

SOURCES

Australian Bishops' Committee for Justice, Development and Peace. 1991. "Christians and Their Duty Towards Nature," In *"And God Saw That It Was Good,"Catholic Theology and the Environment*, ed. Drew Christiansen S.J. and Walter Grazer, 245-57. Washington, DC: United States Catholic Conference.

Barbour, Ian. 1997. *Religion and Science: Historical and Contemporary Issues.* San Francisco: HarperSanFranciseo.

Brown, Raymond E. 1966. *The Gospel According to John I-XII*. Garden City: New York: Doubleday.

Clifford, Anne M. 1991."Creation." In Systematic Theology: Roman Catholic Perspectives, vol. 1., ed. Francis Schüssler Fiorenza and John P. Galvin, 195-248. Minneapolis: Fortress Press.

"Climate Crisis." 2006. The Independent UK. Accessed from http://www. independent.co.uk/environment/climate-change, October 12, 2010.

Edwards, Denis. 1992. *Made from Stardust*. Victoria: Collins Dove Publishers.

———. 1995. *Jesus the Wisdom of God: An Ecological Theology*. Maryknoll: Orbis Books.

———. December 1999. "The Ecological Significance of God-language." *Theological Studies* vol. 60): 708-9.

———. 1999. *The God of Evolution*. New York: Paulist Press.

———. 2000. "Ecology and the Holy Spirit: The 'Already' and the 'Not Yet' of the Spirit in Creation." *Pacifica* 13 (June): 142-159.

———. (ed.). 2001. *Earth Revealing, Earth Healing*. Minnesota: Liturgical Press.

———. 2004. *Breath of Life: A Theology of the Creator Spirit*. New York: Orbis Books.

Fiorenza, Elisabeth Schüssler. 1992. *But She Said: Feminist Perspective of Biblical Interpretation*. Boston: Beacon Press.

Griffiths, Bede. 1994. *Universal Wisdom: The Scriptures of Hinduism, Buddhism, Taoism, Sikhism, Islam, Judaism and Christianity*. New York: Harper Collins.

Hawking, Stephen. 1988. *A Brief History of Time*. New York: Bantam Books.

Haught, John F. 1993. *The Promise of Nature: Ecology and Cosmic Purpose*. New York: Paulist Press.

———. 1995. *Science & Religion: From Conflict to Conversation*. New York: Paulist Press.

John Paul II, Pope. 1996. "The Ecological Crisis: A Common Responsibility: Peace with God the Creator, Peace With All of Creation." In *"And God Saw That It Was Good:" Catholic Theology and the Environment*, ed. Drew Christiansen S.J. and Walter Grazer, 215-22. Washington, DC: United States Catholic Conference.

Johnson, Elizabeth. 1992. *She Who Is: Mystery of God in Feminist Theological Perspective*. New York: Crossroads.

———. 1993. *Women, Earth, and Creator Spirit*. New York: Paulist Press.

———. 2001 "God's Beloved Creation," *America*, 16 April, 8-12.

McFague, Sallie. 2001. *Life Abundant: Rethinking Theology and Economy for a Planet in Peril*. Minneapolis: Fortress Press.

———. 2008. *A New Climate for Theology: God, the World and Global Warming*. Minneapolis: Augsburg Fortress.

Nash, James A. 1991. *Loving Nature Ecological Integrity & Christian Responsibility*. Nashville: Abingdon Press.

Peters, Karl E. 2002. *Dancing with the Sacred: Evolution, Ecology, and God.* Harrisburg, PA: Trinity Press International.

Rasmussen, Larry L. 1996. *Earth Community Earth Ethics.* Maryknoll: Orbis Books.

Raymo, Chet. 2008. *When God Is Gone Everything is Holy.* Notre Dame, IN: Soren Books.

Teilhard de Chardin S.J., Pierre. 1961. *Hymn of the Universe.* New York: Harper & Row.

The New Jerusalem Bible. 1985. New York: Doubleday.

United States Bishops. 1996. "Renewing the Earth," in *"And God Saw That It Was Good:" Catholic Theology and the Environment,* ed. Drew Christiansen S.J. and Walter Grazer, 223-43. Washington, DC: United States Catholic Conference.

CREATION & COVENANT IN THE
THEOLOGY OF THE APOSTLE PAUL

Annemarie S. Kidder

When Pope Benedict XVI declared 2008 the year of the Apostle Paul, people worldwide began reading Paul's letters with renewed interest. They also began noting that biblical commentators had been trying for some time to retrieve from the letters answers to contemporary problems, such as global warming, the ecological crisis, and world hunger, seemingly joining the efforts of scientists, humanitarians, and peace activists. In recent years, a new way of "hearing" Paul, a new hermeneutic that permits reading the ancient texts in light of contemporary issues has allowed for the making of fresh insights and new connections without the insistence on a concordance.

Does Paul offer answers to the current ecological crisis even though he never makes mention of ecology? Are Christians those in whom "the new creation" has begun by the death and resurrection of Jesus Christ, so that their minds and bodies by virtue of a "new covenant" have become intricately intertwined with the rest of creation? This essay explores Paul's theology of creation and covenant and its consequences for us regarding the well-being and care of God's creation. Reading Paul today through the new hermeneutical lens is not an option. Many biologists believe that we are in the midst of "the greatest mass extinction episode since the disappearance of the dinosaurs 65 millions years ago" (Tibet Natural Environment Conservation Network). This mass extinction of an estimated one million species of plant and animal life by 2050 is caused to a large degree by rising temperatures of Earth resulting from human-induced, outputs of carbon dioxide and other greenhouse gases, leading to global warming.

Changes in the climate cause natural habitats to disappear and many species will no longer have anywhere suitable to live. Currently, every twenty minutes one animal species goes extinct and more than 8,400 plant species and nearly 8,500 animal species around the globe are threatened with extinction. This irreversible loss of biodiversity, the fastest in human history over the past one hundred years, has an impact on the survival of remaining animal and plant species; but it also affects humans: we depend on them for food, clean air and water, fertile soil, and medicine. Apart from utility, there is also the question of moral responsibility toward other living creatures on earth. Among the Pauline exegetes explored on the question of human responsibility for plant and animal life on earth in light of global warming are N.T. Wright, Brendan Byrne, S.J, Joseph Fitzmyer, S.J., Udo Schnelle, and James D. G. Dunn. The main New Testament passage examined is from chapter 8 of Paul's letter to the Romans, vv. 18-27, along with passing reference to 1 Corinthians 15, 2 Corinthians 5, and the potentially Pauline creation hymn of Colossians 1.

PAUL & HIS "MASTERPIECE"

The Apostle Paul has been variously called the earliest interpreter, promulgator, and shaper of Christian thought, and any discussion of his theology will rise and fall on an examination of the epistle to the Romans. Considered by most biblical commentators Paul's masterpiece, the epistle is neither epistle in the strictest sense of the word, nor is it a systematic theology or the summary statement or testament of Paul's lifework in retrospect. It is not the typical epistle addressed to a community he helped found, whose problems or conflicts were familiar to him first hand, or whose members he knew. Despite detailed personal greetings (Rom 16:1-15), the letter appears at first as a general letter of instruction, a doctrinal summary of the Christian faith aimed at a broader audience. As one of the last extant literary pieces of a mature Paul, Romans is written by Paul about a decade before his death between the middle to late 50s of the first century from Corinth or nearby. It is the longest of Paul's letters, the longest letter in the New Testament, and it dazzles for its intellectual, theological, and spiritual depth. Grasping Paul's theology will certainly ride on a proper reading of Romans, as does understanding the reasons the letter was written, its audience, and its general themes.

The letter is addressed to a community of Christians in the capital of the Roman Empire that Paul himself had not founded nor visited. Paul gages this community as composed of both Jewish and Gentile Christians with a Gentile majority and a Jewish minority. The latter probably held important traditional, close ties to the Jerusalem church. This means that Paul's letter could have served as a means of introducing himself to the Roman community prior to his coming, to solicit financial support from them for his intended trip to Spain, and to request their prayers on behalf of the Jerusalem church for whom he has collected an offering which he intends to deliver soon but fears it might be rejected. It is likely that Paul had recognized distinct problems in the Roman community and potential divisions regarding him and his understanding of the gospel in the way he had promoted it during his evangelistic mission in the eastern Mediterranean. This understanding had created distrust of him in the Jerusalem church for its message of justification without the deeds of the law; and the distrust had rippled over into the Roman church with its close ties to Jerusalem. Still, the letter is primarily addressed to the Gentile Christians of Rome to explain to them Israel's problem with the covenant and how Jews and Gentiles fit into the new economy of salvation. To sum up, Paul's letter is written for a distinct audience and "thoroughly conditioned by the situation in which it was written." It is neither "a timeless elaboration of Pauline theology" (Schnelle 2005, 308), nor a "dogmatic treatise or a dialogue with Jews who do not accept his gospel; it is rather a didactic and hortatory letter, intended for discussion by the Jewish and Gentile Christians of Rome, for their understanding and their conduct" (Fitzmyer 1993, 79).

Given the presumed audience of Gentiles and Jews in the capital of the Empire, the themes of the letter become more pronounced. Aware of the division existing between Jews and Gentiles, between adherents of the law and those who see no need of adhering, Paul appeals to their common heritage and equality before God as regards their sinfulness and the possibility of salvation: Just as "all" are "under the power of sin" (Rom 3:9), so all "who call upon the name of the Lord shall be saved" (10:13). Just as Abraham had been considered the father of the Israelites and their faith, now Paul regards him as "the father of all who believe" (4:11), which is Jews and Gentiles alike. Paul also seeks to help both groups understand God's economy of salvation: The Jews were first in receiving the promise of the gospel, not on account of their ethnic heritage but on account of their religious faith and the steadfastness of God's covenant promises to them; the Gentiles were second as they were now being grafted into Israel's stock, thus becoming rightful heirs of God's earlier promises made to Israel and equal members of Israel, the people of God, through Jesus Christ.

The overarching theme of Paul's letter to the Romans is to explain to both Jews and Gentiles how such equality before God had been wrought and what the decisive event or mechanism was that could bring about their "righteousness" in the eyes of God. What could warrant their rightful participation as agents in the salvific promises made to God's people for the sake of the world? Paul sums up the letter's theme by the little and frequently misunderstood word "righteousness." The event and mechanism by which people, both Jews and Gentiles, were considered righteous before God and could serve God's righteousness on earth, was the death and resurrection of Jesus Christ, Israel's long-awaited Messiah. Through him, God would restore fallen creation and sinful humanity and set to right what was wrong with the world. Through him justice and righteousness would be restored to the earth and the wages of sin, namely death, would be ultimately destroyed. Together with those whom God had chosen as his instruments, made "righteous" before God by grace through faith, God would accomplish the redemption of the world, so that "cosmic restorative justice" (Wright 2002, 400) would prevail.

The key motif of Romans is God's "righteousness," "*dikaiosis*." The term has been commonly understood to mean that a just God, who will not and cannot tolerate sin and corruption, will consider believers in Christ not as sinful but righteous. Despite their sinfulness, they stand before God as justified, rather than accursed and condemned. Paul says in Romans that God in Christ Jesus "justified" human beings "by his blood" (Rom 3:25). Believers stand before God's tribunal acquitted and innocent, with the judgment not based on their proper observance of Mosaic Law but the fact that Christ died for their sins and paid the wages for them by his death (Rom 6:23). God had shown his righteousness by putting Jesus Christ "forward as a sacrifice of atonement by his blood, effective through faith" (Rom 3:25). This framework has dominated the interpretation of the word "righteousness" for the past five hundred years or so. Commentators had been largely governed in their interpretations by Martin Luther's view of Romans as a letter on the doctrine of "justification by faith." Luther had asked how to find and stand before a righteous God. The answer came with a powerful conversion experience, which showed him that no amount of good works could make a person righteous in the eyes of God; rather, one's righteousness was a "free gift of God" (Rom 6:23). The righteousness needed that allowed one to stand before a righteous God could come about only when one trusted God's action in Christ "by faith alone" through grace, not works. In recent years, commentators reading the letter as a whole with its implied narrative and argument have questioned Luther's interpretive focus and of those who followed in Luther's footsteps. The letter to the Romans is less about individual salvation and the question "How can I be saved?" It is less about the so-called imputed righteousness that makes a person able to stand before God as justified. Rather, it is about God's covenant faithfulness to his people, the Jews. In an unprecedented and cosmic intervention on God's part, the overall plan of salvation for God's chosen people had come about through the death and resurrection of Israel's long-expected Messiah, Jesus Christ. With this action of God, a new chosen people of God is brought into being, no longer defined by blood lineage but by faith in Christ. Both Jew and Gentile who believe in Christ as Messiah are now chosen for God's plan of salvation. They are charged with acting a certain way and bearing the responsibilities that come with election. Thus, the principles of "theology and ethics" and "salvation and politics" (Wright 2002, 404) were not as far removed from Paul's mind as they were, for example, during the periods of the Reformation and the Enlightenment.

The new people of God whom God had chosen to bring about God's righteousness on earth have been justified before God through Christ. Such justification begs the question: justified for what purpose? God's people Israel had been chosen to be a blessing to all nations, to the entire world. But they had failed in the task by taking a self-serving, self-gratifying pride in their superior position before other nations. Their election had come about not by their own doing, nor the righteousness of Abraham their forefather. It had been granted to them as a gift from God for the world's salvation. Due to sin and undue pride of place, the people of God forgot the responsibility that came with election: election not for their sake but to reveal God's glory and to effect the salvation and redemption of the world. Thus, it would be remiss to say that people have been chosen in Christ through no merit of theirs but

by grace alone through faith without asking the question: chosen for
what purpose, elected to what end?

THE MOTIF OF CREATION & COVENANT
IN THE OLD TESTAMENT

Even a cursory reading of Romans shows that Paul is not primarily
concerned with people's righteousness before God as justified by faith,
not works. It is besides the point to wonder how for so many centuries commentators could have been misreading Paul with such consistency. More important is that the misreading has been balanced and
rectified. In recent years, Pauline commentators in the tradition of the
new hermeneutic have begun to explore the socio-historical influences
on Paul's thought. They asked: Which intellectual and theological
currents shaped Paul's world view? What was his self-understanding?
From which worldviews did he operate and to which did he speak?
Asking these questions will mean reading Paul not piecemeal by highlighting some personally favorite verses and pithy sayings. Rather
it will mean reading the text as a theological narrative that discerns
Paul's interpretation of Torah as a diaspora Jew, his status as a Roman
citizen and member of the world's most powerful empire, his Hellenistic cultural background, and his dramatic encounter with the risen
Christ that had turned him into a powerful Christian witness. All four
worlds of Paul appear in his letters as those shaping his thought. The
two decisive ones, however, are his theological Jewish heritage and his
Christian experience, so that he is prompted to seek to correlate the
two.

Paul's novel theological interpretations in his epistles make for a
veritable explosion of metaphor and imagery. He enlists Jewish theological concepts and metaphors that would have been well understood
among his audience, even among Gentiles, in order to re-fill them with
new meaning. A familiar (Jewish) metaphor is picked up by him and
its core re-charged with a Christ focus. Traditional Jewish metaphors,
such as Adam as the icon of human sinfulness and the fall, the Israelites' Exodus, and the messianic typologies of Abraham and Moses are
reinterpreted in conversation and dialogue with the present new age.
The narratives about the covenant God made with Abraham and the
giving of the law with Moses are retold in light of the Christ event.
While Adam had broken the covenant with God, a new covenant was

established through Abraham as the father of the covenant people of Israel, fortified through Moses and the giving of the law, and brought to its ultimate fruition and fulfillment in Christ. While Adam brought about the dissolution of the created order, the new Adam, Christ, brings about a reversal and creation's restoration. While death was the result of the sin of Adam who is of the earth, eternal life springs from the new Adam who is from heaven. Paul freely mixes and matches images of contrast and comparison taken from the Jewish creation myth and the patriarchal narratives by employing simile and allegory, parallel structure and dialectical thought, all in an effort to show the continuity of the covenant promises that God had given Israel and their climactic fulfillment in Christ. Of several images and theological concepts, two stand out in particular for the frequency with which Paul employs them in framing his argument in Romans: the Jewish concepts of creation and covenant, an idea that "remains at the heart of Judaism...and was always central for Paul" (Wright 2005, 21).

The Jewish idea of creation and covenant inherent in Paul's theology comes quickly to the fore when one reads Romans with the help of a reference Bible. Paul repeatedly quotes from the Psalms, Deuteronomy, and the book of Isaiah, particularly Isaiah 40-55, but not in order to proof-text the claim that Jesus is the long-awaited Messiah, "but in order to reground the controlling narrative, the historical story, of God, the world, humankind and Israel" (Wright 2005, 25). This underlying narrative, which unfolds in the book of Genesis, says "that God has called Abraham and his family to undo the sin of Adam, even though Abraham and his family are themselves part of the problem as well as the bearers of the solution" (ibid. 23). The consequences of Adam's sin are despoiled creation. Undoing Adam's sin through the covenant that God makes with Abraham and his descendants is aimed at creation's restoration.

Four Old Testament references shall suffice to illustrate the integral connection in Jewish thought between creation and covenant. The first is Psalm 19, which divides into two distinct and nearly equal movements and is a paean—a praise hymn or poem to God. The Psalm's first six verses celebrate the fact that creation praises God and is able to powerfully declare God's glory even without human language or speech. The day and the night declare the knowledge of the Lord's glory, and their voices and words, though inaudible, go out to the end of the world. The sun, likewise, rises "like a strong man" running "its

course with joy" and "nothing is hid from its heat" (vv .5-6). The second half of the praise poem (vv. 7-14) celebrates Torah, the law of the
Lord, which does in human life what the sun does in creation: by its
light and power it searches out God's presence in the deepest crevices
of the human heart, bringing joy to the heart and "enlightening the
eyes" (v. 8). Thus, the Torah is understood as Israel's covenant charter
by which the Lord establishes the people of Israel and uniquely binds
them to himself so that they may be a living witness to the Lord's glory
and power, as is the rest of creation.

The second reference connecting creation and covenant is Psalm 33.
Here, the Israelites are first asked to give praise for God's law as "the
word of the Lord" (v. 4): "Sing to him a new song; play skillfully on
the strings, with loud shouts. For the word of the Lord is upright, and
all his work is done in faithfulness." Then God is praised for speaking
creation into being: "By the word of the Lord the heavens were made
and all their host by the breath of his mouth. He gathered the waters
of the sea as in a bottle; he put the deeps in storehouses" (vv. 6-7).
Consequently, "all the inhabitants of the world" and the entire "earth"
(v. 8) are to stand in awe of the Lord and his word. And now the poem
switches back as if to ask who above all else and above all other nations has received these commands and the "counsel" of the Lord that
"stands forever" (v. 11a). It is "the nation whose God is the Lord, the
people whom he has chosen as his heritage" (v. 12).

A third reference joins creation and covenant in terms of the land,
based on Deuteronomy 27-30. Before the Israelites are allowed to
enter the Promised Land under Joshua's leadership, God asks them
through Moses whether they would accept the terms of God's covenant made with their forefather Abraham: "See, I have set before you
today life and prosperity, death and adversity. If you obey the commandments of the LORD your God that I am commanding you today,
by loving the LORD your God, walking in his ways, and observing his
commandments, decrees, and ordinances, then you shall live and become numerous, and the LORD your God will bless you in the land
that you are entering to possess. But if your heart turns away and you
do not hear, but are led astray to bow down to other gods and serve
them, I declare to you today that you shall perish; you shall not live
long in the land that you are crossing the Jordan to enter and possess"
(Deut 30:15-18). If Israel obeys the Lord's voice, the Promised Land
will be yielding much return: "Blessed shall be the fruit of your womb,

the fruit of your ground, and the fruit of your livestock.... Blessed shall be your basket and your kneading bowl" (Deut 28:4-5). However, if Israel disobeys, the land will turn against them and ultimately be driving them into exile: "Cursed shall be your basket and your kneading bowl. Cursed shall be the fruit of your womb, the fruit of your ground, the increase of your cattle and the issue of your flock.... The LORD will make the pestilence cling to you until it has consumed you off the land that you are entering to possess. The LORD will afflict you with consumption, fever, inflammation, with fiery heat and drought, and with blight and mildew; they shall pursue you until you perish. The sky over your head shall be bronze, and the earth under you iron. The LORD will change the rain of your land into powder, and only dust shall come down upon you from the sky until you are destroyed" (Deut 28:17-24). When the Israelites violate the covenant, creation itself will turn against them and the land will deny them their existence.

A fourth instance connecting creation and covenant comes from Isaiah, chapters 40-55. This portion of the prophetic book, often called Second Isaiah, along with chapters 56-66, is set against the historical backdrop of Israel's late time in Babylonian exile, a time of hopelessness and defeat. Chapters 40-55 chronicle a drama of creation, history, and redemption with God as its protagonist. More than any other prophetic book, Second Isaiah praises God as creator; the verb "to create" appears sixteen times in chapters 40-55, compared to only one time in Isaiah 1-39 and only eleven times in all of Genesis. The oracles in the book promise God's creative powers becoming active: Israel's status as God's chosen people would be restored in the sight of all nations, the covenant promises be renewed, and the people be returning to their homeland, signaling the establishment of God's universal rule of righteousness and peace (as also indicated in the Servant passages of the book): "Listen to me, you that pursue righteousness, you that seek the Lord. Look to the rock from which you were hewn, and to the quarry from which you were dug. Look to Abraham your father and to Sarah who bore you; for he was but one when I called him, but I blessed him and made him many. For the LORD will comfort Zion; he will comfort all her waste places, and will make her wilderness like Eden, her desert like the garden of the LORD; joy and gladness will be found in her, thanksgiving and the voice of song" (Isaiah 51:1-3).

The God of creation is the God of the covenant, and vice versa: "You have forgotten the LORD, your Maker, who stretched out the heavens and laid the foundations of the earth. You fear continually all day long because of the fury of the oppressor, who is bent on destruction. But where is the fury of the oppressor? The oppressed shall speedily be released; they shall not die and go down to the Pit, nor shall they lack bread. For I am the LORD your God, who stirs up the sea so that its waves roar—the LORD of hosts is his name. I have put my words in your mouth, and hidden you in the shadow of my hand, stretching out the heavens and laying the foundations of the earth, and saying to Zion, 'You are my people'" (Isaiah 52:13-16).

To sum up the relationship between creation and covenant in Jewish thought, one can say the following: First, the covenant that God made with the people of Israel "is there to solve the problems with creation. God called Abraham to solve the problem of evil, the problem of Adam, the problem of the world" (Wright 2005, 24). Through Israel, God will solve the problems of the world, thus bringing about righteousness, salvation, and peace, even if it is yet unclear how this is to come about. Secondly, creation is summoned to solve the problems with the covenant. When the covenant promises are threatening to come unraveled, the people cry out to God as creator and Israel finds itself returning to Genesis 1 and to the vision of the Promised Land. They "trust that YHWH will do again what, as creator, he has the power and the right to do, and what as the covenant God he has the responsibility to do, namely to establish justice in the world and, more especially, to vindicate his people when they cry to him for help" (Wright 2005, 24). The covenant was established to right what had gone wrong with creation. When the covenant goes wrong, creation itself is remembered and invoked. Through the mnemonic tool of invoking the creator God, it is as if creation jumps into action to restore the broken covenantal relationship between the people and God. Certainly, the Israelites are keenly aware that something had gone terribly wrong with creation and with the covenant, not only during their time of exile but also after they returned to the land and rebuilt the destroyed temple on a much smaller scale. It is during this so-called period of Second Temple Judaism, and against this backdrop, that we must hear the words spoken by the Apostle Paul.

PAUL & THE THEOLOGY OF
CREATION & COVENANT

We can now approach Paul's writings to see how he enlists the Jewish concepts of creation and covenant in his formulation of the Christ event. First, though, it might prove helpful to ask how a first-century Jew would interpret the breaking into human history of God become flesh, the God who died as a convicted criminal at the hands of the Romans and rose victoriously from the dead to draw all people unto himself. How to interpret this mind-boggling event of a few decades earlier? How to explain that the God of the universe submitted to the frailty of human existence so people can grasp this event's bearings upon them, world history, the cosmos?

One of the most ingenuous ideas of Paul is to view Christ as the one in whom creation and covenant meet. God in Christ is both the creator of the world and its redeemer, the new creation and the fulfilled covenantal promise. He is both the word and counsel of the divine covenant's precepts as well as its human representative who ably and perfectly keeps it. Christ is the ground of creation's being and the aim toward which it moves. Three scriptures passages can illustrate Paul's argument, which will set the stage for Romans 8. The first is the so-called Christ hymn of Colossians 1:15-20.[1] The similarity with Psalm 19 is striking, including the hymn's division into two halves. The first portion (Col 1:15-17) views Christ as "the firstborn of all creation" in whom and through whom and for whom "all things in heaven and on earth were created" and in whom "all things hold together." Christ is the *creator* God, the ground of all being and its origin. The second portion (vv.18-20) views Christ as "the firstborn from the dead," implying his suffering on the cross and his redemptive sacrifice, which gives him "first place in everything," including the work of the world's and creation's redemption. Now, Christ is also the redeemer and *covenant* God, through whom God reconciles the world, reconciling "to himself all things, whether on earth or in heaven, by making peace through the blood of his cross" (v. 20).

Another passage illustrating Paul's view of creation and covenant coming together in Christ is 2 Corinthians 5. Those in Christ are able to know him not "from a human point of view" but in a new

---

1   Even if Colossians is not generally regarded Pauline, one could reasonably argue that it contains select pieces of Paul's writings.

way. Through the eyes of faith, they are able to see all things as "a new creation," where "everything old has passed away; see everything has become new!" This new seeing of creation entails foreseeing its redemption, the newness that can be anticipated and has already begun. However, where did it begin? It began in those who have been reconciled to God through Christ and who, consequently, are carrying out "the ministry of reconciliation: for "in Christ, God was reconciling the world to himself, not counting their trespasses against them, and entrusting the message of reconciliation to us" (2 Cor 5:19). Thus, believers are his "ambassadors" who are entreating others to "be reconciled to God" in order that all "might become the righteousness of God" (vv. 20-21).

The third passage relates to Paul's view of the created body compared to the resurrected body. How is this earthly human body going to look when fully redeemed? How is the earthly creation going to look in its redeemed and transformed state? Paul answers the question in 1 Corinthians 15 by evoking Genesis 3: "[S]ince death came through a human being," Adam, "the resurrection of the dead has also come through a human being; for as all die in Adam, so all will be made alive in Christ" (1 Cor 15:21-22). The old creation will be replaced by the new; the former mortality will be transfigured into immortality. The heavenly transformed body will, in Paul's view, differ from the earthly one, just as there are differences between the bodies of humans, animals, birds, fish, and the stars—such as sun and moon. The resurrected body will be a new creation, no longer steeped in its own gradual decay, dishonor, weakness, and physicality; it will no longer be shaped by the earth or created from dust as "a man of dust," but will be transfigured "as is the man of heaven." The result is the creation of a new type of human being made in the image of God but, more specifically, fully transformed now into the image of the risen Christ. For Paul, this transformation process has already been set in motion and its fulfillment is certain because God "gives us the victory through our Lord Jesus Christ" over the "sting of death" and "the power of sin" that the law makes so plain to us (vv. 56-57).

In all these references, Paul interprets Christ as the climactic coming together of creation and covenant. He does so most pointedly in the letter to the Romans and more specifically in Rom 8:18-27 with its reference to all creation groaning as it participates in the process of transformation and redemption. How are we to understand this "apparently singular appearance of a passage dealing with human interaction with 'creation' in the flow of Romans" and "its apparent oddity" (Byrne 2000, 194)? What is the distinctive relationship of humans and creation, and what their interaction with each other?

THE GROANING OF ALL CREATION

For Paul, the covenant that God made with Abraham and the people of Israel to rectify the world's wrong is fulfilled in Christ. This covenant is now extended to include both Jews and Gentiles, all those who believe in Christ, are "in Christ," and have received new "life in Christ Jesus." Believers, though still in this earthly life with their mortal bodies, are already united with Christ "in a death like his," are "baptized into his death," and have "crucified" their "old self." Since "The death he died, he died to sin, once for all; but the life he lives, he lives to God," it is only reasonable to say that "you also must consider yourselves dead to sin and alive to God in Christ Jesus" (Rom 6:3-11). Just as formerly you, Jews and Gentiles alike, presented your human bodies and their "members as slaves to impurity and to greater and greater iniquity, so now present your members as slaves to righteousness for sanctification" (v. 19). Paul acknowledges the existing tension: though believers "have been freed from sin and [are] enslaved to God" in the process of "sanctification," whose end is "eternal life" (v. 22), they, like Paul himself, continue to be tempted by "the law of sin that dwells in my members" (7:23). The temporal remedy in this struggle is to surrender to the Spirit repeatedly "since the Spirit of God dwells in you": "If the Spirit of him who raised Jesus from the dead dwells in you, he who raised Christ from the dead will give life to your mortal bodies also through his Spirit that dwells in you" (8:9,11). Paul acknowledges that sanctification involves suffering. This suffering and travail is eased by the Spirit, so that whenever we cry out "Abba! Father!" the Spirit assures us of our "adoption" as "children of God," "heirs of God and joint heirs of Christ" (8:15-17). The tension is real; we "who have the first fruits of the Spirit, groan inwardly while we wait for adoption, the redemption of our bodies" (8:23). The tension and pain are also eased by the anticipated hope of glory and the fact that it is shared by the rest of creation: "We know that the whole creation has been groaning in labor pains until now" (8:22). God's children are not alone in their labor pangs: creation participates in their pain, and, hence is easing their lot.

Creation co-suffers and shares in our travail toward redemption. Such sharing also brings a new responsibility, that of easing creation's pain, in turn. The paradox of suffering is its redemptive value. The more aware we are of the suffering of others, the more willing we may be to endure our own. And, the more aware we are of others' pain, the more motivated we are to practice compassion, reduce pain, and offer healing. As Matthew Fox insists, compassion makes us "vulnerable to the struggle for the liberation of all creation" and "is the working out of our interconnectedness" in praxis (1991, 36). Thus, creation eases our pain because it participates in ours, while eliciting in us compassion toward it, resulting in acts of mercy.

Ideally, compassion results in acts of mercy, but acts of mercy can only be carried out by those in a position to act. Paul says that human beings are in that position, more especially those in whom the redemptive process has begun. Creation looks with "eager longing for the revealing of the children of God" (Rom 8:19). Paul's vision personifies creation as being full of wistful longing. In a vivid imagery, James D. G. Dunn compares creation's posture to a fan club of supporters who are "craning their necks forward to see whether their favorite son has crossed the finish line and their common celebration is assured" (1988, 487). Creation is craning its neck to see whether the children of God are so filled with the "new creation" that this invites for a joint celebration. Here we have the workings of Genesis 1-3 in reverse: God had given the earth to Adam so as to exercise "dominion"—not for a self-serving purpose leading to its ruthless exploitation—but so both humans and creation, each in their own way, can praise their creator and point to his glory. With Adam's sin, however, creation had become "subjected to futility, not of its own will but by the will of the one who subjected it" (Rom 8:20), namely fallen human beings. Humans had appropriated the place reserved for God, an act known in scripture as idolatry. Through their idolatrous behavior and sin's ramifications of evil, creation was fast losing the ability to point to God's glory. By idolatry and the excesses of greed, sloth, and power, humans had "subjected" creation to an existence without a purpose. They had disfigured creation and exploited the earth, thereby robbing it of the capability to carry out its God-given role: praise of the creator. With the reversal of this disastrous development by the Christ event, humans can now exercise a new type of dominion over creation: not one that exploits but one that redeems and allows creation to become what it is meant to be—praising God. Creation itself is hoping that it "will be set free from its bondage to decay and will obtain the freedom of the glory of the children of God" (8:21). Creation longs for what humans in Christ are promised and it "waits with eager longing for the revealing of the children of God" (8:19). As co-redeemers, humans have been empowered to engage with Christ, alive in them by the Spirit, in God's redemptive cosmic work. No longer subjected to the same death, decay, and futility that characterize creation, humans of the new creation are privileged to cooperate with the Spirit of God, indwelling them in Christ as the new creation's "first fruits." They can bring God's righteousness to bear. They can work in the world as ambassadors of Christ, as reconcilers and peacemakers who reveal God's righteousness made manifest in them and through them.

Paul's idea that humans and the rest of creation share a common fate is not new. Both share the same creator God, both have the same responsibilities of pointing to and praising God, both are created for God's pleasure and delight. What is new is that Paul interprets their common destiny in terms of salvation. Both will be set free from the "bondage to decay," both will obtain "the freedom" of displaying God's glory, and both are in the process of being fully birthed under the present "labor pains."

For Paul, there is a common end point and final climax toward which all of history, that of humans and creation, is headed. History is not an endless series of repetitions and circular movements; history has an aim, a purpose, a direction. This directional aim can be explained as originating with Jewish apocalyptic thought, which maintained that a climactic, cosmic transition would occur from this created order to the next, from this age to the age to come in which God's final purpose would be realized and made complete. Paul enlists such apocalyptic imagery while radically reinterpreting its timeline: the age to come is already here and has come to pass with the Christ event. The powers of evil and death have been defeated. Victory for the entire cosmos has been achieved through the ultimate sacrifice, the death of the God-man, the Messiah, and the resurrection of the one in whose body, representative of the earth and all creation, death is overcome and the new creation has begun. Paul's apocalyptic thought is that of an "inaugurated eschatology," as N. T. Wright calls it (2005, 57), one in which the new age, the *eschaton*, has already occurred in the present moment, even if its visible manifestations are lagging behind as a "not yet": the new age will be realized fully only in the future. Or, to cite James Dunn, the two "ages already overlap" and "believers can be confident in the certainty (and imminence) of the eschatological climax because the work of the eschatological liberation and renewal has already begun" (1988, 487). With that, the tension exists not so much between the present and the future age since the future has already been brought forward into the present. Rather, the tension exists between the believer living in the flesh and living in the spirit, dying to the old self in baptism and feeding its selfish ambitions, clinging to the old creation when the new is already at work. This is the tension that causes such suffering, says Paul, making us groan in labor pains together with the rest of creation.

In Paul's view, the future expectation of the full redemption of people and creation is brought forward into the present. That is why he can say "that the sufferings of this present time are not worth comparing with the glory about to be revealed to us" (Rom 8:18) or "in us"— as preferred by Fitzmyer, Wright, Dunn, and Byrne. The foretaste of glory *in us* presently gives us hope for the complete realization and manifestation of the new creation in the distant future. And, because believers have this hope, they can wait for the final fulfillment "with patience" (8:25). All of creation is already caught up in history's progressive movement of God's salvific plan, and God in Christ has made sure that this movement is an ultimate success. All creation is moving toward the final climax in which the righteousness and justice of God will become fully revealed. Given this prospect, believers can live and lean into hope. Already justified before God, they no longer need return to the slavery and bondage of sinful human nature and the law's accusations. Already freed unto the spiritual freedom of the children of God, they have died with Christ and been raised with him into the Spirit's rule. Though their bodies remain subject to temptation, with wills tested and minds riddled by ambiguity, believers carry in them the advance pledge of the Spirit's witness that testifies to them that the future glory is underway.

For Paul, the future glorification of believers and all creation is born by prayer. We recognize "in our weakness" and in the face of concrete evidence that "we do not know how to pray as we ought" (Rom 8:26.). But here again, the very Spirit that makes us "instruments of righteousness" (6:13) and gives us "minds [set] on the things of the Spirit" (8:5) "intercedes" for us "with sighs too deep for words" (8:26). The

Spirit of God "intercedes for the saints" (8:27) to help them along in the full realization of the new creation. The future restoration of creation and the earth, its redemption from the ravages of sin, its salvation from human greed and exploitation rests with no one but God in the power of the Spirit. This is not the work of human ingenuity and a consolidated intervention on the earth's behalf. This is God's work. Certainly, the efforts of environmentalists and all those working for ecojustice are laudable and much needed.

However, from a Christian perspective derived from Paul, the salvation of earth is not a materialist endeavor aimed at advancing its health in hopes of ushering in more speedily the rule of God's righteousness and justice. Earth's health does not result, as in a Marxist perception, "from the necessary working out of socio-economic forces towards a determined future" (Bryne 1986, 171). Rather, Paul's perspective is that the manifestation of the new creation on Earth will hinge on the extent to which believers are made aware, in prayer, of Christ's work in their own bodies. Do they see Christ at work in them as "the hope of glory" (Col 1:27), the foretaste of a future glorification of both humans and creation, so they "'live out' his gift of righteousness" and "allow his saving justice to 'be fulfilled in them' through the Spirit" (Byrne 1986, 171)? Do they recognize that their own sanctification is invariably bound up with that of the earth since both have the same creator? Do they acknowledge their own impotence in light of the visible brokenness of creation, so they gladly surrender to God for the task of undoing it? Do they fathom the cosmic significance of the covenant that God was pleased to make with Israel for the salvation of all creation and its climactic fulfillment in the death and resurrection of Christ, so that now the new creation stands as inaugurated even while awaiting its final glory? Certainly, if this is the cosmic picture that Romans 8 reveals, Paul offers here an "inaugurated-eschatological ethic" (Wright 2005, 151), one that begins with our keen awareness of who lives "in us," in this small part of creation, our body, over which we alone have been given advance responsibility and by which we can serve as "instruments of righteousness" in the world and on creation's behalf.

CONCLUSION

To sum up, Paul's view of creation and covenant governs his theological perspective. For him, the crown of creation is the human Adam who, due to the fall, puts creation in jeopardy and destines it to corruption, futility, and death. The covenant is established as a way of rectifying the problem of creation. When the covenant breaks down because God's chosen people become part of the problem instead of the solution, creation itself is invoked. The images of Eden and the prospect of the Promised Land serve as a reminder of God's goodness and glory. In this escalating dialectic within the Old Testament narrative of God, creation, and the people of God, Paul situates the inconceivable moment in history. God enters human history in bodily form, at once becoming in Christ the first fruits of the new creation and the fulfillment of the new covenant. In him the problem with creation is rectified and the righteousness of God made manifest. With the help of this interpretive lens, Paul develops an economy of salvation. This time, the economy of salvation progresses not only to the Jew and also to the Gentile. Rather, the economy of salvation sees God's glory manifested in human beings first and in creation second. Underlying Paul's thought is, however, that humans realize they are intricately linked and intertwined with the rest of creation. After all, humans are part of creation. When humans sin, creation suffers; when humans are redeemed, creation rejoices. Since both have the same creator, both have the same redeemer. This is Paul's view as illustrated by the biblical passages linking creation and covenant.

However, since the redeemer assumed human form, not that of a mountain or a tree, humans have a primary role to play in creation's redemptive process. While Paul offers no concrete ethical guidelines, he suggests that we can physically experience the wistful groaning of creation in tandem with our own, the deep sighs of the Spirit at work in us, and the welling up of human compassion for inanimate and animate creation as it anxiously waits for us to "get it." The current human-forced climate changes are rapidly threatening and decimating the natural habitat of over sixteen thousand plant and animal species. With one animal species going extinct every twenty minutes, we are not only robbing its members of life. We are depriving ourselves of the very "cheerleaders" that God has appointed in the spectacle of global and cosmic redemption. We depend on animal and plant species not just for food, clean water, air, and medicine; we depend on them, in a theological sense, for our salvation. By potentially eliciting in us compassion and consideration, they assist us in the process of redemption, hence making us more human, more transformed into the image of God's son, Christ Jesus our Lord.

Both animals and plant life are craning their necks to witness our winning the race. Will we allow the Spirit that indwells both to help us pray for what is "according to the will of God" (Rom 8:27)? Will the Spirit in prayer convict us of having placed obstacles and death traps in creation's way that keep it from joining in our own redemption? If that is the case, we are likely to realize our theological interdependence with the thousands of plant and animal species currently endangered by our prideful and reckless disregard. These species look to us as those whom God has appointed creation's crown. The same Spirit who created them will suggest to us the next urgent steps to take.

SOURCES

Byrne, Brendan. 1986. *Reckoning with Romans: A Contemporary Reading of Paul's Gospel.* Wilmington: Michael Glazier.

————. 2000. "Creation Groaning: An Earth Bible Reading of Romans 8:18-22." In *Readings from the Perspective of Earth,* ed. Norman C. Habel, 193-203. Foreword by Archbishop Desmond Tutu. Sheffield: Sheffield Academic Press.

Dunn, James D. G. 1988. *Romans. Word Biblical Commentary.* Dallas: Word Books.

Fitzmyer, Joseph A. 1993. *Romans: A New Translation with Introduction and Commentary. The Anchor Bible.* New York: Doubleday.

Fox, Matthew. 1991. *Creation Spirituality: Liberating Gifts for the Peoples of the Earth.* San Francisco: Harper Collins.

Schnelle, Udo. 2005. *Apostle Paul: His Life and Theology.* Translated by M. Eugene Boring. Grand Rapids: Baker Academic.

Tibet Natural Environment Conservation Network. 2008. "Endangered Species." Accessed from http://www.tibetnature.net on August 21, 2010.

Wright, N. T. 2002. "The Letter to the Romans: Introduction, Commentary, and Reflection." In *The New Interpreter's Bible.* Nashville: Abingdon.

————. Wright, N. T. 2005. *Paul in Fresh Perspective.* Minneapolis: Fortress Press.

3

AN ECOLOGICAL HERMENEUTIC OF
COL 1:15-20: A PANENTHEISTIC PROPOSAL

Vincent A. Pizzuto

The purpose of this essay is to offer an ecological reading of Col 1:15-20 in response to human forced climate change. The global climate crisis presents a relatively recent but urgent hermeneutical context in which exegetes must grapple with the anthropocentric biases in the Bible that have contributed to the human exploitation of the environment. We will begin with a review of Lynn White's now classic accusation that the *imago Dei* reference to humanity in Gen 1:26-28 has played a key role in promoting Christian anthropocentric attitudes which have led to ecological exploitation in Western society. Members of the Earth Bible Project, employing a "hermeneutic of resistance," attempt to counter these anthropocentric tendencies through the application of six "ecojustice principles" to their readings of the Bible. We will critique that interpretive process and explore an alternative ecological hermeneutic rooted in the Christology of Col 1:15-20. In doing so, we will proposes that a Christian panentheistic reading of Col 1:15-20, in which Christ is radically identified with all of creation, may provide a lens through which to expand the *imago Dei* reference in Gen 1:26-28 beyond a strictly human identification. We will propose then, that Col 1:15-20 provides a biblical basis for an eco-theology that could help move the church beyond the anthropomonism of the traditional Christian notion of salvation, to a Christological affirmation of the sacredness of the entire cosmos.

GENESIS 1:26-28 & THE PROBLEM OF
CHRISTIAN ANTHROPOCENTRISM

Lynn White's 1967 article, "The Historical Roots of our Ecological Crisis" indicted Christianity for its desacralization of nature, which he claimed was among the root causes of environmental destruction in Western society (1203-07). He argued that with its triumph over pagan animism, Christianity fostered a sense of indifference toward—and dominance over—nature, thus paving the way for environmental exploitation. His article produced a storm of controversy, but over the past forty years his conclusions have been rightfully mitigated by many critical responses, including Al Gore who addressed this controversy in his book, *Earth in the Balance* (1992, 242-65). Nevertheless, as a premise to our investigation of Col 1:15-20, two of White's observations will claim our attention because their veracity has become increasingly apparent in light of the growing climate crisis. The first is that the health of our biosphere will continue to worsen until we reject the notion that nature exists for the sole purpose of serving humanity—a posture which he claims is inherent in Christianity's anthropocentrism: "Especially in its Western form, Christianity is the most anthropocentric religion the world has seen. As early as the 2nd century both Tertullian and Saint Irenaeus of Lyons were insisting that when God shaped Adam he was foreshadowing the image of the incarnate Christ, the Second Adam. Man [*sic*] shares, in great measure, God's transcendence of nature. Christianity, in absolute contrast to ancient paganism and Asia's religions…not only established a dualism of man and nature but also insisted that it is God's will that man exploit nature for his proper ends" (White 1967, 1205).

As White sees it, the anthropocentrism inherent in Christianity is the culprit that has generated a dualistic and exploitative relationship between humanity and creation. The key text which lies at the heart of Christian anthropocentrism is found in Gen 1:26-28, where humans alone are predicated as the *imago Dei* while the rest of creation is destined to be subdued and mastered by them. This supremacy of humanity over all of creation becomes closely tied with the humanity of Christ in Christian incarnational theology, thus paving the way for an even greater glorification of humanity over-against nature. Thus, White's second observation rightly points to the fact that since the roots of our environmental crisis are largely religious, so also must the remedy be religious, even if not explicitly so (1967, 1207).

In response to these concerns it is imperative that Christianity find new ways to articulate a viable theology which expresses the interconnection between humanity and creation, because such a spiritual consciousness can go far in motivating faith communities to reevaluate humanity's relationship with Earth. While it is clear that much of traditional Christology has indeed contributed to Christian anthropocentrism, it is not necessary to revert to pagan animism in service of a sustainable theology of creation. Rather, new ecological hermeneutics of the Bible must be developed, which will help to deconstruct the anthropocentric assumptions that White has rightly identified in Christianity. This in turn will open ways to explore new theological horizons. The danger however, is that well meaning attempts to eradicate Christian anthropocentrism could be taken too far toward what Denis Edwards calls a "leveling view" of creation, which, when pushed to the extreme "a human person can be understood to have no more value than a worm, or a fungus" (1991, 27). This kind of leveling view is ultimately incompatible with Christian faith because it fails to take into account the unique human capacity for self-consciousness and the ethical responsibilities that ensue as a result of this awareness. We will be better prepared to discuss these issues further toward the end of our essay.

AN ECOLOGICAL HERMENEUTIC OF RESISTANCE: THE EARTH BIBLE PROJECT

In response to the environmental crises we now face, the past decade has seen a growing movement of ecological hermeneutics. The methods vary considerably, but among the most prominent is the Earth Bible Project which was initiated in Adelaide, Australia in 1996 by international scholars, including those who approached the subject from the perspective of indigenous societies. Contributors to the Earth Bible Project developed a set of six "ecojustice principles" which are variously applied to an ecological hermeneutic of the Scriptures, designed to resist the inherent anthropocentrisms in the Bible in the interests of ecology. Adopting a hermeneutics of suspicion, an approach first named by feminist biblical scholars, the writers of the Earth Bible Project examine particular biblical passages in light of one or more of these principles. The point is not to rescue a perceived original meaning of the Bible which has been otherwise distorted by misinterpretations within Christian tradition, but to rescue Earth from the endemic anthropocentrisms of the Bible itself by exposing, resisting and ultimately breaking free of them (see Horrell et al. 2010, 21-22). The principles that serve as tools for this purpose are:

1. The principle of intrinsic worth: The universe, Earth and all its components have intrinsic worth/value.

2. The principle of interconnectedness: Earth is a community of interconnected living things that are mutually dependent on each other for life and survival.

3. The principle of voice: Earth is a subject capable of raising its voice in celebration and against injustice.

4. The principle of purpose: The universe, Earth and all its components are part of a dynamic cosmic design within which each piece has a place in the overall goal of that design.

5. The principle of mutual custodianship: Earth is a balanced and diverse domain where responsible custodians can function as partners with, rather than rulers over, Earth to sustain its balance and a diverse Earth community.

6. The principle of resistance: Earth and its components not only suffer from human injustices but actively resist them in the struggle for justice (Habel 2008, 2).

Among the exegetes of the Earth Bible Project, these principles are variously taken up and used as a basis for ascertaining the various anthropocentrisms that underlie any given biblical text. Careful to distinguish between the meaning of the terms anthropocentric (human centered), anthropogenic (a text originating from human beings) and anthropotopic (a text in which humans claim center stage), the members of the Earth Bible Project attempt to unmask all of these hidden or unacknowledged perspectives that can be identified in the production, reception and interpretation of the Bible, in solidarity with Earth (Habel 2008, 3-4). The goal is to radically reorient the interpreter's view of the Bible through a hermeneutics of suspicion, identification and retrieval, so that Earth is no longer approached as a topic to be explored or an object to be exploited, but as a "subject with a voice" (2). The predominance of biblical anthropocentrism is thus rightfully unmasked as ethically and theologically problematic, in large part because it contributes to human attitudes and behaviors that continue to escalate our current ecological crisis by ignoring, objectifying, or victimizing Earth.

AN ILLUSTRATION: RESISTING THE
ANTHROPOCENTRISM OF GEN 1:26-28

A brief summary of Habel's reading of Gen 1:26-28 in light of the above six principles will illustrate how members of the Earth Bible Project might engage in an ecological hermeneutic of resistance through a process that involves suspicion, identification and retrieval. First, Habel notes that the application of the principle of intrinsic worth would immediately raise suspicion about the anthropocentrisms of Gen 1:26-28 because humans alone are depicted as the *imago Dei*, and thus set apart from the rest of creation to "rule over" and "subdue" it. Secondly, he takes up the principle of interconnectedness in order to identify with the non-human characters in the story so as to hear the passage from the perspective of those who are said to be "ruled over." From this perspective it becomes easier to understand how Gen 1:26-28 has been used to justify human exploitation of the earth. Finally, Habel suggests that the application of the principle of voice can serve as a means to retrieve the voice of Earth in this Genesis narrative. From this perspective, Earth is recognized as a partner with God, from whom arises both the flora and fauna of creation. Earth takes on a positive and active role in the creative process, rather than remain a silent passive object (Habel 2006, 38-39).

A CRITIQUE OF THE
HERMENEUTIC OF RESISTANCE

In preparation for the ecological reading of Col 1:15-20 to be presented below, we submit two critiques of the hermeneutic of resistance which demonstrate the limitations of that approach. Firstly, the six principles that have been adopted by the members of the Earth Bible Project are not themselves rooted in the Bible, and thus remain somewhat arbitrary and peripheral to Christian tradition itself. Certainly, biblical exegesis always takes place in a particular context, and the six principles of the project are arguably a reasonable distillation of key elements to consider when seeking to unmask and resist biblical anthropocentrism with a view to the climate crisis. But because they are extrinsic to the Christian canon of Scriptures, these six principles cannot carry the weight of revealed truth with the authoritative power to motivate and inspire Christians to reexamine their relationship with Earth. For example, in the exegesis of Gen 1:26-28 summarized above, Habel's application of several of the project's principles certainly advances new perspectives on how Earth is objectified by the anthropocentrisms in the Bible, but it does not ultimately address whether and to what extent various biblical texts themselves might serve as the foundations for a new Christian ecological theology. While it is necessary to resist some of the anthropocentric aspects of the Bible, we cannot go so far as to resist the Bible itself. In other words, an authentic Christian ecotheology must ultimately incorporate the Scriptures not reject them.

Secondly, while it is laudable that the members of the Earth Bible Project seek to highlight the intrinsic value of Earth as subject, an ecological hermeneutic of resistance may be ultimately counterproductive because it inadvertently maintains the "dualism of man and nature" which White warned about, and thus does not go far enough in modeling the deep, seamless interconnection that humanity shares with the rest of the biosphere. In particular, the principles of "voice" and "resistance" carry with them certain dangers which may jeopardize the potential hermeneutical advances intended by the remaining four principles because they run the risk of advancing an unacknowledged anthropocentrism or worse, of adopting a well-intended but ultimately disingenuous 'leveling view' of creation noted earlier. In particular, we see a potential contradiction between these two principles and that of the second: the principle of interconnectedness. For example, to seek the "voice" of Earth in our sacred texts as "members of Earth community in solidarity with Earth" (Habel 2008, 3), inadvertently establishes an I-thou relationship with Earth as an *other*. That is to say, the "principle of voice" reinforces the illusion that humanity is somehow above, or at the very least, separate from nature, thereby erroneously segregating ourselves from the biosphere of which we are, in fact, *extensions*. The objectification of Earth which members of the Earth Bible Project attempt to identify and overcome, is actually a symptom of a much deeper spiritual crisis; namely, the self-alienation of humanity from our own ecological body. What is called for instead is the deeper spiritual awareness that the biosphere we have been treating as object is, in fact, an extension of our own subjectivity. Or rather, that Earth and humans are members of one body, and thus possess a single subjectivity. Casting Earth as an *other* to whom we must *listen* essentially creates a division between humanity and creation, when in fact both are seamlessly interconnected, as the second principle would suggest. The distinction here is subtle and thus warrants reiteration: the language of "relationship" implies two-ness (one subject in relationship with another subject), whereas the language of "interconnection" points to the idea that distinctive members together comprise a single subjective unity. The Pauline understanding of the church as the Body of Christ is a clear example (see 1 Cor 6:15; 1 Cor 12:27; Col 1:18). Thus, the six ecojustice principles risk an inner contradiction. We turn then to Col 1:15-20, in preparation for an alternative ecological reading of that passage which goes beyond a hermeneutic of resistance in favor of a Christian eco-theology rooted in the Bible itself.

THE SEMANTIC AUTONOMY OF COL 1:15-20

Paul Ricoeur has observed that all scriptural texts posses a degree of semantic autonomy from their original authors, which is the result of a disconnection between the original intention of the biblical authors and the verbal meaning their texts have for later audiences (1976, 30). Throughout the history of Christianity hermeneutical lenses have continually been developed and applied to biblical texts in order to uncover novel meanings of those texts for new readers and their circumstances. Because the environmental crisis which has given rise to recent ecological readings of the Scriptures is thoroughly modern, the original concerns of the Colossians author bear little relevance for our present interests. Thus, while the ecological focus and spatial limitations of this essay will not permit a thorough exploration of the authorial, literary and historical complexities of Col 1:15-20, it is sufficient to establish for our purposes that in its canonical context this passage demonstrates a Christological vision of the universe in which Christ's role in both creation and redemption extend to the entire cosmos.

Colossians 1:15-20 may be faithfully rendered in English as follows:[1]

[15] Who is the image of the invisible God
firstborn of all creation;
   [16] for in him were created all things
   in the heavens and on the earth,
   the visible and the invisible,
      whether thrones or dominions,

---

1   For an in depth study of the authorship, structure, provenance and Christology of Col 1:15-20, see Pizzuto 2006, especially 203-5.

whether rulers, or powers;
all things, through him and for him, have been created;
And he is before all things,
And all things in him hold together,
And he is the head of the body, the church;
Who is the beginning,
firstborn from the dead,
so that he might come to have first place in everything;
    [19] for in him all the fullness was pleased to dwell,
    [20] and through him, to reconcile all things to himself,
        by making peace through the blood of his cross,
through him [*to reconcile*] whether the things on earth,
or the things in heaven.

A FORM CRITICAL ANALYSIS OF COL 1:15-20

In 1832, Friedrich Schleiermacher was the first to isolate Col 1:15-20 from its epistolary context, based on his observation of parallel passages between vv. 15-16a and 18b-19 (1832, 497-537). However, it was not until 1913 that the philologist Eduard Nordon first published a detailed formal analysis of the pericope in his *Agnostos Theos*, in which he identified two strophes beginning with the relative pronouns ("who is") in vv. 15 and 18b respectively, and which he classified as a "hymn to God the Father and the Son" (1913, 252-253). Based on Norden's findings, there is now widespread agreement among Pauline scholars that the structure of Col 1:15-20 presents a two-fold emphasis on Christ's role in creation (vv. 15-17) and redemption (vv. 18-20) with a central focus on the unity of "all things" (*ta panta*) held together in Christ (vv. 17b). However, many of the problems around the formal designation of Col 1:15-20 as a hymn are closely tied to other questions about the provenance and authorship of the passage, as well as the structure of the verses themselves. For example, some of the criteria frequently cited for the classification of these verses as a "hymn" are the presence of parallelisms reflective of ancient Jewish or Hellenistic poetry; the presence of perceived literary seams before and after vv. 15 and 20 respectively, indicating that they may reflect traditional material predating the literary context in which they presently appear; the use of words in the passage which are atypical of the presumed author, suggesting an independent composition; and poetic, lofty, or doxological language praising the person or mission of Christ

(Pizzuto 2006, 103-111).[2] While it may be argued that Col 1:15-20 reflects any number of these characteristics, the dearth of conclusive evidence regarding the authorship, provenance, and redactive elements of these verses has resulted in the fact that the designation hymn is neither universally accepted nor uniformly employed.

Nevertheless, for the purposes of this essay, we designate Col 1:15-20 as a hymn to signal that these verses bear both confessional and doxological material which serve to provide a clear Christological foundation for the overall letter. The literary context of Col 1:15-20 suggests that the author likely used this hymn as the Christological cornerstone of his epistle intended to convince the nascent churches of the Lycus Valley to resist an unnamed heresy, or perhaps more generally to denounce any syncretistic practices, beliefs, or tendencies (Col 2:6-19) which would situate Christ within a pantheon of angels, demigods, or 'elemental spirits' of the universe (Col 2:8, 20; Hooker 1973, 315). While the semantic autonomy of Col 1:15-20 unmoors it from this original application, the fact that in its epistolary context it expressed a cosmic Christology, opens the way for new ecological readings of the hymn.

THE COSMIC CHRISTOLOGY OF COL 1:15-20

The thematic content of the hymn suggests that the most proximate literary parallels are to be sought in Second Temple Jewish motifs, such as that of Jewish psalmody, and the intermediary figure of Wisdom (i.e., Prov 3:19; 8:30; Wis 7:22; 8:4-6), filtered through the Pauline tradition of the Cross (e.g., 1 Cor 1:23), and that of Christ as the Second Adam (i.e., 1 Cor 15:45-49; Rom 5:14). While references to the church (Col 1:18) and the Cross (v. 20) are sometimes regarded as glosses, they undoubtedly bear Pauline influence. Structurally, the twin themes of creation and redemption which dominate the first and second half of the hymn respectively, also reflect the tendency in Second Temple Judaism to emphasize the continuity between creation and salvation (see Dunn 1980, 180). References to these dual themes abound in Old Testament literature, as can be observed in many psalms (e.g., 146:5-6), the prophets (e.g., Isaiah 42-43; 51:9-17) and in the emphasis on creation in Gen 1-11, followed by that of redemption in Gen 12-50. However, these dual roles of Creator and Redeemer,

---

2   Other texts in the New Testament that are frequently regarded as hymns or "hymn-like" include John 1:1-18; Phil 2:6-11; Tim 3:16; 1 Pt 2:22-25.

attributed exclusively to YHWH in ancient Judaism, are predicated of Christ in the Colossians hymn. Central to the hymn's structure is the affirmation that 'all things' *cohere* in Christ (v. 17b). Thus, Col 1:15-20 serves to depict all of creation as being *of* Christ from its first origin to its final consummation, and held together even now *in* Christ.

The predication of Christ as "the image (*eikon*) of the invisible God" (v. 15a) which begins the hymn, echoes the *imago Dei* reference to Adam in Gen 1:26-28, yet, as "first born of creation" (v. 15b) he is depicted as One who is greater than Adam. Christ is also "the beginning," the One in whom God has carried out the act of creation (v. 16a), and the instrument through whom God continues to create and sustain the universe (vv. 16f, 17b). The elaborate enumeration of things visible and invisible in 16b-e, leaves nothing unaccounted for in the cosmic order (*ta panta*), nor is there anything excluded from the creative enterprise of Christ. More striking still, this hymn is unique among New Testament literature in that it further predicates Christ as the *telos* in whom the entire universe is summed up and given ultimate meaning (see 16f, "*for* him" [*eis auton*]). Because of this, no other New Testament Christological hymn portrays as intimate a relationship between Christ and creation as does Col 1:15-20. This all-inclusiveness prepares a way within the hymn for moving Christianity beyond a myopic anthropocentric soteriology of the cross, to an expansive pan-cosmic soteriology in which "all things" (*ta panta*) are reconciled to Christ "by the blood of his cross" (20b). Thus, F. B. Craddock observes that in the very structure of the Colossians hymn, "The doctrines of creation and incarnation agree in affirming one truth: there is no separation of a spiritual realm from a material realm. There are not two worlds; there is but one 'in him'" (1968, 106; see Col 1:16; 3:11). In other words, the very structure of Col 1:15-20 conveys that in and through the creative and redemptive roles of Christ all things spiritual and material cohere *in him* (v. 17b).

The implications of this cosmic soteriology for the early church are made explicit, in vv. 18a where Christ is predicated as the "*head of the body, the church.*" Given the close identification that Col 1:15-20 has established between Christ and creation in the first half of the hymn, it is all the more significant that Christ is similarly depicted as the beginning (*arche*) or first principle, that is to say the very One in whom salvation 'happens,' also at the start of the second half of the hymn in vs 18b. The body of the church is thus constituted by those who are saved under the cosmic lordship of Christ, and who are radically dependent on him and sustained by him as their 'head' (*kephale*). While muted because of its ecclesial associations in Colossians, this head-body analogy reflects more distant Greco-Roman literary parallels which identify the cosmos as the 'body' of a deity who governs as its head. Nowhere, however, is the analogy used in such a direct and literal sense as we see in Col 1:18a, nor is the reference to Christ as head of the cosmos ever entirely lost in Christian tradition. For example, in a homiletic reflection on Colossians, John Chrysostom interprets the ecclesial headship of Christ to extend to the cosmos (literally, "fullness"), which God has wrought in an act of profound friendship and connection with creation" (1889, 271).[3]

Based on these observations we conclude that both the structure and content of the hymn make clear that humanity is not saved *from* the world, rather the universe itself (*ta panta*) is saved in the redemptive and reconciling work of Christ. Thus, the saving power of the Cross is not restricted to humanity but rather has pan-temporal and pan-cosmic implications (Pizzuto 2006, 258). This does not imply that the particular role of humanity in the history of salvation has been replaced with a leveling view of creation which Edwards warned about earlier. The question is, how might we best define and understand that role. As Lukas Vischer sees it, the problem is not with anthropocentrism, but with "anthropomonism," which postulates that human beings are the sole beneficiaries of redemption, and that only human well being has any value or importance in salvation history. From an athropomonic perspective, Earth is at the disposal of human interests. By contrast, Vischer argues that a modest anthropocentrism, purified of anthropomonism, acknowledges that humanity "is called to fulfill a special and specific role in [the] world," without excluding the rest of the world from the redemptive work of Christ (2004, 22; see Horrell et al. 2010, 260, n. 10).

The Colossians hymn makes clear that the fate of Earth is intimately tied with human destiny because both are grounded in the

---

3   The translation of Chyrsostom's homiletic reflection on Col 1:18 reads: "*He is*," says [Paul], "*the Head of the body, the Church.*" And he said not "*of the fullness*," (although this too is signified) out of a wish to show his great friendliness to us, in that He who is thus above, and above all, connected himself with those below. For everywhere He is first; above first; in the Church first, for He is the Head; in the Resurrection first."

cosmic Christ, who has radically identified himself with the whole of creation through the Incarnation and who has reconciled all things by his cross. Rather than limit ourselves to an ecological hermeneutic of resistance, by which all aspects of biblical anthropocentrism must be unmasked and rejected, an ecological reading of Col 1:15-20 can help to develop the Christological foundations necessary to address the distinctive role of humans in the world. To this end, we will first examine how Col 1:15-20 contributes to the theological language of a Christian "panentheistic" paradigm. Secondly, we will argue that this paradigm, rooted in an ecological reading of Col 1:15-20 serves as a more appropriate lens through which to interpret Gen 1:26-28, than Habel's application of the ecojustice principles that were produced by the Earth Bible Project. Recall that the roots of Christian anthropocentrism are often identified with this text in Genesis because of the *imago Dei* reference attributed exclusively to humanity, and the divine command for humanity to 'subdue' and 'master' Earth. Therefore, our final concern will be to suggest how an ecological hermeneutic of Col 1:15-20 might offer an expanded understanding of the *imago Dei* reference in Genesis to include all of creation. We will propose that the headship of Christ with reference to the ecclesial body in Col 1:18, indicates something of the particular role of humanity in the ecological body, while avoiding a 'leveling view' of creation. As we will see, the foundational issue which underlies our entire ecological hermeneutic is that of *interconnection*. What insights might an ecological reading of Col 1:15-20 yield about the interconnection between humanity and the biosphere we call Earth?

COL 1:15-20 AND CHRISTIAN PANENTHEISM

The Colossians hymn tells us that the universe is *in* Christ, insofar as Christ is the One in, through and for whom God created the universe, and in whom 'all things' (*ta panta*) continue to be held in existence. Indeed, we are assured that in Christ, all 'fullness' (*pleroma*) was pleased to dwell (1:19). Paradoxically, the epistle itself speaks also of the interpenetration of Christ in whom the "fullness of divinity (*theotetos*) dwells bodily" (2:9), and who "is all and in all" (3:11), indicating that we can speak with equal veracity of the universe being *in Christ*, and of the divine Christ being fully present *within* the universe. Thus, as Chrysostom had already hinted in his homilies on Colossians, the hymn opens up the possibility of understanding something of a perichoretic relationship between the church as the mystical body of Christ and that of the cosmic body of Creation. In other words, both are mutually interpenetrated by the other. As Gregory of Nyssa comments in his late 4th century *Catechetical Orations* 25, the Incarnation renders at once a deified humanity and a theophanic universe:

> The fact that God should have *clothed himself with our nature* is a fact that should not seem strange or extravagant to minds that do not form too paltry an idea of reality. Who, looking at the universe, would be so feeble-minded as not to believe that God is all in all; that he *clothes himself with the universe*, and at the same time contains it and dwells in it? What exists depends on him who exists, and nothing can exist except in the bosom of him who is. If then all is in him and he is in all, why blush for the faith that teaches us that one day God was born in the human condition, God who still today exists in humanity? Indeed, if the presence of God in us does not take the same form now as it did then, we can at least agree in recognizing that he is in us today no less than he was then. Today, he is involved with us in as much as he maintains creation in existence. Then he mingled himself with our being to deify it by contact with him, after he had snatched it from death...(cited in Clement, 1995, 39-40, my emphasis).

Whether Gregory explicitly intended an kind of "Christian panentheism" here is doubtful, but Patristic texts such as his already begin to pave the way for panentheistic readings of Scripture. At the very least, Gregory's language points implicitly to a panenchristic notion of Incarnation in that Christ is said to "clothe himself" not only with human nature, but with the universe itself. Here again, is the implication that the locus of the Incarnation is not limited to human nature, but includes the entire universe as the embodiment of Christ. Raimon Panikkar notes similarly that throughout church history, "the idea that God had created the world out of love for the Church was considered virtually synonymous with the idea of the mystical body. This meant that God created the world for the purpose of divinizing his creation by making it become his own body, with Jesus as head, and we the members" (2009, 177). While Christian faith cannot permit an unqualified pantheist equation between the cosmic body of creation and the mystical Body of Christ, the Incarnation nevertheless renders them inseparable. Within a Christian framework this remains an important distinction. Pantheism confuses God with creation, thus requiring any notion of union with the divine to be somewhat mechanical or automatic. Christian panentheism, by contrast, sees all of creation through the lens of the Incarnation, which reveals a world pregnant with the presence of God. Thus, God's union with "the all" in Christ is rendered through a unilateral divine initiative accomplished with sovereign freedom and motivated by unconditional love. In short, for the Christian panentheist God "is all and in all" (Col 3:11) but is not reducible to the universe.

COL 1:15-20 AS A FOUNDATION FOR ECOLOGICAL READINGS OF SCRIPTURE

Our examination of Lynn White's article at the beginning of this essay focused on two of his expressed concerns: First, that Christianity has contributed to our current ecological crisis through the desacralization and exploitation of nature, and second, that since the roots of this crisis are religious, we must search for religious solutions to our current predicament. White locates the problem in the radical anthropocentrism (Vischer's "anthropomonism") that has dominated Christian tradition. In search of a remedy, White proposed a retrieval of Franciscan tradition because St. Francis was a spiritual revolutionary who "… tried to substitute the idea of the equality of all creatures, including man [sic], for the idea of man's limitless rule of creation" (1967, 1207). While Christianity might certainly benefit from Franciscan spirituality as a corrective to its otherwise overly anthropocentric biases, it is not the case that Francis was preaching the "equality of all creatures" as much as "God in all things," or what we have defined above as "panentheism." Thus, it is not equality but *interconnectedness* that underpins Franciscan spirituality, as becomes evident for example, in the way that Francis is thought to have related even to abstract concepts, like "sister death."

However, this panentheistic emphasis in Franciscan spirituality is already a retrieval and magnification of an authentic strand of Christian tradition which can be traced back to the biblical canon itself, as is exemplified in the Christology of Col 1:15-20. At the very least there is a nascent panentheism in the Colossians hymn already recognized in Patristic writings, like that of Chrysostom in the 4th century. Thus, the anthropomonism which has dominated Christianity for centuries is, in part, the result of a failure to take New Testament revelation seriously enough when interpreting Old Testament texts like Gen 1:26-28. In other words, an authentic Christian ecological reading of the Bible must interpret all of the Scriptures Christologically, as is evidenced in the church's writings since Paul. The failure to do so is tantamount to a misappropriation of Old Testament scriptures within a Christian context. Thus, while the ecojustice principles of the Earth Bible Project are helpful in resisting anthropocentric biases in the Bible, the fact that these principles are extraneous to the Bible itself, and thus not necessarily Christological, becomes problematic.

What does it mean to read the Scriptures *Christologically* when there are multiple Christologies presented in the New Testament? Through which of the numerous Christologies of the New Testament should we read the notoriously anthropomonic text of Gen 1:26-28, for example? We suggest here, that a panentheistic model of the universe rooted in the Christology of Col 1:15-20 may help to provide an ecological hermeneutic upon which to develop a Christian anthropocentrism that is morally, theologically and ecologically responsible. Sandra Schneiders speaks eloquently of this Christian interpretive process: "[T]he christological meaning of the Old Testament for the Christian is not an extraneous meaning 'read into' the Old Testament by Christian piety but the effect of the inclusion of the Jewish scriptures in the Christian Bible, part of which is the witness to the historical occurrence of the Christ-event in Jesus of Nazareth that the Jewish scriptures in themselves (i.e., prior to their inclusion in the Christian Bible) viewed as unfulfilled promise" (1991, 88).

AN ILLUSTRATION:

READING GEN 1:26-28 THROUGH THE

CHRISTOLOGICAL LENS OF COL 1:15-20

As Lewis Donelson has noted, the initial verses of the Colossians hymn function as a Christian creation story "told through the lens of Jesus" (1996, 24, 26)[4]. For this reason, Col 1:15-20 is particularly well suited to provide a Christological lens for an ecological reading Gen 1:26-28. The roots of Christian anthropocentrism lie in the creation myth of Gen 1:26-28 where humans are created in the "image of God" (*imago Dei*), and are given "dominion" to fill the earth and "subdue" it. Christians can hardly read the *imago Dei* reference in Gen 1:26-27

---

4   Also see Horrell et al. 2010, 102-5

without calling to mind the predication of Christ as "image of God" and "Firstborn of creation" in Col 1:15. However, R. F. Collins noted that in Colossians "the phraseology suggests that the prototypical humans were in fact a kind of antitype for Christ. God was thinking of the humanity of Christ when he formed Adam. It is not so much Adam who is the summit of creation as it is the beloved Son" (1988, 192). In other words, by placing salvation in the context of a universal claim of Christ's lordship, the pinnacle of creation is shifted from Adam to Christ, in whose image Adam was created, and in whose image we are also being conformed (Col 3:10). This interpretation of the Colossians hymn is the antithesis of the theology that has predominated Christian thought since the second century writings of Tertullian and Irenaeus (see White 1967, 1205). For them, Adam foreshadowed Christ, but as Collins has pointed out, Col 1:15 posits Christ as "firstborn of all creation" and therefore it is Christ who prefigures Adam. From the radical identification of Christ with the cosmos that follows throughout the hymn, we can begin to construct an ecological hermeneutic which decentralizes humanity as the sole locus of the Incarnation and opens the way for a new understanding of humanity's place within the cosmic body of Christ.

For example, when reading Genesis through the lens of Colossians, the *imago Dei* reference in Gen 1:26 can be expanded from a strictly human designation, to one that includes the entire cosmos, indeed all of created order (see Col 1:15-17, 19). If Christ is "all in all" as the epistle goes on to affirm (Col 3:11), then Gen 1:26 is not incorrect to say that humans reflect the *imago Dei*, it is simply inadequate in light of an ecological reading of Col 1:15-20. The entire universe can rightfully be said to reflect the *imago Dei* because the bodiliness or physicality of Jesus is both deeper and broader than his humanity. In fact, we cannot affirm Christ's humanity, without *de facto* affirming his bodiliness. This opens the way to an ecotheology based on humanity's interconnection *with* Earth, rather than merely our relationship *to* it. In light of this ecological reading of Col 1:15-20, humans can no longer claim to be the exclusive expression of the *imago Dei*, but rather, as members of the cosmic Body of Christ, an instance of the *imago Dei*. Thus, humanity's reconciliation with God in Christ is a participation in a cosmic reconciliation in which humanity is inextricably caught up and which the church is called to embody in history.

In Colossians, the ecclesial body—of whom Christ is the head (Col 1:18)—takes on cosmic significance. As Hammerton-Kelly observed: "The application of cosmic terms to the church [signifies that] the church is being understood as a cosmic, metaphysical entity. The missionary penetration of the world is the penetration of the cosmos by the reconciling power" (1973, 175). In Colossians, the cosmic scale of Christ's reconciling power is rendered through the peculiar Greek term *apokatallosso* used in Col 1:20, 22 to convey the idea of a comprehensive "reconciliation." Outside Colossians, this term appears only in Ephesians 2:16 (which has close literary parallels to Colossians), but nowhere else in ancient Greek literature. The term *katallasso* (reconciliation) used elsewhere in Pauline literature lacks the prefix *apo-* which in Colossians is used to intensify the extent of the reconciliation— that is, "to reconcile *completely*" (Abbott-Smith 1936, 51; see Horrell et al. 2010, 100). The reconciling power of Christ then, is not limited to people, cultures, and societies, but to every aspect of creation; to the very relationship between humanity and creation itself. For this reason, the all-inclusive notion of creation as the *imago Dei*, and the expansive notion of redemption which includes "all things" in the reconciling work of Christ, are to be manifested in the theology, work and ministry of the church (see Col 1:1-17), whose "missionary penetration of the world" can no longer be thought of in mere anthropocentric terms (Hammerton-Kelly 1973, 175).

This cosmic perspective offers a key insight for awakening a deeper ecological consciousness within Christian spirituality and serves as a cornerstone of an eco-theology that must inform every aspect of the churches mission in the world. In light of what we now know about the interdependence of all creation, the extent of the environmental crisis, and the potential for ecological annihilation which rests in human hands, our fundamental doctrines of faith cannot just be occasionally adapted or applied to new or emerging "green" theologies. Rather, our reconceptualization of the very core aspects of Christian faith must be radical, pervasive, and permanent. Inasmuch as Christian scholastic tradition has come to understand creation as a *creatio continua* (ongoing creation) so too might we recognize the entire cosmos as an *incarnatio continua*, (ongoing incarnation) whereby the doctrine of an interventionist God must be abandoned in favor of a new theological paradigm which reinterprets Christian dogma, life and praxis through a panentheistic perspective (see Panikkar 2009, 128). God, in other words, cannot be said to act upon creation from the "outside," because in God there is no outside.

AN ECOLOGICAL READING OF THE
HEAD-BODY RELATIONSHIP IN COL 1:18

Karl Rahner has coined the notion that in an evolutionary universe humanity represents the cosmos come to self-consciousness (1996, 183-92). Humans are neither above nature, nor merely in relationship to it. Rather, we are interconnected with it as members of the cosmic body of Christ. Furthermore, our capacity for self-reflection, moral decision-making and self-transcendence situates us in a unique position to reflect Christ's role as "head of the body" (Col 1:18) vis-à-vis Earth. In other words, taking our cue from the heady-body reference to Christ and the church in Col 1:18, we might suggest analogously that the particular way in which humanity reflects the *imago Dei* is in Christ's headship vis-à-vis his cosmic body of creation. This manner of reflecting Christ in the world is not intended to endow humanity with yet another anthropocentric privilege but to recognize a spiritual and ethical vocation which demands that humans live in balance with the biosphere. As instances of the universe come to self-consciousness, humanity does not speak *for* Earth but as an extension *of* Earth. Thus, when humans listen to the "voice of Earth," as members of the Earth Bible Project would encourage us to do, it can only be ourselves that we hear. That is to say, it is not an "other" that we hear, but "another member" of the Body of Christ, of which we ourselves are a part.

By way of analogy, what would it mean for a person to listen to their hand or foot or eye as though it were a separate entity, a distinct "voice"? In contrast to the "principle of voice," we are suggesting that the head-body analogy of Colossians opens the way to an even more radical association between Earth and humanity. As it is, the "principle of voice" cannot really introduce a fair hearing of all voices in the biosphere without moving humanity toward a "leveling view" of creation about which Denis Edwards had warned. In other words, unless we are going to advance such a leveling view, some voices of Earth will be given more weight than others as determined by human bias, leaving the process unavoidably anthropocentric. By contrast, a panentheistic model of the universe which sees the whole universe as a reflection of the *imago Dei* and humanity as particular reflection of Christ's headship, certainly advances a moderate form of Christian anthro-pocentrism, but at the same time reconciles the dualism established by ecological readings of Scripture that would segregate the voice of Earth from that of humanity.

An exaggerated attempt to overcome anthropocentrism altogether by placing the value of every living thing on equal footing, for example, could not permit the eradication of the polio virus in favor of the human lives it is capable of destroying. What "voice" does polio have in the human struggle to eradicate this disease? Are we to suggest that it has a voice? If not, why not? But if so, what is polio saying about its right to survive? Indeed, to *thrive*? Why do the trees of the Amazonian rainforests have a voice in the wake of human destruction, but not HIV? We must recognize that while humanity is not the only species intent on survival, we are the only ones who have the capacity to single-handedly destroy the entire planet, and thus the only species who has an ethical vocation to preserve and protect it. The key then, is for Christians to realize a spirituality which reflects a morally, theologically, and materially viable approach not only to a sustainable human civilization but more fundamentally, to a healthy planetary biosphere.

In light of Col 1:18 then, the specific task of humanity with respect to the ecological body of Christ is to participate in Christ's headship, not over-against creation, but with reverential respect for it as a continuous expression of the bodiliness of Christ. Therefore, while we can speak of the whole universe as the *imago Dei*, the "head" best reflects the particular vocation of humanity to govern Earth wisely and reverently, because in humanity the universe has evolved the capacity for self-reflection. This insight confronts us with the spiritual realization that in the ongoing degradation of the environment perpetrated by human hands, it is the cosmic body of Christ whom we continue to crucify—indeed the body of which we ourselves are members. We must resolve then to ask, "how big is our 'we'?" (Bruteau 1990, 510), that is to say, where do we draw the line between "us" and "them"? An ecological reading of Col 1:15-20 invites us to recognize that, in fact, no line can be drawn to ultimately separate humanity from the rest of creation. In a panentheistic universe of which Christ is the personal center, we are only differentiated members of a unified cosmic body. In such a universe, there can be no "them."

CONCLUSION: A COLD LOOK AT GLOBAL WARMING

Humans are the only species in the history of our planet to have chosen to live out of balance with our ecosystems and the greater biosphere. This fact must be acknowledged, unobscured by religious doctrine, political ideology, or anthropocentric bias. If we listen carefully and honestly, human clamors to "save the world" are most often thinly veiled clamors to "save human civilization." While a sense of deep spiritual connection to and harmony with Earth is endemic to many indigenous cultures, contemporary thought disconnects humanity from the created order. Thus, with the exception of species that have been in some way domesticated (e.g., farm animals, household pets) or inadvertently advanced by human populations (e.g., rats, viruses), from a strictly biological view, the extinction of the human species would be advantageous for virtually every other species on the planet. As the human population nears seven billion worldwide, no amount of human waste, pollution, or carbon emissions is really "acceptable" because Earth is limited in its resources. And, the more the human population explodes, the more we approach the outer limits of those resources to the detriment of every living entity on the planet.

As the climate grows warmer under the impact of carbon emissions that are required to grow and sustain human civilization, our polar ice caps are melting with alarming speed, more powerful hurricanes rage off our shores, forest fires burn longer and hotter and with more frequency, while entire plant and animal species are racing toward the brink of extinction. The prophets of our day are not religious zealots or social activists, but the global community of environmental scientists who, with their instruments, calculations and climate models, are able to read with great precision these new "signs of the times." With virtual unanimity, they are raising a collective voice of dire warning. It is therefore imperative that Christian theologians and religious leaders also confront human forced climate change by uncovering within the Christian scriptures and tradition a theology of the sacredness of creation.

We have proposed in this essay that a panentheistic Christology derived from an ecological reading of Col 1:15-20 provides new ways for Christians to think about their interconnectedness with all of creation when addressing the climate crisis. While anthropocentrism underpins the traditional Christian emphasis on the humanity of Christ, a deeper emphasis on the bodiliness of Christ opens new avenues for understanding the vast material cosmos, not merely "humanity," as the focal point of the Incarnation. An ecological hermeneutic of the Colossians hymn offers a biblical basis for just such an emphasis and, thus, can serve as a catalyst to generate a biblical, theological and ethical mandate for action on behalf of the environment. In a universe held in existence by Christ (Col 1:17), the reconciling power of the Cross is not limited to the human milieu but given a cosmic power and significance to reconcile "all things" (Col 1:20). Redemption, the hymn assures us, is not about being saved from the world, but about the world itself being saved in and through Christ.

In a panentheistic universe, Earth and humans are understood to be extensions of one another, as both are members of Christ's cosmic body. Thus, it makes no sense to harm Earth, because in doing so it is Christ whom we persecute, indeed, members of our own body that we inflict harm. As reflections of the *imago Dei*, we can metaphorically think of humanity as reflective of Christ's head, because as an instance of the universe come to self-consciousness we possess a moral agency that demands we act with justice on behalf of Earth. Thus, there is a common human vocation to act toward Earth, not as a separate subject, but as a continuation of our own bodily subjectivity—much as Christ is the head of his body, the church. As the cosmic Christ, Jesus is radically identified with all of creation which has both its first origins and final redemption in him. Thus, as this same creation travails under the impact of changes humans are forcing on the global climate, we see before us the new wounds of Christ crucified to which we are urgently called to attend.

SOURCES

Abbott-Smith, G. 1936. *A Manual Greek Lexicon of the New Testament*. Edinburgh: T&T Clark.

Bauckham, Richard. 2010. *The Bible and Ecology: Rediscovering the Community of Creation*. Waco: Baylor.

Bruteau, Beatrice. 1990. "Eucharistic Ecology and Ecological Spirituality." *Crosscurrents* (Winter): 499-514.

Clement, Olivier. 1995. *The Roots of Christian Mysticism*. New York: New City Press.

Collins, R. F. 1988. *Letters that Paul Did Not Write: The Epistle to the Hebrews and the Pauline Pseudepigraphia.* Wilmington: Michael Glazier.

Craddock, F. B. 1968. *The Pre-existence of Christ in the New Testament.* New York: Abingdon Press.

Deane-Drummond, Celia. 2008. *Eco-Theology.* London: Anselm Academic.

Donelson, Lewis R. *Colossians, Ephesians, 1 and 2 Timothy, and Titus.* Westminstser Bible Companion. Louisville, KY: Westminster John Knox, 1996.

Dunn, J. D. G. 1980. *Christology in the Making: A New Testament Inquiry into the Origins of the Doctrine of the Incarnation.* London: SCM Press.

Edwards, Denis. 1991. *Jesus and the Cosmos.* Mahwah: Paulist Press.

Fragomeni, Richard N. and John T. Pawlikowski, eds. 1994. *The Ecological Challenge: Ethical, Liturgical and Spiritual Responses.* Collegeville: Michael Glazier.

Gore, Al. 1992. *Earth in the Balance: Ecology and Human Spirit.* New York: Houghton Mifflin.

Gregory of Nyssa. 1967. *Catechetical Orations* 25. In Jacques-Paul Migne. *Patrologia Graeca* 45, [1875], New Jersey: Gregg Press.

Greeley, Dolores, R.S.M.B.A. 1977. The Church as 'Body of Christ' According to the Teaching of Saint John Chrysostom. Ph.D. diss., University of Notre Dame.

Habel, Norman C., ed. 2000. *Readings from the Perspective of Earth.* Cleveland: Pilgrim Press.

Habel, Norman C. 2006. "Playing God or Playing Earth? An Ecological Reading of Genesis 1.26-28." In *"And God Saw that it was Good." Essays on Creation and God in Honor of Terence Retheim,* ed. Frederick Gaiser and Mark Throntveit, 33-41. Word and World Supplement Series 5. St. Paul: Luther Seminary.

Habel, Norman C. and Peter Trudinger, eds. 2008. *Exploring Ecological Hermeneutics.* Atlanta: SBL Symposium Series 46.

Hammerton-Kelly, R. G. 1973. *Pre-Existence, Wisdom & the Son of Man: A Study of the Idea of Preexistence in the New Testament.* Cambridge: University Press, 1973.

Hooker, M. 1973. "Were There False Teachers in Colossae?" In *Christ and Spirit in the New Testament,* ed. B. Lindars and S. Smalley, 315-31. Cambridge: University Press.

John Chrysostom. 1889. *Homilies on Colossians* 3. Translated by John A. Broadus, *Nicene and Post-Nicene Fathers, First Series*, Vol. 13. Edited by Philip Schaff. Buffalo: Christian Literature Publishing Company.

Kiley, M. 1986. *Colossians as Pseudepigraphy*. Sheffield, England: *JSOT* Press.

Lightfoot, J. B. 1904. *Saint Paul's Epistles to the Colossians and to Philemon*. London: MacMillan and Company.

Lohse, E. 1968. *Die Briefe an die Colosser und an Philemon*. Göttingen: Vandenhoek & Ruprecht.

McFague, Sallie. 1993. *The Body of God: An Ecological Theology*. Minneapolis: Fortress.

Moule, C. F. D. 1957. *The Epistles of Paul the Apostle to the Colossians and to Philemon: An Introduction and Commentary*. Cambridge: University Press.

Norden, Eduard. 1974. *Agnostos Theos: Untersuchungen zur Formengeschichte religiöser Rede*. Stuttgart: B.G. Teubner [1913].

Northcott, Michael, S. 2007. *A Moral Climate: The Ethics of Global Warming*. New York: Orbis.

Panikkar, Raimon. 2009. *Christophany: The Fullness of Man*. Translated by Alfred Dilascia. New York: Orbis Books.

Pizzuto, Vincent. 2006. *A Cosmic Leap of Faith: An Authorial, Structural, and Theological Investigation of the Cosmic Christology in Col 1:15-20*. Dudley, MA: Peeters.

Rahner, Karl. 1996. *Foundations of Christian Faith An Introduction to the Idea of Christianity*. New York: Crossroads.

Ricoeur, Paul. 1976. *Interpretation Theory: Discourse and the Surplus of Meaning*. Fort Worth: Texas Christian University Press.

Schleiermacher, F. 1832. "Über Koloss. 1:15-20." *Theologische Studien und Kritiken* 5: 497-537.

Schneiders, Sandra M. 1991. *The Revelatory Text: Interpreting the New Testament as Sacred Scripture*. San Francisco: HarperSanFrancisco.

Vischer, Lukas. 2004. "Listening to Creation Groaning: A Survey of Main Themes of Creation Theology." In *Listening to Creation Groaning: Report and Papers from a Consultation on Creation Theology Organised by the European Christian Environmental Network at the John Knox International Reformed Center from March 28 to April 1ˢᵗ 2004*, ed. Lukas Vischer, 11-31. Geneva: Centre international réformé John Knox.

White, Lynn Townsend, Jr. 1967. "The Historical Roots of our Ecological Crisis." *Science* 155, March 10, 1203-07.

PERSPECTIVES INSPIRED BY
MEDIEVAL THEOLOGIANS

4

BONAVENTURE'S FRANCISCAN CHRISTOLOGY

A RESOURCE FOR ECO-CONVERSION TOWARD HALTING HUMAN-FORCED GLOBAL CHANGE

Dawn M. Nothwehr, O.S.F.

Sallie McFague asks three ultimate questions that lie at the heart of the dilemma concerning if and how humans will act to halt human-forced climate change: "Who are we? Who is God? How shall we live?" (2008, 2). A vast interdisciplinary literature exists, linking the causes of human-forced global climate change and other environmental devastation to some appropriations of Christianity (McFague 2008, 27-40).[1] Often the loss of life's meaning, alienation, and a spiritual malaise are specifically named as the *root* causes (Sagan et.al., 1998, 2-6).[2] Furthermore, science has shown with great certainty that the consequent break down of the planet's major ecological cycles has forced unprecedented life-threatening global warming, bringing us to the brink of a shutdown of Earth's vital climatic functions (Intergovernmental Panel on Climate Change 2007).

Our hope resides in the reality that things were not always this way. Indeed Christianity and other religions offer healthy modes of being and acting. We must recover those life-giving resources, again lay hold of moral meaning, and act immediately to halt and reverse human-forced global warming. One Christian resource, ripe for retrieval of a renewed and ecological ethical vision, is the Christology of Bonaventure of Bagnoregio (1217-1274).

---

1   Also see Attfield 2009, 31-50; Northcott 2007, 45-80; Boff 1997, 63-85; and Toolan 2001, 41-74.

2   Also see Raven 2006, 4-5; and Doyle 1981, 72.

Bonaventure's Franciscan cosmic Christology provides a framework for a hope-filled theological and morally compelling response to McFague's three questions. In Jesus, we discover the divine clue to the structure and meaning not only of humanity, but of the entire universe. Integral to this Christology is its intimate and necessary integration with an incarnational spirituality. That characteristic of Franciscan Christology places it in a unique position to respond to the moral and spiritual malaise that undergirds today's anthropogenically generated climate change (Hayes 1996, 16-17; Delio 2008). In Bonaventure's Franciscan cosmic Christology, the Incarnation is integral to the possibility of creation itself; one is integral to the other. Christ is not accidental or an intrusion in creation but the inner ground of creation and its inner goal. Thus, Franciscan theologians hold that "a world without Christ is an incomplete world, that is, the whole world is structured Christologically" (Hayes 1996, 6). Though solidly anchored in the Christian tradition, Bonaventure presents a cosmic vision with universalist implications that invites dialogue with others who seek after truth, a quality so necessary when dealing on a planetary scale (Hayes 1996, 6).

In this three-part essay, I claim that Bonaventure's Franciscan cosmic Christology is a resource for eco-conversion toward halting forced global warming. First, I briefly situate Bonaventure's thought within the Franciscan Theological Tradition. Second, I outline the relevant theological tenets articulated by Bonaventure, showing the hope-filled vision of the intimately related and interrelated reality: God—God-Human—All Creation. Bonaventure identifies the Triune, fully relational God as the one whom, out of overflowing love, created a thoroughly related, relational, and connected world. The Incarnation is God's complete self-revelation that opens the fullness of life to humans and the entire universe. In such an interrelated world, humans are co-creators with the divine and are called to take on the mediating role as guardians of creation. At the heart of Bonaventure's understanding of God, the human person, and ethics is his Christology. His speculative cosmic Christology also grounds his historical and spiritual writings concerning Christian discipleship and ethics, namely the *"imitatio Christi."* In the last part, informed by Bonaventure's Christology as well as the IPCC's "Summary for Policy Makers," I address McFague's question, "How shall we live?" Here I incorporate Pope John Paul II's call for "ecological conversion" (2001) and social scientific studies illuminating Christian moral living toward halting anthropogenically generated climate change.

SITUATING BONAVENTURE'S CHRISTOLOGY IN THE FRANCISCAN THEOLOGICAL TRADITION

The Franciscan theological system is characterized by a particular set of core values and beliefs namely: conversion, peacemaking, the emphasis on divine love and freedom, the primacy of Christ, the centrality of the Incarnation, Christ crucified, the sacramentality of creation, the goodness of the world, the human person in the image of God, emphasis on poverty and humility, and the development of *affectus* (Delio 2002, 1-19.) The "common thread" sustaining Franciscan theology across time is the person and witness of Francis of Assisi, "the Patron of Ecology" (Ioannes Paulus II 1979). Francis' authority (*ex beneficio*, not *ex officio*) as a "vernacular theologian" (McGinn 1983, 6-7; 1998, 21) originated in the gracious gift of his experience of God (CDF 1990, 119; USCCB 1989, 101). Three major themes in Francis' vernacular theology irrevocably link Franciscan spirituality and theology: the humanity of Christ (Jn 14:6-9), the mystery of God as generous love, and the sense of creation as family (Francis of Assisi 1982a, 25-26).

Significantly, Francis asserted the life-sustaining virtue of mutual obedience among all the creatures. The obedient one "is subject and submissive to all persons in the world, and not only to human beings, but even to all beasts and wild animals so that they may do whatever they want with that person, in as much as it has been given to them from above by the Lord" (Francis of Assisi 1982b, 151-52). Furthermore, as reported in the "Mirror of Perfection," Francis denounced abusive human relationships with others (1972). Humans use other creatures to meet daily needs, yet they are often ungrateful to them, failing to recognize the Creator of such blessings. Bonaventure develops these linkages between Christ, creation, creatures, and humans, solidifying an inseparable connection between Franciscan theology, spirituality, and ethics (Hayes 1996, 6).

ST. BONAVENTURE'S CHRISTOLOGY: A HOPE-FILLED VISION

Bonaventure is the founder of the Franciscan theological tradition (Osborne 1994, 7-9; Hayes 1980, 62-62), and the best interpretive Christological voice of St. Francis of Assisi (Johnson 2001, 12; Daniel 1975, 48). Augustine, Pseudo-Dionysius, and Richard of St. Victor (Bougerol 1964, 23-49) ground and support his evangelical theological synthesis that is faithful to the intuitions of Francis and a rigorous intellectual structure and method (Delio 2002, 8; 1998). Bonaventure's academic career was truncated when he was elected the General of the Franciscan Order. This mixed blessing profoundly shaped the maturing of his theology (Hayes 1980, 62-62). The result was an integration of a refined theology and a powerful spirituality that gives Bonaventure's work great promise for our time (Leo XIII 1879).[3]

Bonaventure's Method and Metaphysics

The metaphysical structure supporting Bonaventure's thought consists of the notions of emanation, exemplarity, and consummation (Hoebing 2002, 273-75). Simply put, in Bonaventure's world view we come from God, we live in God, and we return to God. These three dimensions are also interrelated. Thus, it is not easy to discuss any part of Bonaventure's theology without also referencing other dimensions because it is a carefully interwoven whole. Concerning *emanation*, the mystery of divine, infinite self-diffusive goodness is the basis of the life *ad intra* of the Trinity. The inner relational life of the Trinity overflows externally to creation and humans in creation. Thus God, creation, and humans are radically related.

Concerning *exemplarism*, God is the prototype of everything that exists, and that expresses the divine in creation, thus revealing the Creator. Exemplarism can be defined as "the doctrine of the relations of expression between God and creatures" (Bowman 1975, 184). Using his modified notion of Plato's divine ideas, Bonaventure held that the Word expresses the Father's ideas, and thus, the Word is the Exemplar of all that exists. Further, the Exemplar is the active side of that expression, namely that which expresses itself. Correlatively, that in which the Exemplar is expressed is called an *image*. As Delio puts it: "God is the exemplary cause of all things, because God in knowing himself

---

3   See also McCool 1989, 163, 171-72, 190, 196-97.

expresses his most perfect idea, the *rationes* or the external patterns of all possible things. Everything that exists, therefore, is in some way related to God as a copy or imitation. Bonaventure distinguishes three different degrees of resemblance: the vestige, image and similitude. The exemplar is the pattern or original model in whose likeness all things are made. It is the basis of imitation. Because the Word expresses the Father's ideas, the Word is the Exemplar of all that exists" (Delio 2001, 200-01).

The third major component of Bonaventure's metaphysics is *consummation* or *reduction*. Here Franciscan spirituality and Bonaventure's theology converge. While all creation images and reveals God in Christ the Exemplar, its purpose is to lead humanity back to God. "The First Principle created this perceptible world as a means of self-revelation, so that like a mirror of God or a divine footprint, it might lead [humans] to love and praise [their] Creator" (Bowman 1975, 188). In fact, Bonaventure saw that the world in its ascending levels of reflecting God was like a ladder for [humans] to ascend in their return to God. Certainly Bonaventure grasped that nature can be admired, misunderstood, or abused. However, he insisted that the true seekers of wisdom (philosophers) will try to comprehend something by first understanding its causes. The goal of understanding is to know the essential structure of a thing, a structure constituted by a thing's relationship to its causes. This process of analysis (*plena resolutio*) is a goal of wisdom and the mirror image of exemplary causality. Though limited by sin, humans have the capacity of *contuition*—the ability to see God through things and things in God which is the ultimate significance of things in the economy of exemplarism.

Bonaventure's Christology opens a horizon to a renewed image of God intimately concerned with creation, understanding of the human person, and "earth friendly" moral living relevant for our world that is threatened by changes we are forcing on the global climate. His theological vision grounds a moral normativity of mutual relations that compels us to halt forcing these climate changes. Thus, his insights are useful in responding to McFague's three questions.

"Who is God?" Bonaventure's Trinitarian Theology

Bonaventure's starting point was the Trinity and he ended up with a highly relational theological system that centered on a cosmic Christology. Thus, his Christology is best understood in the context

of his Trinitarian theology. He followed the questioning of the Cap-
padocians trying to understand the relationship of the Christ of the
Gospels and the God of creation (Delio 2008, 55; LaCugna 1991,
53-81). Thus, Bonaventure focused on the divine *persons* of the Trinity. Implicit in the notion of *persons* is the experience of relationship;
to be a person is to be social, a person-in-relationship. Bonaventure
described this relationship of Trinitarian persons as *circumincessio* (to
move around one another in a communion of love) (Delio 2001, 41;
LaCugna 1991, 272).

The Trinity ad intra: Bonaventure first cited Jesus' statement to
the young man (Mk 10:18; Lk 18:19) that "No one is good, but God
alone" (1978, 94-95). From this valuation, Bonaventure concluded that
God's truest name is "Good." He also noted that Pseudo-Dionysius
called God "the self-diffusive goodness." So not only is God, "Being"
(Ex. 3:14), God is "that Goodness which is beyond Being, and which
gives rise to being" (Delio 2001, 52). He also utilized the Aristotelian
axiom—to the degree that something is prior, it is fecund. Therefore,
because Bonaventure understood God the Father as the Unoriginate
One, he asserted that the Father is the absolutely fecund source of the
entire cosmos (Delio 1999, 233-37; La Cunga 1991, 164). Thus understood, God becomes Bonaventure's "Fountain Fullness" or "Font of
All Goodness." God is absolute Goodness, therefore God is *necessarily*
uniquely *relational*. Yet, God also *chooses* to love (Gilson 1965, 163).

To further explain the Trinitarian relations, Bonaventure turns to
Richard of St. Victor (Hayes 1979, 3-40). Richard of St. Victor calls
God's goodness charity (love). By its very nature, love is dynamic and
requires a second person; a lover requires a beloved. If a love relationship is not turned in on itself, then a perfect love requires a third person for the relationship with whom the lover and the beloved can each
share their love. Thus, Richard of St. Victor provides the model for
the Trinitarian God: the totally gratuitous love (Father); absolute love
(Spirit); and both gratuitous and receptive love (Son). For Bonaventure therefore, God is the personal, self-communicating, generous,
overflowing Goodness and Source of all reality who desires to express
the divine self *ad intra* and *ad extra*. Like an infinite intensely flowing
fountain this self-diffusive primal and eternal Goodness generates the
Son *ad intra* and the entire cosmos *ad extra* in a *kenotic* expression of
that goodness.

The Father: Bonaventure sees the generation of the Son by the Father as the personal diffusion of the love of the Father (*per modum naturae*) (Delio 2001, 52). This is important for our purposes because: "The Son therefore, is everything the Father is in One other than the Father. The relationship between the Father and the Son in Bonaventure's thought is the very ground of all the other relationships. An understanding of this relationship allows us to comprehend the deeper meaning of creation—the centrality of Jesus the Christ" (Delio 2001, 45).

The Son is also the "Image of the Father" in the sense that, like the Father, the Son is the source of others. Indeed, the Son is the true and perfect expression of the Father. Bonaventure preferred the title "Word" for the Son because it best encompasses the reality of the Son as the Father's expression. More precisely, the Son is the *causal word* (Bonaventure 1989). In language usage, we distinguish between the *mental word* or internal thoughts of a person and their actual expression of their ideas or thoughts. Similarly then, "the eternal self-thinking of the Father, that is, all the eternal ideas of the Father, are expressed in the Word" (Delio 2001, 46). Bonaventure thus calls the Son the "Art of the Father." Understanding the Word as the Father's self-expression and self-diffusive goodness gives rise to a set of relationships *ad extra*. The speaker is the Father; the audience is all of creation; the sound or expression is the Incarnation; and, the knowledge attained by others through the mediation of the Word is found in the Scriptures (Delio 2001, 53).[4]

As the Father's self-expression, the Word though singular, is however the one through whom all of the multiple divine ideas are articulated. Thus, the vast diversity that is creation reveals something about the Father and the Son, but it is not God (Hayes 1996, 11-12). "When the Word becomes incarnate, the eternal Word enters into union with creation through humanity, and the mystery of the Father is uttered in history and time" (Delio 2001, 48). This connected understanding of God's expression in the world also holds vast implications for the moral meaning of forced global climate change.

The Holy Spirit: In Bonaventure's understanding of the relational Triune God, the Holy Spirit is the communication of the Father's goodness by the will (*per modum voluntatis*). The Spirit proceeds from the free will of the Father and the Son. The Father is the primal source

---

4   See also Hayes 1992, 55-63 and 1979, 51-52.

of love and the Son is the emanation of love proceeding from pure liberality; their sharing of love as an act of the will, spirates forth the Spirit. The Spirit is the fruit of mutual love. It is the gift of the Spirit in creation that unites humanity with the Father and the Son. And, most significantly, the mutual Trinitarian relations signal the normative mode of all moral relationships.

Bonaventure understands the Word as the *medium* between the Father and the Spirit. The Son with the Father is the source of others (all creation, Holy Spirit); the Father and the Son together spirate the Spirit. The Holy Spirit and the Son share the characteristics of receptivity in so far as they are both generated from the Father's self-diffusive goodness. The Word is at the center of Trinitarian relations *ad intra* and as we shall see, *ad extra*. This holds broad implications for global climate change—in so far as the inspirited, sacramental, mutually related, created world draws in, inspires, heals, and refreshes humanity.

"Who is God?" Bonaventure's Doctrine of Creation

The Triune God is the God who creates the entire cosmos. Using modified Platonic models, Bonaventure explained that the world is one where God freely created everything as ordered, oriented, and directed toward its goal, in a manner that all of creation is fully interrelated and interdependent. (Hayes 2002b, 249-67). He sees the community of creation as the product of the overflowing self-diffusive and self-expressive love of the community of the Trinity. Based on hylomorphism (everything is open to change) and the platonic doctrine of seminal principles (everything has potential), creation is dynamic and open to transformation (Delio 2001, 56-57; Hayes 1994, 72-79). Simply put, Bonaventure held that it was part of God's plan that all creation would desire to be perfected. The Word is the Art of the Father, and the creation is the expression of the Artist. Of all in creation, human beings are those most capable of participating in and manifesting God's glory (Hayes 1994, 65).

"Who is God?" Bonaventure's Christology

Bonaventure began his Christological discussion with the faith claim that "the Word became flesh" (Jn 1:14), and he asked "why?" He then proceeded to probe this question from the God side and from the human side (Hayes 1994, 87). Bonaventure found the answer rooted

in the ways God reveals godself. The union of human and divine natures is grounded in the possibility of God as the Creator. God creates
out of self-diffusive and self-communicative love. God's self-revelation
is given in its broadest sense in all of creation. The Divine is revealed
more specifically in the human nature (*imago Dei*). But, the most
complete revelation of God is in the human and divine, Jesus Christ
(Hayes 1994, 83).

As we have said, in Bonaventure's Logos-centered theology of revelation, the Word, God's revelation is given in the entirety of the universe. Also, within creation, God's revelation is found in humans as
the *imago Dei*. And, recall that in Bonaventure's Trinitarian theology,
the Son is also known as the Image. "Thus, in the mystery of the incarnation, the created image is filled with the eternal exemplary Image. In this way humanity reaches its fullest participation in the divine
archetype and thus, the deepest fulfillment of its potential" (Hayes
1994, 86). Humans are unique among the creatures (as is each creature) in that by virtue of their spiritual dimension, they have an inner
ordering to immediacy with God. When this capacity is brought into
act by divine initiative, "the created order finds its highest form of fulfillment. This is what Bonaventure understands the hypostatic union
to involve" (Hayes 1994, 87). Here we see in Bonaventure a kind of
evolutionary world view that later theologians would name "Christogenesis" (Boff 1997, 174-86).

The cosmic dimensions of Bonaventure's Christology also have significance for our purpose. He "perceive[d] the possible relations between the story of Jesus and the larger picture of the world" (Hayes
1996, 6). The Incarnation must be thought of as God's intent from
the very moment of creation as part of a broader cosmic vision; it
was not an afterthought subject to human sin. Thus, the entire world
was transformed at the moment of the Incarnation—when the divine
becomes part of the material world: "All things are said to be transformed in the transfiguration of Christ. For as a human being, Christ
has something in common with all creatures. With the stone he shares
existence; with plants he shares life; with animals he shares sensation;
and with the angels he shares intelligence. Therefore, all things are said
to be transformed in Christ since—in his human nature—he embraces something of every creature" (Hayes 1989, 13).

"Who Are We?" Bonaventure's Christian Anthropology

Humans are "the ones in creation who freely choose God, and in choosing God, become like God—perfect in love" (Delio 2001, 59; Hayes 1976, 318). Because humans as the *imago Dei* are both spirit and matter, they have the potential to unite everything with God. Thus, humans stand in the middle of creation as the creatures who can lead it to perfection (or dominate and oppress it). In Bonaventure's view, that humans are in the center, indicates a position of mediation. Humans and other creatures of creation serve one another in relationship. Humans advocate for the other creatures, giving them voice before God and the world (Nothwehr 2005, 38-39; Cousins 1992, 152-55). The other creatures inspire and thrill humans, arousing their awareness of the magnificence and generosity of God, and moving them contemplation. From the Uncreated Lover comes an unlimited variety of ways in which love is communicated, namely all of creation; and in Christ Incarnate the entire cosmos is united and transformed (Bougerol 1964, 40).

Yet humans are also contemplative, ethical beings, according to Bonaventure. His speculative Christology served to ground and integrate a spirituality and agapistic virtue ethics for humans, the spiritual and ethical beings at the center of creation. In *The Soul's Journey into God*, Bonaventure extends Francis' insight of cosmic union. Not only did Bonaventure see union with God reflected in the *external* world, but also, *within the human person*, in the inner act of sensation, in aesthetic experience, in the activities of memory, understanding, and will, and in the contemplation of Being and self-diffusive Goodness, or God" (Cousins 1992, 151). In *Collationes in Hexaemeron* 12, 14 Bonaventure explains: "[T]he entire world is a shadow, a road, a vestige, and it is also a book written without. (Ex2:8; Ap.5:1) For in every creature there is a shining forth of the divine exemplar, but mixed with the darkness. Hence creatures are a kind of darkness mixed with light. Also they are a road leading to the exemplar. Just as you see a ray of light entering through a window is colored in different ways according to the colors of the various parts, so the divine ray shines forth in each and every creature in different ways and in different properties; it is said in Wisdom: *In her ways she shows herself* (Wis 6:17). Also creatures are a vestige of the wisdom of God. Hence creatures are a kind of representation and statue of the wisdom of God. And in view of all of this, they are a kind of book written without" (trans. Cousins 1992, 152).

Creation was executed by God like an artist who conceives an idea in her imagination and then expresses it externally. Beginning with time, God created everything from nothing, setting the cosmos on a journey that moves forth from God and moves toward a return to God, revealing God-self along the way. The role of the material and natural world is to arouse the human consciousness to praise and love God, and in turn, humanity is to serve nature by giving it a voice that which would not otherwise be heard in the cosmos (Hayes 1994, 67-68). Bonaventure thus demonstrates how there is an intimate relationship between God and each creature of the cosmos (including humans), and of each of them with one another through Christ.

If humans ignore this relationship, they fail to comprehend their position in the world and the cosmos! As Bonaventure puts it: "Therefore whoever is not illuminated by such great splendors in created things is blind. Anyone who is not awakened by such great outcries is deaf. Anyone who is not led from such great effects to give praise to God is mute. Anyone who does not turn to the First Principle as a result of such signs is a fool. Therefore open your eyes, alert your spiritual ears, unlock your lips and apply your heart, so that in all the creatures you may see, hear, praise, love and adore, magnify and honor God, lest the entire world rise up against you" (Bonaventure 2002, 61). Humans created in the *imago Dei* with the faculties of memory, intellect, and will are drawn by the goodness in creation to comprehend and value it as sacred in the panentheistic sense as a revelation of the Creator. In so doing, humans find their role as the mediators between God and creation and as guardians of the multiple manifestations of God's self-revelation in Christ (Cousins 1992, 152-55).

HOW SHALL WE LIVE?
BONAVENTURE'S CHRISTOLOGY
IMITATIO CHRISTI, VIRTUE ETHICS, &
HUMAN-FORCED CLIMATE CHANGE

The *imitatio Christi* has both an internal and external dimension. Humans as spiritual beings seek union with God while on a journey in our world that is utterly bursting forth God's self-revelation. Because Jesus Christ is the full revelation of God in the flesh, he is also the concrete source for humanity's comprehension of the moral life. Therefore, citing the gospels (especially Luke and John) and the experience of St. Francis, Bonaventure asserted that the whole life of Jesus Christ is instructive (Hayes 1992, 27). The vision is one of a loving Father who gave his Son to sinful humanity, and a Son who, though divine, became our human brother in Jesus of Nazareth.

However, Jesus is far more than a mere ethical model. The *imitatio Christi* moves beyond a simplistic mimicking of Jesus' activity, to the transformation of an individual's interior disposition, grounded in a loving relationship with God in Christ. Indeed, we must distinguish between example and Exemplar. There is a constant interplay between the ethical (external) and the spiritual (internal) that is illuminated by God's grace and through which the human soul is "rendered deiform" (Hayes 1992, 42). Christ mediates this deification by purging human guilt, enlightening people through his example and teaching, removing punishment, and conferring grace and glory (Hayes 1992, 44-45). The spiritual life directs the moral life of the Christian. Central to this transformation are the virtues of humility, poverty, obedience, and love.

Humility is the virtue most central to the God-human relationship. Humans must revel (not grovel) before God who loved so much as, in the Incarnation, to humbly bend down and lift the lowliness of our nature into unity with his own person (Bonaventure. 1989). This is the child-like humility of those "greatest in the kingdom of heaven" (Matt 18:4). Yet, the depth of moral maturity is exemplified in Jesus' eucharistic washing of the disciple's feet (Jn 13:1-15). Jesus, knowing full well who he was, consciously chose to serve those not his equal, and then commissioned them to do likewise. Imitating Christ's humility, requires a personal self-identity in integral balance; gift/needs or strengths/limits. People need to see themselves intimately loved by God, yet creatures amid others beloved of God in the cosmos (Bowman 1975, 187).

Poverty of spirit is nearly synonymous with humility, though Bonaventure focused on material poverty while stressing the integration of the two virtues in Christ. Jesus's life and ministry began in a manger, thrived in relationships with the marginalized, and culminated in his death with common criminals on the cross. In moments of material poverty, God's grace and presence can be most tangible and desirable, complementing human limitations. Many who are materially wealthy know only a false poverty (material), the illusion of never having enough of anything, yet always ultimately craving something beyond what they thought would satisfy (spiritual poverty). Key to the virtue of poverty is being receptive to God's gifts while not *possessing* them, ready to give away what has been given as gift (Matt 10:8).

Obedience (Latin *oboedire* means "to pay attention") is the virtue Christ enacted by his constant attention to his Father's will (Heb 10:7, 9; Jn 14:31; Jn 5:30) and to the needs of human and non-human others (Hayes 1992, 37). This attentiveness to relationships enabled him to embrace the cross and open the post-resurrection new creation.

Love drives the Christ-mystery in Bonaventure's speculative theology and spirituality (Hayes 1992, 38-39). God's love overflows to humankind and creation, reaching perfection in the Incarnate Christ an in all of his acts. Humans respond by mediating that love among themselves and all of creation (Jn 13:34-35). Because human efforts always fall short, love needs to join with justice utilizing the discipline of law. Justice is able to restore beauty to the world; making upright what was deformed by sin. Citing Aristotle's axiom, "Virtue consists in the middle way," Bonaventure held that the Christ, as the center and archetype of the world and all virtue, is the ultimate norm and arbiter of justice (love) (Hayes 1992, 202-03).

Christ: Paradigm and Norm

The spiritual life in Christ is a journey deeper into the ontological realities of the world. Understanding those realities also shapes Christian ethics. For Bonaventure, "To perceive the life of Christ as a paradigm is to accept its fundamental values as normative for human life. The fundamental attitudes and values of Christ must be so personalized in one's life, that they truly define one's relationship to reality" (Hayes 1992, 39).

For our purposes, we find Jesus' life, ministry, and teaching radically relational and intimately connected with Earth. Beginning with his inaugural sermon (Lk 4:16-22) he shows the vital bond between care for Earth and care for the poor. That was grounded in the theological understanding alive in Israel and expressed in Leviticus 25, particularly in the notions of Sabbatical and Jubilee Year that mandate debt forgiveness and return to one's own land (economics), restoration of the land and rest for animals (ecology), and liberation of all enslaved (ecumenicity).

Imitatio Christi: *Virtue Ethics & Human-forced Climate Change*

As we have seen, Bonaventure's cosmic Christology is integral to a positive and radically relational view of God, creation, redemption, and ethical praxis (Boff 1997, 174-86).[5] "Bonaventure consistently claims that Christ belongs to the very structure of reality—as the Word, to the reality of God; as the Incarnate Word, to the reality of the universe created by God. It is Christ who reveals to the world its own meaning" (Delio 2003, 313). All creation is loved by God for its own sake. Humans have the capacity to read the signs of the divine in creation and then take up their proper relationship to God, creation, and fellow humans.

Bonaventure's cosmic Christology flows from the Gospel of John: "and the Word became flesh (*sarx*) and dwelt among us, full of grace and truth" (Jn 1:14). The Johannine writer uses the Greek word "*sarx*" to join the Incarnation with the contingent, material, perishable, fragile and finite—conditions that characterize the created cosmos and human flesh (Edwards 2006, 58-60).[6] The incarnation unites humans with the whole cosmos, and Jesus Christ incarnate embodies and exemplifies the norm for all ethical relations. By following the example and teachings of Jesus, humans can participate with Christ as co-creators and co-redeemers of the cosmos. Bonaventure insisted that there was no imitation of Christ without humility, poverty of spirit, austerity of life, and genuine charity (Hayes 1992, 39). He was equally emphatic that those virtues must be lived in compliance with Jn 13:34-35: "Just as I have loved you, you also must love one another. By the love you have for one another, everyone will know that you are my disciples." This is the way of the *imitatio Christi* and the basis for healing in the world that is offered to humanity in Christ.

Bonaventure's Christology also opens the way to an "inclusive Christology," that is, "what Christians call God, grace, and salvation is in the some way universally accessible to human experience even though it may be differently understood and articulated at different points in history and in different cultures" (Hayes 2002, 53-55). Practically, Jesus models this inclusiveness as the meaning of *agape* through his life, ministry and teaching. Quite explicitly, our relationship with God brings us into relationship with the world.

---

5   See Jn 1:3, 14; Heb 1:2; Col 1:15-20; Eph 1:3-14; Rev 1:8 and 21:6.

6   See also Reid 2000, 69-83; Gregersen 2001, 205.

Considering his grasp of the cosmos and openness to science (Hayes, 2002b, 249-67) and arguing from his *agapic* virtue ethics, Bonaventure would require our halting activities that are forcing changes to the global climate and threatens all earthly life (IPPC 2007). Jesus's engagements with the marginalized, sick, and poor model what is currently (and what will be) required to care for the needs of the poor who are and will be victims of the perils of desertification and flooding, famine and disease that are predicted to immediately result from unchecked human-forced global warming (Baum et al. 2007; Johnson 2009, 27-30).

A Story

Susan Strauss tells of Gerry who was walking with a Native American friend amid noon hour traffic in Washington, DC. Roaring engines, honking, and jabbering crowds made it difficult to hear anything else: "In the middle of the traffic, Gerry's friend stopped and said, 'Hey, a cricket!' 'What?' said Gerry. 'Yeah, a cricket,' said his friend. 'Here, look,' and he pulled aside some bushes that separated the sidewalk from the government buildings. There in the shade was a cricket chirping away. 'Wow,' said Gerry. 'How did you hear that with all this noise and traffic?' 'Oh, said the Native man. 'It was the way I was raised...what I was taught to listen for. Here, I'll show you something.' The Native man reached into his pocket and pulled out a handful of coins...nickels, quarters, dimes...and dropped them on the sidewalk. Everyone who was rushing by stopped...to listen" (Strass 1996, 9). Dominantly econonmistic Western culture will fail to listen and to respond ethically to the need for halting anthropogenically generated changes to the global climate unless we learn to value and attend to things beyond only money (Brandt 1995). The IPCC reports the litany of about a dozen major catastrophic changes that are already occurring and makes clear recommendations for personal and public policy changes.

"Ecological Conversion"

The late John Paul II called for "ecological conversion:" "[I]f one looks at the regions of our planet, one realizes immediately that humanity has disappointed the divine expectation.... [Humans have] unhesitatingly devastated wooded plains and valleys, polluted the waters, deformed the earth's habitat, made the air unbreathable, upset the hydrogeological and atmospheric systems, blighted green spaces, implemented uncontrolled forms of industrialization, humiliating... the earth, that flower-bed that is our dwelling. It is necessary, therefore, to stimulate and sustain the "ecological conversion," which over these last decades has made humanity more sensitive when facing the catastrophe toward which it was moving" (John Paul II 2001). Some people have successfully stepped back from the abyss. Yet many Christians languish in apathy and continue "business as usual." What can motivate us to change? The social sciences can shed light on the harmful ecological behaviors of otherwise good and faithful people.

Eco-philosophy

Eco-philosopher Joanna Macy shows that when facing human-forced global warming facts straight up, people (especially Westerners) tend to become paralyzed and move to denial and inaction (1998). The resolution to human-forced climate change requires values, attitudes, and skills that have become foreign to us in the developed world. We repress the overwhelming pain of ecocide and gaiacide because expressing pain is thought dysfunctional! We ignore the complexities of forced global climate change because we fear looking intellectually inadequate to develop solutions to the problem—so we "leave it to the experts." In the U.S. we think of ourselves as the "good guys." The facts tell otherwise! We are 20% of the world's population living in opulence, consuming 80% of the wealth! We can't claim or cope with the guilt (personal or collective).

Following Robert J. Lifton's findings from Hiroshima survivors, Macy holds that the way out of our paralysis is to realize that the pain is not only real. It is healthy! While in times past, people were not sufficiently concerned about human-forced global warming, there is renewed hope in our experience of pain. Once aware, we can restore and renew ourselves and Earth by doing the opposite of what much of Western culture trained us to do. By facing fears, acknowledging pain, and choosing to live within limits, we break the isolation and begin the healing. This is done best in relation to the web of life and in community with other persons. All of these moments of awakening are opportunities to live out humility, poverty, obedience, and love.

Motivational Studies

Recent motivational studies show some common factors that assist and empower people to act toward halting forced global warming.

Noreen Allossery-Walsh found four motivational trends contributing to ecological responsibility that are common to the people studied: (1) emotional connections with nature were expressed through care and concern; (2) support was found from social groups; (3) ecological literacy and knowledge of God as incarnational was discovered; and (4) a sense of the ability to make a difference (creative ideas and strategies) was realized (Allossery-Walsh 2009, 99). Additional factors motivating responsibility include: visions of future possibilities, making verbal commitments, cultivating a sense of individual responsibility and competence in bringing about change. Guilt and fear are the least effective motivators; in fact they deter proactive behavior.

Allossery-Walsh's research on Catholic lay Earth Ministers shows that "the Christian tradition does appear to offer tangible motivation for ecological responsibility in the case of lay leaders" (ibid. 100). She notes that the emergence of the "ecological self," the self understanding that expands beyond individualism to embrace communities, the rest of the natural world, our evolutionary past and the distant future, is a positive on-going motivation. Catholic lay Earth Ministers named two major motivating factors leading to ecological responsibility and action. Earth Ministers found cognitive dissonance in an individualistic self-identity but were drawn to something more:

> Consciousness of an ecological self involving a widening identity and self-interest due to three converging developments.
>
>   a. The dangers that are threatening to overwhelm us and a loss of certainty that there will be a future.
>
>   b. The emergence of the new paradigm in science, replacing old assumptions about a distinct, separate, continuous self with processive 'self' seen as an open, self-organizing system whose principal functions arise in interaction with the environment.
>
>   c. The resurgence of non-dualistic spiritualities...An expanded self is emerging, arising in the grief, the anger, the fear we experience before the decimated ecosystem and our deconstructed world (Allossery-Walsh 2009, 101-02).[7]

The second motivating factor was found in the conflict produced by a utilitarian understanding of Earth that led Earth Ministers to a "Consciousness of a deep ecology. A vision of life which believes that we cannot deal with nor transform nor solve ecological conflict by coming up with an endless series of ad hoc remedies to each catastrophe"

---

7   See also Ludwig 1995, 192; and Fuller 1992, 32.

(ibid.). Allossery-Walsh concludes: "Ultimately then, ecological responsibility motivated by elements of Christian tradition is determined by consciousness of the ecological self, where the expansion of the self acknowledges the centrality of relationship with the Creator. And where caring for all Creation is understood as a call to participate communally in the divine life where justice flourishes" (ibid. 103).

Public Health Studies

Cindy L, Parker and Steven N. Shapiro describe forced global warming as the ultimate public health catastrophe (2007). They show that experience with controlling epidemics and natural disasters stipulates two dimensions of action: immediate care for victims and preventative measures. Both dimensions require a connected vision of the past, present, and future in order to be effective. In terms of personal immediate actions and larger national and international policies, the current state of affairs concerning human-forced global climate change is the quintessential example of what results from living in a false reality. The climatic cycles are globally interdependent and any change anywhere affects everything else, yet we have lived as though we could control it all in one place (U.S. Global Change Research Program 2009). Health requires a return to behaviors that connect people to each other in sustainable ways and support them in living within the limits of Earth's carrying capacity.

CONCLUDING OBSERVATIONS

Motivational studies confirm what Bonaventure knew long ago—we live in an interdependent, interconnected world for which we need to care in ways that respect the interrelationship of all. In reply to McFague's questions, "Who are we?" and "Who is God?" we must acknowledge our creaturely status as loved and sustained by a generous, fecund God who desires relationship with us and the entire cosmos (humility). God is utterly present in the cosmos that reveals God to us. Answering her question, "How shall we live?" we must let go of our arrogance, our fears, and our constant "need" to consume more (poverty). We have to learn the 13.7 billion year story of the cosmos and the new science that reveals the chaos, diversity, and evolutionary order of the cosmos that is held in the hopeful promise of God (obedience). Grounded in that Great Love, we must act to love our neighbor as ourselves by reducing our carbon footprint, listening for the chirp of the crickets, and living sustainably upon Earth.

SOURCES

Allossery-Walsh, Noreen. 2009. Christian Ecological Responsibility: Intimations of Prophetic Witness for the Church in the New Millennium. Unpublished D. Min. Thesis, Catholic Theological Union, 14 May 2009.

Anderson, Vinton R. 1998. "Statement by Religious Leaders at the Summit on Environment." In *Ecology and Religion: Scientists Speak*, eds. John E. Carroll and Keith Warner, 7-12. Quincy, IL: Franciscan Press.

Attfield, Robin. 2009. "Social History, Religion, and Technology: An Interdisciplinary Investigation into Lynn White's 'Roots.'" *Environmental Ethics* 31: 31-50.

Baum, Gerald, Monika Hellwig, and W. Malcolm Byrnes. 2007. "Global Climate and Catholic Responsibility: Facts and Responses." *Journal of Catholic Social Thought* 4: 313-401.

Boff, Leonardo. 1997. *Cry of the Earth, Cry of the Poor*. Ecology and Justice Series. Translated by Phillip Berryman. Maryknoll: Orbis Books.

Bonaventure of Bagnoregio. 1978. *The Soul's Journey into God*, 5: 1-2. In *The Soul's Journey into God-The Tree of Life-The Life of St. Francis*. Translated and Introduced by Ewert Cousins. New York: Paulist Press.

———. 1989. "Sermon II on the Nativity of the Lord." In *What Manner of Man? Sermons on Christ by St. Bonaventure*. Translated by Zachary Hayes. Chicago: Franciscan Herald Press.

———. 2002. *Works of St. Bonaventure: Itinerarium Mentis in Deum*. Translated by Zachary Hayes. St. Bonaventure, NY: The Franciscan Institute.

Bougerol, J. Guy. 1964. *Introduction to the Works of Bonaventure*, Vol. 1. Translated by José de Vinck. Patterson, NJ: St. Anthony Guild Press.

Bowman, Leonard J. 1975. "The Cosmic Exemplarism of Bonaventure." *The Journal of Religion* 55: 181-98.

Bragg, Elizabeth Ann. 1997. "Ecological Self: An Invitation on a Shamanic Journey," paper presented at the cameo session of the Environment Stream - 'Thinking Like a Mountain,' World Futures Studies Federation XV World Conference, University of Queensland, Brisbane, Australia, September 28-October 3, 1997. Accessed from http://www.rainforestinfo.org.au/deep-eco/shaman.htm on May 25, 2009.

Brandt, Arbara. 1995. *Whole Life Economics*. Gabriola Island, BC: New Society.

Congregation for the Doctrine of the Faith (CDF). 1990. "Instruction on the Ecclesial Vocation of the Theologian." *Origins* 20: 119.

Cousins, Ewert. 1992. *Christ of the 21ˢᵗ Century*. Rockfort, MA: Element, Inc.

Daniel, E. R. 1975. *The Franciscan Concept of Mission in the High Middle Ages*. St. Bonaventure, NY: The Franciscan Institute.

Delio, Ilia. 1998. *Crucified Love: Bonaventure's Mysticism of the Crucified Christ*. Quincy, IL: Franciscan Press.

————. 1999. "Bonaventure's Metaphysics of the Good." *Theological Studies* 60: 228-46.

————. 2001. *Simply Bonaventure: An Introduction to His Life, Thought, and Writings*. Hyde Park, NY: New City Press.

————. 2002. "The Franciscan Intellectual Tradition: Contemporary Concerns." In *The Franciscan Intellectual Tradition*. CFIT/ESC-OFM Series No. 1. Washington Theological Union Symposium Papers, 2001, ed. Elise Saggau, 1-19. St. Bonaventure, NY: The Franciscan Institute.

————. 2003. *A Franciscan View of Creation: Learning to Live in a Sacramental World*. The Franciscan Heritage Series Vol. 2. St. Bonaventure, NY: The Franciscan Institute.

————. 2008. *Christ in Evolution*. Maryknoll: Orbis Books.

Doyle, Eric. 1981. *St. Francis and the Song of Brotherhood and Sisterhood*. New York: Seabury Publishers.

Edwards, Denis. 2006. *Ecology at the Heart of Faith*. Maryknoll: Orbis Books.

Francis of Assisi. 1982a. Admonition 1:1-4. In *Francis and Clare*, eds. Regis Armstrong and Ignatius Brady, 25-26. Mahwah, NJ: Paulist Press.

————. 1982b. Salutation to the Virtues, 14-18. In *Francis and Clare*, ed. Regis Armstrong and Ignatius Brady, 151-52. Mahwah, NJ: Paulist Press.

Fuller, Robert C. 1992. *Ecology of Care: The Interdisciplinary Analysis of the Self and Moral Obligation*. Louisville: Westminster/John Knox Press.

Gilson, E. 1965. *The Philosophy of Bonaventure*. Patterson, NJ: St. Anthony Guild Press.

Gregersen, Neils Henrik. 2001. "The Cross of Christ in an Evolutionary World." *Dialog* 40: 192-207.

Hayes, Zachary. 1976. "Incarnation and Creation in St. Bonaventure." In *Studies Honoring Ignatius Charles Brady, Friar Minor*, Theology Series No. 6. eds. Romano Stephen Almagno and Conrad L. Harkins, 309-339. St. Bonaventure, NY: Franciscan Institute.

————. 1979. "Introduction to the Disputed Questions on the Mystery of the Trinity." In *Works of Saint Bonaventure*. vol. 3, 3-40. St. Bonaventure, NY: The Franciscan Institute.

————. 1980. "The Life and the Christological Thought of St. Bonaventure." In *Franciscan Christology: Selected Texts, Translations and Introductory Essays*, Franciscan Sources No. 1, ed. Damian McElrath, 62-64. St. Bonaventure, NY: The Franciscan Institute.

————. 1992. *The Hidden Center: Spirituality and Speculative Christology in St. Bonaventure*. Franciscan Pathways. St. Bonaventure, NY: The Franciscan Institute.

Hayes, Zachary. 1976. "Incarnation and Creation in St. Bonaventure." In *Studies Honoring Ignatius Charles Brady, Friar Minor*, Theology Series No. 6., eds. Romano Stephen Almagno and Conrad L. Harkins, 309-339.

————. 1994. "Bonaventure: Mystery of the Triune God." In *The History of Franciscan Theology*, ed. Kenan B. Osborne, 72-79. St. Bonaventure, NY: The Franciscan Institute.

————. 1996. "Christ, Word of God and Exemplar of Humanity." *Cord* 46: 16-17.

————. 2002a. "Bonaventure: Paradigm for Franciscan Theologians?" In *The Franciscan Intellectual Tradition*. CFIT/ESC-OFM Series Number 1/ Washington Theological Union Symposium Papers, 2001, ed. Elise Saggau, 53-55. St. Bonaventure, NY: The Franciscan Institute.

————. 2002b. "The Cosmos, a Symbol of the Divine." In *Franciscan Theology of the Environment: An Introductory Reader*, ed. Dawn M. Nothwehr, 249-67. Quincy, IL: Franciscan Press.

Hoebing, Phil. 2002. "St. Bonaventure and Ecology." In *Franciscan Theology of the Environment: An Introductory Reader*, ed. Dawn M. Nothwehr, 273-75. Quincy, IL: Franciscan Press.

Intergovernmental Panel on Climate Change. 2007. "Climate Change 2007: Synthesis Report—Summary for Policymakers." Accessed from http://www.ipcc.ch/pdf/assessment-report/ar4/syr/ar4_syr_spm.pdf on May 19, 2009.

Ioannes Paulus Pp. II. 1979. Litterae Apostolicae. "Inter Sanctoss, Franciscus Assisiensis Caelestis Patronus Oecologiae Cultorum Eligitur." Accessed from http://www.vatican.va/holy_father/john_paul_ii/apost_letters/1979/documents/hf_jp-ii_apl_19791129_inter-sanctos_lt.html on March 18, 2009.

Johnson, Elizabeth A. 2009. "An Earthy Christology: 'For God so loved the Cosmos.'" *America* 200/12, Whole No. 4852: 27-30.

Johnson, Timothy J. 2001. "Lost in Sacred Space: Textual Hermeneutics, Liturgical Worship, and Celano's Legenda Ad Usum Chori." *Franciscan Studies* 59: 109-131.

La Cugna, Catherine Mowry. 1991. *God for Us: The Trinity and the Christian Life*. New York: HarperCollins Publications.

Lee, Kaman. 2009. "Factors Promoting Effective Environmental Communication to Adolescents: A Study of Hong Kong," *China Media Research* 4 (3): 28-36. Accessed from http://www.chinamediaresearch.net/vol4no3/04Kaman%20 Lee-final.pdf on May 25, 2009.

Ludwig, Robert A. 1995. *Reconstructing Catholicism for a New Generation*. New York: Crossroad Publishing Company.

Macy, Joanna. 1998. "The Greatest Danger: *Apatheia,* the Deadening of Mind and Heart." In *Coming Back to Life: Practices to Reconnect Our Lives, Our World*, 23-38. Gabriola Island, BC: New Society.

McCool, Gerald A. 1989. *From Unity to Pluralism: The Internal Evolution of Thomism*. New York: Fordham University Press.

McFague, Sallie. 2008. *A New Climate for Theology: God, the World, and Global Warming*. Minneapolis: Fortress.

McGinn, Bernard. 1983. *Meister Eckhart and the Beguine Mystics*. New York: Continuum.

———. 1998. *The Flowering of Mysticism: Men and Women in the New Mysticism-1200-1350*. New York: Crossroad.

"Mirror of Perfection." 1972. In *Omnibus of Sources*, ed. Marion A. Habig, 1236. Chicago: Franciscan Herald Press.

"Motivating and Sustaining Pro-environmental Behaviors." 1998. A Review of the Literature. Accessed from http://www-personal.umich.edu/~sarhaus/courses/NRE530_F1998/kjlawren/motivate.html on May 25, 2009.

Northcott, Michael S. 2007. *A Moral Climate: The Ethics of Global Warming*. Maryknoll: Orbis Books.

Nothwehr, Dawn M. 2005. *The Franciscan View of the Human Person: Some Central Elements*. The Franciscan Heritage Series, Vol. 3. St. Bonaventure, NY: The Franciscan Institute.

Osborne, Kenan B., ed. 1994. *The History of Franciscan Theology*. St. Bonaventure, NY: The Franciscan Institute.

Parker, Cindy L. and Steven N. Shapiro. 2008. *Climate Chaos: Your Health at Risk*. Westport, CT: Praeger Publications.

John Paul II, Pope. 2001. General Audience Address, January 17, 2001. Accessed from http://www.vatican.va/holy_father/john_paul_ii/audiences/2001/documents/hf_jp- ii_aud_20010117_en.html on May 26, 2009.

Leo XIII, Pope. 1879. "*Aeterni Patris*, Encyclical on the Restoration of Christian Philosophy (1879)." Accessed from http://www.vatican.va/holy_father/leo_xiii/encyclicals/documents/hf_l-xiii_enc_04081879_aeterni-patris_en.html on March 12, 2009.

Raven, Peter H. 2006. The Sustainability of the Earth: Our Common Responsibility. A paper presented at "Without Nature: A New Condition for Theology," University of Chicago Divinity School, October 26-28.

Sagan, Carl, et.al. 1998. "An Open Letter to the Religious Community." In *Ecology and Religion: Scientists Speak*, ed. John E. Carroll and Keith Warner, 2-6. Quincy, IL: Franciscan Press.

Reid, Duncan. 2000. "Enfleshing the Human." In *Earth Revealing-Earth Healing: Ecology and Christian Theology*, ed. Denis Edwards, 69-83. Collegeville, MN: Liturgical Press.

Strauss, Susan. 1996. *The Passionate Fact*. Golden, CO: North American Press.

Toolan, David. 2001. *At Home in the Cosmos*. Maryknoll: Orbis Books.

United States Conference of Catholic Bishops (USCCB). 1989. "Doctrinal Responsibilities: Approaches to Promoting Cooperation and Resolving Misunderstandings Between Bishops and Theologians." *Origins* 19: 101.

U.S. Global Change Research Program/U.S. Climate Change Science Program. 2009. "Climate Literacy: 'The Essential Principles of Climate Sciences' A Guide for Individuals and Communities." Accessed from http://www.climatescience.gov/Library/Literacy/ on May 26, 2009.

5

THOMAS AQUINAS, THE COSMIC COMMON
GOOD, & CLIMATE CHANGE

Daniel P. Scheid

O ne of the most promising avenues for applying Catholic social thought to ecological ethics and particularly to the threat of human-induced climate change is the principle of the common good. A central component of Catholic ethics, the common good is the telos of justice and of all human laws, making it central to debates about environmental justice. Thus, the principle of the common good offers a positive vision for social relationships, and the ecology question at its root challenges our notions of sociality and the boundaries of community. As Merle Longwood explains: "Our conception of the common good must obviously include the whole biotic community, since the quality and health of human life is integrally tied to the quality and health of the lives of all the other members of the biosphere. There is, after all, only one ecology" (1973, 479-80).

Because human well-being and the health of the planet are intimately linked, the U.S. Catholic bishops and popes have invoked "the common good" when addressing environmental issues. More dramatically, in their 1992 pastoral letter *Renewing the Earth*, the U.S. bishops urge all people to pursue the planetary common good (USCCB 1996, 232). This document represents one of the first instances in which an episcopal conference has applied common good language to include Earth of which humanity is only one part. The Bishops reiterated this language and accorded it a more central position in *Global Climate Change: A Plea for Dialogue, Prudence, and the Common Good* (USCCB 2001). In this document, they call attention to "the universal common good" and its ethical priority when addressing the climate crisis.

The common good is a natural fit for discussing human-forced climate change because it corresponds well to the key ecological principle of interdependence. The actions of one affect the whole, and each individual flourishes only when remaining committed to the well-being of the whole of which she is a part. Climate change ranks as perhaps the most devastating ecological threat to face the human family, and it constitutes a clear example of a common problem that transcends any particular state, nation, continent, or hemisphere. Climate change therefore offers a salient opportunity for articulating what a shared planetary common good could entail. In *Global Climate Change*, the US Catholic bishops urge all people to acknowledge the inescapable reality of social and ecological interdependence and the fact that the choices made by one nation or even one individual have consequences in other parts of the world: "Individual nations must measure their own self-interest against the greater common good and contribute equitably to global solutions" (USCCB 2001, 8). Indeed, the Bishops' language suggests that the planetary and universal common good includes more than just human beings: "Working for the common good requires us to promote the flourishing of all human life and all of God's creation" (USCCB 2001, 11). By invoking the planetary common good, the U.S. Bishops implicitly recognize that interdependence extends far beyond the human community. The climate crisis alerts us to our radical interdependence with other animals and plants.

While the intention to slow changes humans are forcing on the climate and to mitigate its effects is evidently praiseworthy, the bishops implicitly raise a pivotal question: How do we correlate the human common good with the planetary common good? Does the planetary common good primarily or even solely exist to serve humans? Or, does humanity fit into a broader purpose that God intends for all creation? Is a dramatic shift in the Earth's climate just bad for humans, or for the planet itself and God's intentions for it? I propose another and even broader aim of justice—the idea of a *cosmic common good*—as a way of reconciling this tension. Articulating a cosmic common good that incorporates more than human well-being can clarify the relationship between the human common good and the planetary common good. A helpful way of elucidating what is meant by planetary and cosmic common goods may be to probe the theology of creation in the Catholic theological tradition. It is well poised to develop a robust understanding of the common good because the tradition offers a coherent worldview that unites God, the universe, and humanity into a holistic paradigm. When the U.S. bishops and others voice principles like the planetary common good and universal common good, they generally presume a theology of creation.

Thomas Aquinas (1225-1274) offers an excellent resource for providing greater detail to the planetary and cosmic common good. Though he certainly affirms humanity's dominion over non-human creation and its privileged position over all other Earthly creatures, he also envisions a cosmos in which all creatures, including human beings, contribute to a glorious cosmic end centered on God.

Therefore, I underscore three key concepts in Aquinas that can contribute toward articulating a cosmic common good. The first concept is his understanding that the whole universe surpasses in excellence any individual creature. Second, the most valuable feature of the universe is the order among its various parts; the ordered interconnections between creatures represent the cosmic common good to which all should contribute. Recognizing and contributing to the order that God has instituted acknowledges God's goodness and wisdom. Third, Aquinas teaches that this cosmic common good glorifies God, and to glorify God is the ultimate purpose of every creature and of the universe as a whole. Finally, I will apply these basic themes to climate change.

THOMAS AQUINAS' THEOLOGY OF CREATION

Key to understanding Aquinas' theology of creation are his reflections on the goodness of the universe as a whole, God as the absolute cosmic common good, and the order and unity of the universe as the temporal cosmic common good. These teachings are discussed in the following sections.

The Goodness of the Universe as a Whole

Despite a steadfast presumption of humanity's dominion over nature, human superiority is not the only theme in Aquinas' theology of creation. Indeed, multiple passages envision a cosmos in which all creatures, including humans, share a common end of giving glory to God. Though humanity retains a privileged position among creatures, the universe as a whole possesses its own intrinsic goodness and worth.

Aquinas argues that creation in its totality is a better reflection of God's goodness than any species of creature. Only the entirety of the universe deserves the title of most good, because the perfection of the universe requires a composition of various parts: "For goodness, which in God is simple and uniform, in creatures is manifold and divided; and hence the whole universe together participates the divine goodness more perfectly, and represents it better than any single creature whatever" (Aquinas 1948, 1.47.1). He recognizes that no single creature, in its finitude and partiality, could best represent the goodness of God, who is infinite and the source of all good. The various creatures that populate this vast cosmos fulfill a purpose that no species or creature alone could fulfill, however remarkable its achievements. Aquinas sees no contradiction between the idea that all non-rational creatures are directed for human use and the idea that all things represent the divine goodness better than any single creature.

From his medieval understanding of the world, Aquinas taught that God, the omnipotent and sole cause of creation, imparts to the universe its present form in all its details. A diversity of creatures is intended by God: "The form of the universe is intended and willed by God.... Now the form of the universe consists in the distinction and order of its parts. Therefore the distinction of things is not from chance" (Aquinas 1924, 2.39.6). He affirms that this intricate system did not emerge accidentally, as if differentiation and order were not due to God's providential wisdom. Furthermore, the distinction of things in nature is not a result of sin, as Origen postulated, nor does a variety of creatures detract from the superior goodness of the human person. Alone among Earth's creatures, humans are made in the image and likeness of God, so only humans have the power of reason to know or understand (Aquinas 1948, 1.93.2) and a "capacity for the highest good" (ibid. 1.93.2 ad 3). For this reason other creatures are considered "less noble" than humans and exist for humanity's sake (ibid. 1.65.2).

However, differences among creatures are integral to the universe's form and are necessary for the universe's goodness: "As the divine wisdom is the cause of the distinction of things for the sake of the perfection of the universe, so is it the cause of inequality. For the universe would not be perfect if only one grade of goodness were found in things" (ibid. 1.47.2). Though humans are more noble than other creatures because they have the power to act rationally, the universe still requires a diversity of creatures to best manifest God's goodness. Consequently, a universe composed of only one kind of creature would not be an improvement, even if that species is of greater worth: "The goodness of the species surpasses the good of the individual, even as the formal exceeds that which is material. Hence multitude of species adds more to the goodness of the universe than multitude of individuals in one species. Therefore it concerns the perfection of the universe, that there be not only many individuals, but that there be also different species of things, and consequently different degrees in things" (Aquinas 1924, 2.45.6). He thinks primarily in terms of species before individuals, and the cosmos before individual species. Similarly, in his commentary on Peter Lombard's *Sentences*, Aquinas provocatively compares the worth of a universe composed of angels and rocks to one comprising only angels. Once again, Aquinas affirms that the universe best imitates the divine goodness when it contains a variety of species: "Although an angel taken absolutely may be better than a rock, still both natures taken together (*utraque natura*) are better than either one alone: and hence a universe in which there are angels and other things is better than where there would be angels only, because the perfection of the universe is seen essentially according to the diversity of natures, by which diverse degrees of goodness are filled, and not according to the multiplicity of individuals in one nature (1 *Sentences* 44.1.2 ad 6 quoted in Blanchette 1992, 125-26). That Aquinas perceives a need for a multiplicity of types of creatures in the universe, rather than a mere multiplicity of one kind, affirms his thinking that no creature is superior to the entirety of creation. Angels, who surpass the human in nobility due to their superior intellect, do not suffice for the perfection of the universe. On the contrary, only a universe composed of many kinds of creatures of varying complexity and nobility could best demonstrate God's goodness.

Aquinas also writes about the importance of diversity in the universe in terms of conditional necessities or debts. Debt is a term that is linked to justice, because justice seeks to give each its due, and what it is due is a form of debt (Aquinas 1948, 1.21.1). Justice that corresponds to what is due must be distinguished from the virtue of justice, which orients humans to the human common good. The virtue of justice implies equality (ibid. 2|2.58.2) whereby one renders to another what is due to the other according to a "certain proportion of equality" (ibid. 2|2.58.10). That is, all humans are treated equally as they receive what is owed to them as a matter of right according to their nature. Since humanity's rational nature elevates them above other creatures, the virtue of justice cannot direct humans to treat lesser creatures as they would human beings. Likewise, the powers of reason and will enable human beings to have a love of friendship with other human beings, marked by mutual love and benevolence, while humans can never love other creatures with a such love (ibid. 2|2, 25.4).

Thus Aquinas did not recognize any moral duty to non-human creatures per se. In addition, in absolute terms, God does not owe a debt to any creature. God creates and orders the universe "to the fulfillment of his purpose" (Aquinas 1924, 2.28-29.17), not on behalf of any creature. Still, Aquinas observes an appropriateness in the universe's composition, and he discusses this in terms of conditional necessities. The universe was created by God in a certain way, and this must be maintained in order to fulfill what God intends for it. So while human beings may not owe a debt to other creatures, upholding the order that God has established may be seen as a kind of justice to God's wisdom. Aquinas discusses two kinds of these conditional necessities. In the first conditional necessity, the universe as a whole is due each individual part that contributes to its perfection. If the distinction of parts and their varying dignity is essential for the universe's perfection, then the universe is conditionally indebted to contain all of those creatures: "[There is a conditional indebtedness] on the part of the whole universe of things in relation to each part thereof that is necessary for the perfection of the universe. For if God willed such a universe to be made, it was due that He should make the sun and moon, and such-like things without which the universe cannot be" (ibid.). The entire universe has a conditional debt in which certain creatures exist so the universe might be perfect. God has willed such a universe to be, so each part of the universe that is necessary for the perfection of the universe must not be lacking. This point will become clearer in the next section when explaining the cosmic common good and again in relation to the climate crisis.

The second conditional necessity that Aquinas identifies is that certain creatures need to exist according to other creatures that depend on them. If God desires that humans should exist, then He has to make plants and animals and the other things that humanity requires for its existence. In these cases, God can be said to do these things by necessity, not because God owes it to the creatures but because God wills them to exist to fulfill God's purpose (Aquinas 1924, 2.28-29.17). These two forms of conditional indebtedness reflect Aquinas' dual understanding of creation: all lower creatures like plants and animals that lack the capacity for rational action exist to serve human beings, and the universe as a whole has a composite goodness that exists to serve God's purposes.

In his understanding of creation, Aquinas consistently focuses on the whole of the cosmos and its relationship to God, not solely on its use for humans. The universe as a whole composed of myriad parts surpasses in greatness any individual part. While some creatures are less noble and exist for humans, the distinction of things in nature is due to God's providence and has a central role in God's own purposes. To unpack this goodness of the cosmos, we now turn to the cosmic common good.

God as the Absolute Cosmic Common Good

As we saw above, each creature is necessary for the universe to be perfect. In turn, each part has a duty to contribute to the well-being of the whole. The concepts of debt and duty are both ways of reflecting on the importance of justice and its ultimate aim of the common good. Aquinas' concern for the goodness of the universe as a whole suggests the viability of thinking about justice and the common good in terms of the entire cosmos rather than just in terms of human wellbeing. Aquinas' theology of creation thus offers an important resource for articulating a cosmic common good.

In his understanding of creation, Aquinas indicates two facets to the cosmic common good. One can be seen as an absolute or ultimate common good that exists eternally, while the second is a finite and created good that obtains only in this temporal existence. For example, Jame Schaefer distinguishes these two aims in different ways: she speaks of the proper function of the universe as its internal common good (2005, 795-796) or the created common good (799) while God is the absolute common good (2005, 796) or the uncreated common good of creatures (ibid.). I shall use the terms absolute common good to refer to God, and the temporal common good to refer to the order and unity of the universe.

For the first, Aquinas emphasizes that God remains the absolute common good of the cosmos. "God is the universal good, and under this good both man and angel and all creatures are comprised, because every creature according to its being naturally belongs to God" (Aquinas 1948, 1.60.5). God remains the primary end of all creatures and of all their inclinations because all beings belong to God. Thus, the absolute cosmic common good transcends the universe itself.

Just as every being has its end in God, so too does the cosmic common good absolutely and ultimately aim towards God. All creatures belong to God and move towards God as their natural end through a kind of love, and to be human means to act and move towards this end using the power of reason and free will. This common love shared by all beings in turn contributes to a common good: "Now the fellowship of natural goods bestowed on us by God is the foundation of natural love, in virtue of which not only man, so long as his nature remains unimpaired, loves God above all things and more than himself, but also every single creature, each in its own way, i.e. either by an intellectual, or by a rational, or by an animal, or at least by a natural love, as stones do, for instance, and other things bereft of knowledge, because each part naturally loves the common good of the whole more than its own particular good" (Aquinas 1948, 2|2.26.3) The common good of the whole, absolutely speaking, is God, who is the foundation of natural love. Aquinas insists that God is the common good of all (*bonum commune omnium*) (ibid.), and, while humans and angels can pursue this good in a rational manner and even continue this relationship in eternity, all creatures are ordered to love the absolute common good of all who is God.

The Order & Unity of the Universe as the Temporal Cosmic Common Good

Aquinas teaches that while God in Godself as the source of all good is the absolute cosmic common good, there is also a temporal cosmic common good to which all creatures ought to contribute. This he identifies as the order of interconnected parts: "Now the highest good existing in things is the good of the order of the universe, as the Philosopher clearly teaches in the Metaphysics. Therefore the order of the universe is properly intended by God (Aquinas 1948, 1.15.2). There is a parallel, therefore, between the goodness of the cosmos and cosmic order. As we have seen, only a universe consisting of a variety of creatures best imitates the divine goodness. Similarly, the order of creatures in the universe is the highest temporal good for all creatures. If it requires the entire universe to best imitate God's goodness, then the proper functioning of the universe best glorifies God.

In a finite and temporal sense, therefore, the order of the universe is the end of creation. While God remains the ultimate final end, the temporal end of the universe is its order and interconnectedness: "That which is good and best in the effect is the end of its production. But the good and the best in the universe consist in the mutual order of its parts, which is impossible without distinction: since by this order the universe is established as one whole, and this is its best. Therefore the order of the parts of the universe and their distinction is the end of the production of the universe" (Aquinas 1924, 2.39.7).

This universe can be considered a single unified entity in two ways. First, all created things are united in the fact that they rely utterly on God for their existence. Only God must necessarily exist; all other creatures exist because they participate in God: "Therefore all beings apart from God are not their own being, but are beings by participation. Therefore it must be that all things which are diversified by the diverse participation of being, so as to be more or less perfect, are caused by one First Being, Who possesses being most perfectly" (Aquinas 1948, 1.44.1). Thus the universe of creatures is unified in its distinction from the creator.

Second, all things are related to one another, according to Aquinas, and the orderly relationship of creatures constitutes a unity of order: "The very order of things created by God shows the unity of the world. For this world is called one by the unity of order, whereby some things are ordered to others. But whatever things come from God have relation of order to each other as well as to God" (Aquinas 1948, 1.47.3). The order that God has established in creation signifies a second way in which creatures are related and linked to each other: "Now the best among all things caused is the order of the universe, wherein the good of the universe consists, even as in human affairs *the good of the nation is more God-like than the good of the individual*" (Aquinas 1924, 2.42.3). In a similar way that the nation's good exceeds the individual's, the integrity of the universe and its form outweighs the well-being of its constitutive parts as God intended. Aquinas indicates that there is a realm of moral value apart from human affairs, namely the good of the universe. The order of the whole universe, which incorporates human goods, retains supreme importance: "Now the greatest good in things created is the perfection of the universe, consisting in the order of distinct things: because in all things the perfection of the whole takes precedence of the perfection of each part" (ibid.). Aquinas likens God to a master craftsman who envisioned and sought to create various kinds of creatures in order to effect the best possible creation: "Again, since the good of the whole is better than the good of each part, it does not befit the best maker to lessen the good of the whole in order to increase the good of some of the parts: thus a builder does not give to the foundation the goodness which he gives to the roof, lest he should make a crazy house. Therefore God the maker of all would not make the whole universe the best of its kind, if He made all the parts equal, because many degrees of goodness would be wanting to the universe, and thus it would be imperfect (Aquinas 1924, 2.44.17). As explained earlier, Aquinas attributes the distinction among creatures to God's providence and not to chance, or, as Origen contended, to God's punishment for original sin. Far from detracting from God's intentions for creation, the order and diversity of creatures is in fact the best aspect of creation: "It seems absurd to assign a defect in things as the cause of what is best in them…[which is] their distinction and order" (ibid. 2.42.4).

Aquinas attributes the order in the universe to God's providence. While humans order their acts to their proper ends through their exercise of prudence, God orders all things to one another through providence. Divine providence can have two different meanings from Aquinas' perspective—either to the type of order of things towards an end or to the type of the order of parts in the whole (Aquinas 1948, 1.22.1). As described above, the order of things is always towards God as an ultimate end. In addition, each part of the universe is ordered to the universe as a whole so it functions properly, and the whole universe best imitates God's goodness. Therefore, the order of the universe is an integral way by which God has ordered all things to God's self.

Thus, the order and the unity of the universe are integral features in God's designs for creation. When entertaining questions about God's power, Aquinas reasons that God could not create in a better manner, for God could not access a greater source of goodness and wisdom. God is goodness itself and perfectly wise. In addition, since God is never compelled to create, or to create just as God has, God could conceivably create a better and more noble species of creatures by imparting more goodness to them. Nevertheless, Aquinas cannot envision God creating a better creature as part of this particular universe that God has in fact created, because he presumes that the order of the current cosmos must emanate from God's will. For Aquinas, this is additional proof that the order and unity exhibited by the universe constitute its greatest temporal good: "The universe, the things that exist now being supposed, cannot be better, on account of the most noble order given to these things by God, in which the good of the universe consists. For if any one thing were bettered, the proportion of order would be destroyed, just as if one string were stretched more than it ought to be, the melody of the harp would be destroyed" (Aquinas 1948, 1.25.6 ad 3). In a rare poetic simile, Aquinas likens the status of creation to a musical instrument, and altering any part of the universe's intrinsic order would shatter the delicate harmonies that God has ushered forth by God's creative act. Thus Aquinas acknowledges that God could create new and other creatures or add perfections to extant creatures, which would elevate the level of goodness in the universe. Yet, in his non-evolutionary worldview, the universe as it appeared to him represents God's desires for creation. "No other order would be suitable and good to the things which now are" (ibid. 1.25.6 ad 3). God has established a coherent and unified cosmos, both for the benefit of rational creatures and for God's own purposes. The well functioning and orderly universe participates in the divine goodness more completely than any single creature because it manifests the tremendous power and providence of God.

Hence the cosmic common good is not an ancillary concern or wholly subordinated to human welfare. The cosmic common good remains pertinent because it relates directly to God: "Now the good of the whole universe is that which is apprehended by God, Who is the Maker and Governor of all things: hence whatever He wills, He wills it under the aspect of the common good; this is his own Goodness, which is the good of the whole universe" (Aquinas 1948, 1|2.19.10). The absolute cosmic common good is God's own goodness, and God orders and upholds the functioning of the universe as the temporal cosmic common good: "A good existing in the universe, namely the order of the universe, is an end thereof; this, however, is not its ultimate end, but is ordered to the extrinsic good [God] as to the end" (ibid. 1.103.2 ad 3). As we saw, the entire universe and its interrelated parts are required to best represent God's goodness and wisdom, which explains why upholding the perceived order of creation and the sundry relationships between creatures is the cosmic common good: "Whoever has an end in view, cares more for what is nearest to the last end: because the other ends are directed to this. Now the last end of God's will is his goodness, the nearest thing to which among created things is the good consisting in the order of the universe: because every particular good of this or that thing is ordained thereto as its end, just as the less perfect is ordained to that which is more perfect: even as each part is for the sake of its whole. Consequently that which God cares for most in created things, is the order of the universe" (Aquinas 1924, 3.64.10). Just as all things are good because they participate in the goodness of God, all things are good because of their potential service to God's will for the universe.

The Glory of God

Finally, the temporal cosmic common good, as it imitates God's goodness and wisdom, culminates in the glory of God. Manifold creatures are necessary to imitate the divine goodness more completely than any one species or groups of species, and this mirroring is important because it enhances and contributes to the glory of God. If the temporal cosmic common good is the order and relationship of the various parts, the absolute common good is God and the glory of God. In the following passage can be seen a combination of the various ideas discussed above: each creature possesses its own intrinsic end and purpose; the human person possesses a superior nobility, resulting in the service of almost every creature to humanity; the cosmic harmony of order among distinct parts that emphasizes every part serving the common good; and, the purpose of creation as God's glory:

> So, therefore, in the parts of the universe also every creature exists for its own proper act and perfection, and the less noble for the nobler, as those creatures that are less noble than man exist for the sake of man, whilst each and every creature exists for the perfection of the entire universe.
>
> Furthermore, the entire universe, with all its parts, is ordained towards God as its end, inasmuch as it imitates, as it were, and shows forth the Divine goodness, to the glory of God. Reasonable creatures, however, have in some special and higher manner God as their end, since they can attain to him by their own operations, by knowing and loving him. Thus, it is plain that the Divine goodness is the end of all corporeal things (Aquinas 1948, 1.65.2).

As the only creature made in the image of God, humanity occupies a privileged role in creation; and, as the only animals who possess the faculties of reason and free will (ibid. 1|2.1.2) humans can participate in God's providential ordering in a way other creatures cannot—by knowing and loving God. Still, in at least two ways humans are contextualized in a greater purpose: humankind is part of a variegated universe whose complexity and interrelatedness is necessary to imitate the Divine goodness; and, the purpose of creation is not a human end but the glory of God. The dominion of human creatures over other creatures remains central to Aquinas, both for scriptural reasons— God's blessing to Adam and Eve to have dominion over the Earth (Gen. 1:28)—and for philosophical reasons—that which is less noble exists to serve what is more noble (Aquinas 1948, 1|2.1.2). Nevertheless, Aquinas accords a central importance to the good of the universe, and all creatures are subordinated to God's purposes for the entirety of creation.

HUMAN-FORCED CLIMATE CHANGE & THE COSMIC, PLANETARY, & HUMAN COMMON GOODS

Thomas Aquinas offers a comprehensive and nuanced theology of creation, but it is important to acknowledge that, as a medieval theologian, his perspective is vastly different than our own contemporary worldview when informed by cosmology and evolutionary biology. Aquinas envisioned a universe created just ten thousand years earlier, whereas modern science now describes a fourteen billion year history of the cosmos. Similarly, Aquinas believed that the order he perceived in the universe had been instituted by God's providence at the very beginning of creation, while contemporary science depicts a cosmos in which order emerges over incomprehensibly long spans of time. While Aquinas conceived of all the various species of plants and animals as fixed and viewed their interrelationships as predetermined by God, evolutionary theory depicts greater fluidity among species. We are beginning to appreciate how delicate and vulnerable Earth and its various ecosystems are to human activity. Modern biology tells us that no species is guaranteed its survival, and the destruction of an ecosystem can mean the annihilation of an entire web of life. Aquinas does not convey in his works any anticipation that the human species or any other species could become extinct, whether through natural actions or through the consequences of human action.

Nevertheless, though Aquinas wrote centuries before ecological concerns and from a vastly different understanding of the world than articulated today by scientists, his theological understanding of the human relationship to the universe still offers various lessons for addressing global climate change. Even in a theologian who so thoroughly emphasized the supremacy of humans and the fact that all creatures are ordered to human use, there is a strong respect for the goodness and integrity of the universe. In addition, his theological reverence for the order of the universe encourages us to approach Earth with a similar posture of humility. Aquinas insists on approaching the universe with "the things that exist now being supposed" (Aquinas 1948, 1.25.6 ad 3). Similarly, we may find his holistic appreciation of the entire universe consisting of various parts a valuable contribution to understanding our present position on Earth.

To conclude, I point to three dimensions of the common good to which Aquinas directs us. First, he suggests a common good of the entire universe in which God is glorified by all—the cosmic common good. Second, Aquinas suggests ways of seeking the planetary common good, which I identify as the Earth's sustainability. Third, a critical component of the planetary and cosmic common good is the human common good, in which concern for the poor and marginalized must remain central.

The Cosmic Common Good

Aquinas helpfully outlines a broader context for understanding climate change because his aim encompasses more than just Earth. His theology of creation aids us in adding substance and detail to what the universal common good could signify. First, a diversity of creatures is required for representing God's infinite goodness. One creature, however noble and wonderful in its capacities, does not suffice. A related point is that God does not will only a multiplication of creatures within one species; God also wills a diversity of various kinds of species. Aquinas thereby offers a poignant defense of biodiversity based not on human needs and desires but simply on the order and functioning of the universe. Because it is better that a multiplicity of species exist rather than a multiplication of one species, humans cannot justify economic development that hastens climate change and the deaths of various plant and animal species merely because it aids in sustaining more human lives.

Second, the greatest aspect of creation is the order and harmony of parts in the universe. Due in part to the fact that some creatures have greater nobility than others, it is fitting and proper that there be a myriad of species that function in harmony with each other to enable each to flourish. The common good indicates not only a multiplicity of kinds of animals but also the ecosystems in which they live and flourish. We must remain aware of the ways in which creatures relate to each other, mindful that their order and harmonious interactions are representative of the greatest good in creation.

Third, maintaining the ordered interconnections between diverse creatures is an essential moral and spiritual task that culminates in glorifying God. Even if we presume that one species is greatest in dignity and worth, God does not subject all things to humans simply for humanity's benefit. Rather, humans must also recognize the ways in which other creatures are also ordered to God and to each other—for God's own purposes. Humans ought to acknowledge the radical interdependence of ecosystems and therefore preserve them and the creatures who inhabit them as a crucial part of our human vocation. The effects that human-caused climate changes have on all creatures of Earth need to be identified as critical threats to the common good God intends for Earth and its multitudinous constituents. Aquinas offers a theological defense for upholding the interdependence of various creatures—an order gravely threatened by climate change.

Moreover, given Aquinas' truly universal scope, we may ponder Earth's important role within the cosmic common good. To date, Earth is the only planet known to contain life, let alone a multitude of living creatures. The loss of a species or whole ecosystems diminishes not only the Earth (and therefore also the human being) but the entire universe. If plants exceed the elements in nobility, and animals plants, then humans have a special responsibility on behalf of the entire cosmos to preserve their existence, to the glory of God. Indeed, if there is a conditional necessity that various creatures exist for the perfection of the universe, then humans may be said to owe a debt to God to preserve the diversity of creatures that God has established. While we cannot presume that the universe exists solely to have produced Earth or to sustain her creatures, we may value our planet's contribution to the cosmic common good as the sole planet, or at least one of the rare places, where such a variety of creatures can flourish.

The Planetary Common Good & the Climate Crisis

Insight drawn from Aquinas' thinking about the cosmic common good facilitates understanding the U.S. bishops' teachings about universal and planetary common goods and helps us reflect on their significance for changes humans are forcing on the global climate. Clearly, our notion of the common good must advance beyond what is good only for humanity. Knowledge of our interconnections with other species and within ecological systems dictates an expansion of our thinking. The common good of Earth with its many varied creatures is more than the "sum total of conditions of social living, whereby persons are enabled more fully and readily to achieve their own perfection," as Pope John XXIII proclaimed in *Mater et Magistra* (1961, 65). Today, when considered from the perspective of human-forced climate changes, the common good is better described as the planetary common good— the sum total of conditions of Earthly living whereby all creatures are enabled more fully and readily to achieve their perfection.

First and foremost, the planetary common good seeks the Earth's sustainability, a term that includes the long-term flourishing of the Earth, of the diverse ecosystems it contains, and the plants and animals they sustain. Schaefer similarly identifies the temporal common good of the universe as its sustainability: "The sustainable functioning of ecosystems and the greater biosphere is indeed a good that humans have in common with other biota and abiota that comprise these systems. When humans will that common good, they are conforming their wills to God's valuation of the physical world" (2005, 803). Schaefer rightly notes that attempts to use any creatures would have to abide by this new notion of the common good, which places an emphasis on the ordered connections of the parts (804). The planetary common good is not simply an application of traditional common good language to the Earth. Whereas all humans are equal participants in the human common good, creatures have varying moral significance in the planetary common good. Killing a tree does not represent the same loss to the cosmic common good as killing a human person because, building on Aquinas, humans are made in the image of God, and their powers of reason and free-will give humans a nobility that other creatures lack. Nevertheless, there is still a need to protect species that are most vulnerable to suffering and becoming extinct because their loss may upset the sustainability of the planet. Promoting the planetary common good means that the sustainability

of Earth must be at the forefront of all ethical decisions, whether political, economic, or personal.

While Aquinas may not have fathomed the need to protect non-
human creatures from extinction, we recognize today that the reality of changes forced on the global climate demands an exhortation
to safeguard the common good of non-human creatures, ecosystems
of which they are parts, and the biosphere. The global climate crisis is the crowning example of human interference with the orderly
functioning of the natural environment in which species evolved and
ecosystems formed and the sustainability of the planet is maintained.
The impact that humans can wield on non-human nature no longer
allows us to presume that God will continue to watch over animals
and plants without the assistance of humankind, as Aquinas trusted
(1948, 2|2.55.6). The temporal cosmic common good of the order
of creation and God's intentions to create myriad species demands
protecting the ecosystems whose creatures are nearest to extinction
and most vulnerable to human intervention in order to safeguard the
Earth's sustainability. Recalling Aquinas' notion of conditional indebtedness, we may describe our actions to preserve other creatures as a
form of justice. While Aquinas limited the virtue of justice to concern for human beings, we may recognize human interference in the
Earth's sustainability as a threat to the order of creation intended by
God's wisdom. We owe a debt to God's goodness to strive to maintain
the sustainable functioning of a variety of creatures and ecosystems.
Moreover, in coping with the inevitable effects that a rise in the Earth's
temperature will prompt, we must seek at all costs to avoid protecting
human welfare at the cost of destroying whole ecosystems of which
humans are parts and in which humans function. We must commit
ourselves to the sustainable flourishing of the variety of creatures that
populate Earth. Whatever damage climate change will inflict, humans
must remember God's ultimate intentions for creation and seek to
maintain and promote contexts in which creatures may thrive so they
can contribute to the sustainability of Earth.

Human Common Good & Climate Change

Though the planetary common good includes and transcends humanity, a Catholic perspective must also uphold the importance of
human life and dignity. The cosmic and planetary common goods
must include the human common good because there is a potential

for mutual love and benevolence between humans that is not possible with other creatures. Aquinas urged his readers to love God and to love their neighbors for God's sake (Aquinas 1948, 2|2.25.1). Moreover, Aquinas taught that we are meant to do good to our neighbor and to give alms to those in need as a matter of precept (2|2.32.5). Aquinas understood that we cannot do good to everyone (2|2.31.3), and we are not " bound to search throughout the world for the needy that one may succor them … it suffices to do works of mercy to those one meets with" (2|2.71.1). Therefore Aquinas cautioned that we should give alms to those who are more closely united to us (2|2.32.9). Yet he also stressed that love of neighbor should extend even to our enemies in case of urgency (2|2.25.9) and that it in certain cases one ought to "succor a stranger, in extreme necessity, rather than one's own father, if he is not in such urgent need" (2|2.31.3). In the context of climate change, love of neighbor and doing good to those in need means being aware of and responding to those most directly affected and especially to the poor, since people who live in poverty are the most vulnerable to human-forced changes to the global climate. For example, the first wave of climate refugees has begun, whether the island nation of Tuvalu steadily relocating to New Zealand (Morris 2009) or the Yup'il Eskimos in southern Alaska (Pilkington 2008). In addition to loss of land due to rising oceans, people are also facing a loss of wetlands and other crucial ecosystems, harsher effects of hurricanes, and an increase in disease due to higher temperatures. All of this will compound an already troubling refugee crisis. The International Organization for Migration (IOM) reports that forecasts for climate refugees by 2050 fall within a broad range, from 25 million to one billion, with 200 million the most frequent (International Organization for Migration 2010). In the case of refugees, we do not need to search for the needy, as Aquinas warned. Rather, they are the people we meet with, and Aquinas calls us to succor their needs if our needs are not so urgent. The human common good means, at its most fundamental, the opportunity for people to live in sustainable ways in which they can provide for their basic human needs.

Furthermore, the good of interacting with other species and the natural environment is not simply a matter of sustaining human communities, though this is obviously essential. Rather, allowing people to experience and contribute to the sustainability of Earth is one of the goods that must be shared by all humans. As David Hollenbach argues when describing the human common good, a life lived with others is not just an instrumental necessity for finite creatures. Communication and social interaction "are not merely extrinsic means to human flourishing but are aspects of flourishing itself. This shared life of communication and interaction with others, in all its aspects, is good in itself" (Hollenbach 2002, 81). Similarly, our dependence on Earth for our survival and our relation to other animals, to plants, and to other living and non-living creatures is not merely a biological or material fact. Though not equivalent to our relationships to other humans, our interactions with the rest of Earth constitute a feature of human flourishing and not merely an aid to that flourishing.

CONCLUSION

Aquinas' theology of creation offers a good opportunity to add substance and detail to the concept of the cosmic common good, which in turn helps to clarify the meaning of both the planetary and the human common good. Identifying the cosmic common good may encourage a rigorous response to the climate crisis. Climate change remains a critical threat not only to humans but also to other creatures and to the well-being of Earth. God desires not just the flourishing of the individual human person or even the prosperity of the human community. God desires a universe comprised of diverse creatures who contribute to the ordered interconnections between them as a way of glorifying God. A vision of the cosmic common good and humanity's part in it may enable us to address the root causes of climate change and make the sacrifices necessary to work for the good of all Earth's creatures with special concern for the poor and marginalized.

SOURCES

Aquinas, Thomas. 1948. *Summa Theologica*. Translated by Fathers of the English Dominican Province. 5 Vols. Chicago: Benziger Brothers.

———. 1924. *Summa Contra Gentiles*. Translated by Fathers of the English Dominican Province. 5 Vols. Chicago: Benziger Brothers.

Blanchette, Oliva. 1992. *The Perfection of the Universe According to Aquinas: A Teleological Cosmology*. University Park: Pennsylvania State University Press.

Hollenbach, David, S.J. 2002. *The Common Good and Christian Ethics*. New York: Cambridge University Press.

International Organization for Migration. 2010. Migration, Climate Change, and Environmental Degradation: A Complex Nexus. Accessed from http://www.iom.int/jahia/Jahia/complex-nexus#estimates on October 28, 2010.

Pope John XXIII. 1961. *Mater et Magistra: Christianity and Social Progress.* In *Catholic Social Thought: The Documentary Heritage*, ed. David J. O'Brien and Thomas A. Shannon, 82-128. Maryknoll: Orbis Books.

Longwood, Merle. 1973. Common Good and Environmental Issues. *Theological Studies* 34.03: 468-480.

Morris, Rachel. 2009. "What Happens When Your Country Drowns?" Mother Jones. Accessed from http://motherjones.com/environment/2009/11/tuvalu-climate-refugees on October 28, 2010.

Pilkington, Ed. 2008. The Village at the tip of the Iceberg. Guardian News and Media Limited. Accessed from http://www.guardian.co.uk/environment/2008/sep/28/alaska.climatechange on October 28, 2010.

Schaefer, Jame. 2005. "Valuing Earth Intrinsically and Instrumentally." *Theological Studies* 66.4: 783-814.

United States Conference of Catholic Bishops. 1996. Renewing the Earth. In *And God Saw That It Was Good: Catholic Theology and the Environment*, ed. Drew Christiansen, S.J. and Walter Grazer, 223-243. Washington, DC: United States Catholic Conference.

———. 2001. *Global Climate Change: A Plea for Dialogue, Prudence, and the Common Good*. Washington, DC: United States Catholic Conference.

6

ON KNOWING ONESELF IN AN AGE OF
ECOLOGICAL CONCERN

William C. French

Aristotle and Aquinas famously argue that a distinctive characteristic of humans is the ability to reason through which we can "master" our own actions. Where animals act, via instinct and raw inclination, they believed, humans by virtue of reason act on intentions and thus are capable of knowing and controlling their own actions. Self-knowledge has long been both an ideal in Western religious and philosophical ethics and a necessary condition for human agency and moral responsibility. Socrates famous charge to "Know thyself" echoes across the centuries along with Jesus' plaintive cry "Father forgive them for they know not what they do." Knowing what we do is rather important. In many countries today, however, it is quite difficult to know what we are doing because our actions and their chain of effects are understood within, and often constrained by, our range of attention. And, in the global economy today, our understanding of many of our ordinary day-to-day actions is mediated partly through and constrained by the information and often disinformation that the pricing of goods and services offers us.

Specifically, I examine the problem of the economics of "externalities." An externality exists when the market price fails to incorporate the full costs of the production and costs for using a good or service. In many cases the full production costs are reflected into the market price, but the cost of negative social or ecological impacts is ignored. Pollution from factories has been a classic case of "external" costs, namely, those real costs or damages that are pushed onto local third-parties, society at large, or future generations but are "hidden" in the sense that such costs are not incorporated into the market price. This failure of the market to accurately reflect the full range of real costs has a direct impact on personal self-understanding, even as it functions to provide a societal-wide incentive for "business as usual" consumption patterns. By allowing real and sometimes massive ecological costs to escape incorporation in the price of coal, oil, and gasoline, we as a society sustain a key obstacle that functions to help block timely shifting to solar, wind, and geothermal energy sources that emit no greenhouse gases. As Al Gore has noted, by systematically excluding significant ecological and social costs that are building damage into the future, we get a highly distorted vision of what counts as a rational economy, as a responsible society, and as smart national policy. We also get a distorted vision of what counts as smart personal choice. "Externalities" are real costs that are imposed on third-parties now and will be imposed in the future on people other than the producer and the purchaser/consumer. By leaving these real but widely-diffused costs out of the price paid in the market transaction, we get an immense privileging of the present interests of some coupled with a structural blocking of necessary concern for burdens on others—other humans or animals and plants in ecosystems—our common future. Short-term concerns by some for profits and power are allowed to swamp long-range responsible planning for the common good. As Gore describes this tragedy, "The future whispers while the present shouts" (1992, 170). Thus, the problem of "externalities" has important implications for religious people's concerns to be compassionate to neighbor needs, including the needs of our neighbors in future generations, and to be thoughtful about how best to respect God's grace embodied in the world of creation all around us.

My key argument is that externalities cut to the core an individual's or a society's ability to know and name our actions rightly in their full complexity of impacts. In short, externalities cut our ability to attend to, in a sustained way, the full range of impacts that our actions engage us in when we drive our cars or when we eat hamburgers at McDonald's or when we purchase many other products or services. Externalities deplete what Aristotle and Aquinas call our "self-mastery," namely, our rational understanding and responsible control over our actions. The way to recover our knowledge and mastery over our lives and our actions, lies, at least in part, by supporting honesty in the information flows that guide what counts as rational choice in our economic life. And in a time of concerns about climate change, this means encouraging our governments to adopt fossil fuel tax schemes to help "internalize" the real costs that the burning of fossil fuels is imposing on natural ecosystems, myriad animal and plant species, present coastal communities, and our children's and their children's generations.

In this essay, I correlate the ecological problem of how failures of moral vision push failures of timely responsible action with Aquinas's understandings of the role ignorance plays in the moral life and how habits can structure intentionality and actions across time in ways that promote virtue or vice. The problem of "externalities" is not just an issue of theoretical economic analysis. Externalities have a significant impact on our individual and collective ability to respond to climate change threats because they function as a key obstacle to timely policies aimed at protecting God's creation. Aquinas's discussions of the complexity of moral intentionality, action, and failure offer, I believe, a way for the one billion Roman Catholics in today's global Catholic community to see how climate change threats should not be seen as distant from the core of Catholic concern and sensibility.

Catholic moral theology originated historically out of the need to help guide Catholic priests in their understanding of sins and in helping them to evaluate the gravity of various sins in their administration of the sacrament of confession. Thus, Catholic moral theology started with action-theory—a sustained effort to elaborate a richly-detailed classification of human acts, both good and bad. This sacrament requires an "examination of conscience" that is surely required in an ecologically concerned age. However, externalities make it difficult for us to attend to the full range of impacts of our ordinary life choices about production and consumption—about what we do day to day and about what we buy and use. What impact do I have on the soils and aquifers, on the dwindling fisheries or maybe on the Amazon or the Greenland Ice Sheet, if I buy and eat this and not that? What effects on Earth, on others, and on the future do I have by driving this kind of car or perhaps flying to Europe? By undercutting our ability to sustain attention to the full range of impacts and costs that flow from our participation in the national and global economy, externalized costs make it difficult to sustain an on-going examination of conscience about many of our basic life responsibilities and the broadening challenges of ecological negligence.

CLIMATE CHANGE, MORAL BLINDNESS, &
THE GREAT DRAMA ALL AROUND US

The last two and a half centuries have been a revolutionary period in human and planetary history. However, because we have been born into this ongoing revolution, most of us on some deep emotional level take it for granted as an historic norm. Thus we have difficulty foreseeing the heights of the remarkable drama unfolding all around us. Since the industrial revolution, we have seen a vast acceleration of history. Humanity's raw technological power has increased exponentially. The harnessing of coal and then oil gave an upsurge of power to whole new industries of production in mining, manufacturing, and agriculture. These new powers helped prompt a vast upsurge of human population growth. In 1750, human population stood roughly at 791 million. By 1804, it crossed one billion, stood roughly at 1.65 billion in 1900, and reached the second billion 123 years later around 1927, the third billion 33 years later in 1960, the fourth billion 14 years after that in 1974, the fifth billion 13 years later in 1987, and the sixth billion 12 years later in 1999 (United Nations 1999, 3-8). The current United Nations' best estimate holds that global population will hit 8.9 billion by 2050 (2004, 3-4). Soaring human numbers have meant huge expansions of production and consumption and, thus, the required transformation of many natural ecosystems into agro-ecosystems to produce crops for human consumption. We have pushed land and sea creatures hard both by habitat destruction and by direct exploitation. Extinction rates have soared in recent decades as scientists have come to speak of our era as witnessing the "Sixth Great Extinction Event," one sadly for which we humans are responsible (Leakey and Lewin 1995).

Hans Jonas has observed how the vast expansion of humanity's range of raw power that promotes ecological destruction and degradation requires of us a corresponding appreciation for a new wide range of direct moral responsibilities (1984, 1, 6-8, 21-22). Where once nature—its mountain ranges, its seas, its herds of mammals and schools of sea creatures, its coral reefs and rain forests, its prairies and myriad other ecosystems, and its ice sheets and great glacier networks—seemed to be vast and stable, a solid and undamageable "given," now we see nature as vulnerable and fragile in the face of humanity's surging numbers and powers.

This is truly a new condition and challenge for humanity. Our expanded powers to destroy require a novel corresponding expansion of our direct moral attention and sense responsibility. As Jonas puts it: "If the new nature of our acting then calls for a new ethics of long-range responsibility, coextensive with the range of our power, it calls in the name of that very responsibility also for a new kind of humility—a humility owed, not like the former humility to the smallness of our power, but to the excessive magnitude of it, which is the excess of our power to act over our power to foresee and our power to evaluate and to judge" (Jonas 1984, 21-22). Therein lies the great tragedy of the market failure of "externalities." Just when we most need to view our lives and actions in a globally expansive frame that is attentive to our individual and societal impacts, the expansion of a global market system, through its vast failures to assess rightly real social and ecological costs, functionally blinds us to the full range of consequences that flow from our actions. It incentivizes the status quo dominant productive and consumptive patterns of society, at a time when we should be mobilizing our energies across the globe to restrain fossil fuel consumption and promote population stabilization, a shift to solar, wind, and geothermal energy production, and vast reforestation efforts. Just when we most need a clear eye, the global economy with its distorting market pricing blinds us to critically important realities.

The three main factors pushing climate change concerns are: (1) the upsurge of human population growth that requires greater conversion of natural ecosystems for farm and ranchland; (2) the consequent wave of deforestation in the last half century; and (3) the vast increase of demand for fossil fuels for humanity's expanding car and truck fleets, commercial and home heating needs, and expanding industrial energy uses. In 2008, for example, roughly "7.9 billion tons of carbon were emitted from the burning of fossil fuels and 1.5 billion tons were emitted from deforestation, for a total of 9.4 billion tons. But since nature has been absorbing only about 5 billion tons per year in oceans, soils, and vegetation," the rest builds up in the atmosphere (Brown 2009, 57).

The global economic upsurge across these two and a half centuries has been largely powered through the burning of fossil fuels—coal and oil—which have been slowly increasing the amount of carbon dioxide in our planet's atmospheric. In many poorer countries of the world, wood remains a key cooking and heating fuel, and it too makes significant contributions to greenhouse gas buildup in the atmosphere. Through core samples of Arctic ice, scientists have been able to gauge that in 1750, prior to the Industrial Revolution, atmospheric carbon dioxide stood at 280 parts per million (ppm). By 2008, atmospheric carbon dioxide has climbed to 386 ppm. This carbon increase when coupled with significant emissions of methane through rice production and cattle raising and increases of a broad range of other greenhouse gases is altering the very chemical composition of the atmosphere and pushing the threat of severe global climate change.

Several threatening aspects of the planetary drama loom before us. The melting of glaciers in all the major mountain ranges of Earth threatens to end their natural function as water reservoirs which offer sustained river flow around the year and provide great benefit to human communities and ecosystems below. The atmosphere contains roughly 5% more water vapor than in recent centuries which is prompting increased extreme precipitation events. As ocean waters heat up due to global warming trends, they increase the energy and wind velocities of many storms and hurricanes. The last decade has given us vivid evidence of the capacity of more extreme storms and hurricanes to cause remarkable damage to human communities, barrier islands, forests, and whole ecosystems which serve as the habitat for many natural animals and plants.

Furthermore, worries mount about the rapidity of increased melting of polar icecaps and of evidence that the Greenland Ice Sheet is also exhibiting some alarming signs of increased melting. As Lester Brown notes, we worry now about two major feedback loops. First, as the North Pole ice melts, dark waters take the place of white ice. Polar ice reflects roughly 70 percent of sunlight back into space, but darker ocean water reflects only 6 percent of sunlight leaving 94 percent to be absorbed to increase the water's warming. Thus, the more ice we lose, the quicker the heat dynamic works to cause even more rapid polar melting. Second, the northern tundra areas contain significant amounts of methane frozen and sequestered in the permafrost. As melting occurs, this methane is released and surges to contribute to the buildup of greenhouse gases in the atmosphere, thus reinforcing the cycle of increased global warming and increased polar methane release (Brown 2009, 57, 61-63).

Current projections are deeply alarming. The Intergovernmental Panel on Climate Change projected in 2007 that global temperatures will rise on average from between 1.1 and 6.4 degrees Centigrade or 2 to 11 degrees Fahrenheit by the end of this century (Brown 2009, 58). One study has estimated that by 2100 a sea level rise of 3-6 feet will result from the expansion of warming waters and the increased melting of icecaps. If, in the future, the West Antarctic ice sheet melts, a sea level rise of 16 feet could occur. If, in the future, the Greenland Ice Sheet melts, the sea level would rise 23 feet. The vast impact of these scenarios on coastal populations and the basic security of nations around the globe is staggering (Brown 2009, 61).

While sea level rise and an increase in wind velocity and rainfall of storms and hurricanes threaten coastal regions around the globe, the inland ecosystems will also experience a jolting impact through both gradual temperature shifting and precipitation shifting. Animal and plant species over the millennium have evolved to adapt to the climate, temperature, and particular resources and food chains of particular ecosystemic niches. Both animal and plant species evolved across the eons into an intimate fittingness with the particular opportunities offered by a specific ecosystem enjoying a specific temperature and climate pattern. When humans push climate change, animal and plant species that have adapted to the constraints of a particular locale are adversely affected and ill prepared for rapidly changing temperature and weather fluctuations. Global warming will so rapidly alter the basic temperature and climatic conditions of many ecosystems that many plant species will simply die off and, in turn, animal species will suffer the loss of their traditional food supply and habitat conditions.

Despite these dire predictions, why does society lack a sense of urgency? Surely we face a double problem. First, the dynamism of human actions today in ordinary "business as usual" production and consumption is degrading global ecosystems, endangering myriad life-forms, both animal and plant, and destabilizing long-standing climate and weather patterns. This is really big. However, the second and perhaps more alarming problem is that many do not recognize this problem and have no sense for the onrush of these global scale threats, threats borne out of certain unsustainable human practices. Most shocking is not so much the real world threat, but our global societies' inability to "get it" and respond in a serious and timely fashion.

Attempting to understand the obstacles—political, economic, cultural, and religious—that retard societies across the globe from mobilizing to respond to these planetary scale threats is indeed important.

Some peoples and nations are already waking up to the need to take serious action via policy change. Others are relatively more asleep regarding the seriousness of the drama swirling all around us.

If we face a double problem, we also face a deep tragedy. First, humanity developed in the twentieth century a new science, ecology, that gives us for the first time an empirical and expansive and indeed awe-inspiring understanding of how the planetary system in its remarkable complexity and beautiful diversity sustains a plethora of buzzing life, skies of beautiful colors, shapes of cloud, magnificent sunsets, animals of remarkable dignity and diversity, and plants and plankton that sustain the entire food chain by their miracle of photosynthesis. Second, the sadness arises from knowing that we can now understand how our planet sustains us while concurrently our collective actions are attacking Earth's life-support systems. Our awe at the vast gift of evolution and the sustaining forces that uphold our families, communities, and nations coincides with our awe and shame for how these life-sustaining systems are so fragile, so vulnerable, and how they are weakening decade after decade through our own collective failures of attention, care, and moral restraint.

MARKETS & THE PROBLEM OF EXTERNALITIES

For a number of years, economists have recognized the problem in which the price of a commodity or service fails to incorporate the full costs of the production and delivery of that good or service and the full costs imposed on others or future generations by the consumption of that good or service. As Barry Commoner describes in his book, *The Closing Circle*, market exchanges typically are marked by mutuality and by free choice (1971, 251). No one compels the buyer to buy or the producer to sell which Milton Friedman and other economists contend maximizes our freedom (Friedman 1962, 7-21). However, as Commoner notes, many market exchanges include negative impacts that are pushed onto unsuspecting third-parties. The costs of these negative impacts are not incorporated into the market price paid by the purchaser and have been dubbed "externalities," costs that lie outside the market price set by supply and demand and profit margin interests. For these third-parties receiving the negative impact, the "externality is neither mutually beneficial nor voluntary" (Commoner 1971, 251). These third-parties receive none of the benefits of the exchange; they receive only damages as a result of these externalities. The exchange imposes a violation and an injustice to third-parties who, through no consent of their own, are harmed, sometimes in significant ways. The paradigm cases of negative externalities are pollution cases in which companies allow toxic effluent to be dumped into a stream or river system or they allow lead, mercury or other air pollutants to go up a smokestack to be diffused by the winds across a region.

Thus, the market often engages in a structural distortion by narrowly focusing on a certain range of costs to be incorporated into the exchange price, while paying no attention to other genuine costs borne by natural ecological systems, by other individuals nearby, by society in general, or by future generations. At the policy level, externalities are important for a number of reasons. First, there is a basic injustice done when a market exchange hurts another third-party who derives no benefit from the exchange nor volunteered in any way to be joined in the flow of impacts from this transaction. These third-parties can be people in the general locale who are negatively impacted, animal and plant species in the local ecosystems, or future generations—our kids, and their children. Second, there are powerful economic rewards for companies that push negative externalities onto third parties. It is in the economic interest of companies to suck up profits and to shed production costs and other financial liabilities. The expense of proper cleanup and treatment of a manufacturing waste-stream can add significant expense to the production process, thereby driving up prices and reducing market share and profit margins. Thus, companies are often tempted to keep costs of production low by sending pollutants up the smokestack or down the river where their negative impact will be diffused across an unsuspecting population or onto a generation not yet born. There are powerful incentives to kick the problem and the costs down the road into the future.

While much of the focus in early ecological problems concentrated on local and regional water and air pollution concerns, today they have been joined by an escalating concern about the truly globalwide impact of the experiment of having surging human population growth coupled with surging global economic growth on a modest sized planet. Today's concerns are framed in planetary wide impacts of global ecosystem transformation and degradation, massive threats to biodiversity, and massive concerns about global climate change. Gore and others began to draw popular attention to the concern about the problem of economic "externalities" by showing the rising challenges of human-pushed climate change and global ecological degradation. Gore wrote powerfully in *Earth in the Balance* about the need to appreciate the gravity of our current market distortions:

> The hard truth is that our economic system is partially blind. It 'sees' some things and not others. It carefully measures and keeps track of the value of those things most important to buyers and sellers, such as food, clothing, manufactured goods, work, and indeed, money itself. But its intricate calculations often completely ignore the value of other things that are harder to buy and sell: fresh water, clean air, the beauty of the mountains, the rich diversity of life in the forest, just to name a few. In fact, the partial blindness of our current economic system is the single most powerful force behind what seem to be irrational decisions about the global environment" (Gore 1992, 182-83).

Rather than empowering a broadening of care and responsibility, economic pricing distortions through the mechanism of externalities pushes a basic ignorance about the wide range of the impacts of our productive and consumptive actions in many spheres of life today. As Gore puts it:

> [O]ur current system of economics arbitrarily draws a circle of value around those things in our civilization we have decided to keep track of and measure. Then we discover that one of the easiest ways to artificially increase the value of things inside the circle is to do so at the expense of those things left outside the circle. And here too, a direct and perverse ratio emerges: the more pollution dumped into the river, the higher the short-term profits for the polluter and his shareholders; the faster the rain forest is burned, the quicker more pasture becomes available for cattle and the faster they can be turned into hamburgers. Our failure to measure environmental externalities is a kind of economic blindness, and its consequences can be staggering (Gore 1992, 189).

Failure to incorporate all costs of a given mode of agricultural or industrial production often arises from a general historic assumption that nature's systems will continue to benefit us in the same ways they always have.

Pricing Nature's Services

To highlight this problem of taking nature for granted, ecologists in the last 30 years have begun to draw analogies from economic language about the value of the services provided in the economy in order to highlight the importance of myriad "nature's services." Often costs of agricultural or manufacturing production that are ignored as "externalities" are the opportunity costs incurred by damage to ecosystems that hinders their ability to perform certain ecosystem "services" that they once did (Daily 1997). In these types of cases, nature has long offered generalized "services" to human and nonhuman communities, and some specific agricultural, forestry, or manufacturing operation comes in and gains direct short-term profits by producing salable goods from nature but at a cost of the loss or attenuation of nature's ability to continue to perform its other historic generalized services. Because these services have historically been long enjoyed, society has never before worried about them or considered their monetary value. Only their loss or attenuation draws society's attention to their critical value.

Today, ecologists are reminding us that if governments and general societies continue to take nature's array of services for granted without intentionally naming their economic value, business interests that focus solely on the monetary value of particular natural goods—lumber, corn, oil, wheat, beef cattle, salmon, coal—will be given free rein to make economic decisions that continue with a "business as usual" approach. Short-term profits via production will be privileged, but only by continuing to ignore the value of longer-term services that might be lost to human and nonhuman communities at large.

Valuing the Services of Trees

Among nature's many services, the broad services of trees and forests have become a significant area of focus for many ecologists. Powerful economic forces have pushed deforestation in the last half century in much of the world. Many individuals, companies, and national governments enriched themselves by felling trees for sale as lumber on the burgeoning global market. The lure of high short-term profittaking pushed the felling of many forested regions to the long-term detriment of local natural and human communities. The impact of the loss of trees and forests has meant the attenuation of the natural flood control function that forests serve by slowing the runoff of rainfall. In addition, forests provide the critical natural service of sequestering atmospheric carbon dioxide, slowing the buildup of atmospheric greenhouse gases, and slowing the rate of global climate change. Unless these critical services can be intentionally monetized and "internalized" back into an assessment of the value of a living tree or a tract of living forest, the customary "business as usual" calculus of value will continue to value dead trees sawed into lumber over living trees—trees capable of catching rain, holding runoff, and breathing out oxygen as they breathe in and sequester carbon dioxide.

In recent years, China, Thailand, and the Philippines have all enacted partial or full bans on logging. China was clobbered in 1998 by record floods, and this woke up government officials to the enormous costs the country was paying due to widespread logging in the country's forests that cut their natural services of rainfall stabilization and flood control. These officials concluded that the "flood control service of trees standing...was three times as valuable as the timber from trees cut." This new information pushed a national policy shift whereby the government started paying loggers not to cut trees, but to plant trees (Brown 2009, 199).

The serious real and projected effects of human-pushed climate change are attracting the attention of governments and societies and leading them to appreciate the critical carbon dioxide sequestering service that trees provide. A major effort of the United Nations Environment Programme supports massive reforestation efforts today in many countries in order to aid climate stabilization (Brown 2009, 200). Wangari Maathai won the Noble Peace Prize for her efforts in founding the Green Belt Movement that has planted forty-five million trees across Kenya in an effort to provide a general replenishment of ecosystem health, a more productive environment for Kenyan villagers, and mitigation of the adverse effects of climate disruption in Africa (Maathai 2010, 22-24, 77-91).

Instead of privileging the future as most parents would, or mayors or national leaders should, economic theory tends to discount the future gravely due to the increasing uncertainty about conditions in the distant future. However, if the needs of the present are consistently prioritized over the needs of the future, we fail gravely to be responsible to our children and our children's children and the world they will inherit. Are we so unable to be moved by the compelling needs of future human and nonhuman generations?

As Gore puts it: "In drawing a circle of value around those things we consider important enough to measure in our economic system, we not only exclude a great deal that is important in the environment, we also discriminate against future generations.... The effect is to magnify the power of one generation to compromise all future generations" (Gore 1992, 190-91).

Green Taxation & Tax Shifting

At a time when the rising threat of significant global warming should push all nations and peoples to mobilize our resources with a sense of war-time urgency to restrict sharply our fossil fuel consumption rates and our policies that push deforestation, the failure to internalize costs borne from both practices is the chief obstacle to shifting to a rapid development of a solar, wind and geothermal power-based economy that would greatly reduce our emission levels of CO_2 and other greenhouse gases. Research and development is burgeoning in all areas of these alternative energy sources, and rapid growth has been occurring in these sectors around the world for the last five years. However, a major retardant to the needed shift away from fossil fuel consumption is the fact that solar and wind are not yet viewed as cost-effective as fossil fuel power generation. What is needed is world-wide action by governments to follow the lead of Europe and Japan which impose high carbon taxes on fossil fuel consumption. As Gore insists: "[E]very one of the solutions to the climate crisis will be more effective and much easier to implement if we place a price on CO_2 and other global warming pollutants.... Once we have a price on carbon, the negative externality that was invisible and not tracked by the market will become visible and will be included in the decisions of the market participants" (Gore 2009, 327). Herman Daly and John Cobb agree with Gore that we need to internalize currently externalized costs back into the market price via governmentally imposed taxes placed on commodities like fossil fuels so their prices reflect the true equivalent of its full range of costs that include both the costs originally included in the market price and an added tax hike seeking to approximate the added burden of social and ecological costs imposed on third-parties, general society at large, and future generations (Daly and Cobb, 1989, 54-58).

By raising the market prices of fossil fuels, tax mechanisms immediately make solar, wind, and geothermal much more cost-effective and help push the needed rapid societal shift to these renewable energy sources. By "internalizing" real costs borne in fossil fuel consumption through "green" taxation schemes, economic prices become more honest. By raising significant pools of revenue, green taxation also gives governments important financial resources to help jump-start and support the solar, wind, and geothermal economy and to push reforestation, species protection, climate change mitigation measures, and mass transit alternatives to private vehicle usage. By raising the costs of fossil fuel consumption green taxation measures send powerful and widespread educational incentives to change societal and individual habits of energy consumption (Brown 2009, 241268).

However, there are significant political obstacles to national leaders who wish to initiate fossil fuel taxation schemes. This is clearly the case in the United States, but Americans are not alone in our dislike for calls for raising taxes. As Brown notes, higher fossil fuel taxes can be coupled with coordinated reductions in income tax levels so a population can be assured of a relatively flat overall tax burden in any given year. Various policy devices can be used to accomplish this goal. For Brown, this "tax shifting" might offer a politically viable way to pass legislation that imposes higher costs on fossil fuel consumption and jump starts the needed energy revolution that will help prevent the worst case scenarios of global warming from coming to fruition (2009, 244-49). As Brown insists: "To create an honest market, we need to restructure the tax system by reducing taxes on work and raising those on carbon emissions and other environmentally destructive activities, thus incorporating indirect costs into the market price" (2009, 243).

Additionally, policies would need to be developed to mitigate any undue impact on the poor and lower middle-class, some of whom live in rural areas and must drive longer distances in their day-to-day schedule. As a matter of justice, higher gas taxes, for example, as a sales tax are "regressive" in that they impose a far greater burden on the poor and middle-income people than on the rich. Some mechanism for reducing this unfair, undue burden would need to be identified. This could take the form of making gas receipts partially tax-deductible on one's income tax or by adding a direct federal payment to families for the first documented spending of $800 in gas receipts across any given year. Justice requires mitigating the regressive impact, but global security and charity owed to future generations requires the bold action of green taxes.

The Costs of Cheap Gasoline

Focusing on the price of gasoline helps us to understand what is at stake when we call for honest pricing that includes external costs of ecological degradation in general and human-pushed climate disruption in particular. In societies across the globe, the price paid at the pump for gasoline is a major issue. Anyone who drives a car engages this price weekly. Whereas Europe has long added significant taxes to raise the price of gas at the pump, the United States since World War II has resisted high gas taxes. The price of gas in Europe has checked the growth of suburban sprawl, supported mass transit and urban concentration, and held the average European person's annual emission of carbon dioxide into the atmosphere at roughly one half of that of the average American citizen's.

While American drivers regularly grumble at higher prices for gasoline, our main problem is not high prices when viewed from an ecological perspective. Our problem is prices that are set far too low by leaving out many genuine social and ecological costs of burning a gallon of gas. Because gas has been priced low in America since World War II, we have allowed ourselves as a society to become addicted to cheap gas by our widespread adoption of motor vehicle transit and our allowance of metropolitan growth to sprawl widely in low-density suburban districts that tend to lock large segments of America's population into daily dependence on long auto and truck commutes. Cheap gas policies have encouraged sprawl development patterns and high rates of auto and truck usage. Our transit mix is disproportionately weighted to car and truck usage, and securing adequate governmental subsidies for public transit alternatives has been difficult politically. Accordingly, the United States today relies most heavily on car and truck driving, and, for longer trips, on air travel. Problematically these two transit options are the highest in fossil fuel consumption per passenger per mile and, thus, the highest in terms of carbon dioxide emission that pushes global warming (Durning 1992, 78-86). These two choices in travel options are among the main reasons why the United States bears such a heavy burden of responsibility for our vast daily emission of carbon dioxide that pushes global climate change.

None of these American policy choices after World War II were inevitable. For example, Europe adopted quite different policies that have led to the average Europeans' far more modest carbon footprint today. Europe placed high taxes on gasoline, and these higher prices functioned as disincentives for long auto commutes and for big heavy cars. Europe imposed strict zoning regulations to protect the density and vitality of their cities and to channel growth along public transit bus and rail routes. And, Europe has always understood that public transit is a public good warranting significant public subsidies and funding. Consequently, for many Europeans walking and bicycling today are much used modes of urban transit, and bus, rail, and trolley are extensive and efficient. Europe has long raised disincentives for private motor vehicle usage and raised incentives for public transit usage. Most European cities are vibrant because of the relatively high density that can sustain a vital array of shops, restaurants, galleries, and parks. In contrast, vast tracts of American suburbia are quite isolating and lonely if one does not have access to a car or if one is unable to drive across the miles to the Mall or shopping area. Europe adopted one set of policies that have shaped its common life for the last half century, and the United States adopted another set of policies. We are living today with the consequences that flow from them (French 1993, 15-16, 19-21).

According to Brown, the combination of the distortions promoted by hidden subsidies and hidden externalities leads to the policy justification for "business as usual" practices of production and consumption. These practices have pushed the United States' chief "occasion of sin" in being roughly 5% of the world's population but consuming roughly 25% of the world's annual use of fossil fuels and thus contributing roughly 25% of the world's annual greenhouse gas emissions. Such distortions and practices are both unsustainable and unjust. Brown and others estimate that if all the externalities, including future costs of global warming, surrounding the production and consumption of a gallon of gasoline were internalized into the price that Americans pay at the pump, we would need to pay roughly $15 per gallon, $12 in various taxes on top of the $3 or so for the current (distorted) market value (Brown 2009, 245). However, a far more modest gas tax of about $1 per gallon coupled with other broader fossil fuel taxes on coal and natural gas would provide a powerful push in the right direction of alternative sources of energy and more fuel efficient motor vehicles, more efficient home heating and cooling, and more efficient corporate energy usage.

ECONOMIC EXTERNALITIES &
CORE CONCERNS OF CATHOLICISM

At first blush, an extended discussion of economic policy and climate change seems quite distant from the core of Christian life and passion. It seems coldly distant from liturgical and parish life, the power of scripture, the call to open our hearts to Christ's grace, or the command of neighborly love. Economics we are told is the "dismal" science, and discussions of taxation schemes seem to be technical and dry. By contrast, discussions of sexual and medical ethics have been readily understood as directly related to core Catholic concerns about dignity, justice, and relational well-being in the intimate zone of personal subjectivity and agency. Though economics and environmental policy analysis often seem unrelated to the sphere of interiority and of personal relations between and among the self, family, neighbors, and God, I will examine the ethical views of Aristotle and Aquinas that offer some suggestions for how the issue of externalities and human-supported climate change are indeed quite relevant for Catholic concern and reflection today.

Aristotle on Fitting Action, Perception, & "Ultimate Particulars"

Thomas Aquinas could not have correlated the best theology of his day to the best scientific understanding of his day had not translation centers at Toledo in Spain and Palermo in Sicily translated Arabic translations of Aristotle's corpus along with important Arabic commentaries into Latin and thus made them available to Europe's rising university system and its scholars (Rubenstein 2003, 12-23). Aristotle's scientific understanding of the world and his view of the moral life powerfully shaped Aquinas's views and those of many other theologians of the Catholic Medieval period.

Therefore, it is helpful to explore some major themes in Aristotle's approach to ethics. His *Nicomachean Ethics* is a profound discussion of human agency, vision, and responsibility. Three themes stand out as important for an age facing global climate change. First, Aristotle is a realist. As he puts it, "the acceptance of a fact as a fact is the starting point" (Aristotle 1962, 1095b, 7). Second, habits of action generate understandings of value. Self-control means that an action is done by one who "must know what he is doing" (Aristotle 1962, 1105a 30, 39). Third, Aristotle's ethic focuses on the "fitting" action. He stresses that there must be a critical responsiveness in the moral life to different problems, and he stresses the virtues that are ideal-types of how we should act along the mean between extremes. However, when stressing an ethics of the appropriate or fitting response, he centers his ethics in practical wisdom, not theoretical reason, for those who have the worldly experience to appreciate the importance of "ultimate particulars." I think he means that we really need to deal with details and specifics. None of this lazy generalization of "Let's just love nature" will do. Our ability to be responsible and moral, for Aristotle, is dependent in good part upon our capacities of attention and discernment to these "particulars." Clearly, ethics is not just an exercise of learning the general principles of ethical theory; ethics is an activity of practical reason about concrete cases involving distinctive circumstances and needs. Ethics requires an appreciation for the significance of these "ultimate particulars" (Aristotle 1962, 1141b 15, 157; 1142a24, 160). Attention, vision, and responsiveness are central to Aristotle's view of the moral life. Experience is crucial. As he puts it: "[U]niversals arise out of particulars. Hence one must have perception of particular facts, and this perception is intelligence" (Aristotle 1962, 1143b 5, 166).

Aquinas: On Creation, the Common Good & Rational Self-mastery

Aquinas remains a distinctively important figure in the Catholic Church and for good reason. His corpus of writings painted a vast canvas of God's love for all creation, especially humanity, and an understanding of grace embodied in the drama of creation and salvation. He serves as a model for Catholics today—all one billion or all one-sixth of humanity—who wish to engage faith concerns with the best science of our day. That is what he did in his day when correlating the best of the Christian theological tradition with the Aristotelian scientific world-picture that was newly made available to Western Europe.

Of course, though Aristotelian science was the best and most sophisticated of Aquinas's day, it is no longer sufficient for ours. Nevertheless, his example of engaging theology with the best philosophy and science of his day can help inspire us to engage the best thinking and science of our day (Curran 1974, 45, 79-80). Aquinas's robust appreciation for God's grace in creation correlates in important ways with the findings and affirmations of today's ecological sciences. Both emphasize the complex relationality of beings who are in themselves good but who support the good of their species, which, in turn, coupled with the good of the diversity of species, supports the common good of the universe and the ultimate common good, namely God. Both Aquinas and contemporary ecological science celebrates the diversity of life-forms on Earth (Schaefer 2009, 22-24, 78-79; Jenkins 2008, 115-151). In contrast to so much of dominant streams of both Catholic and Protestant theology, Aquinas attends to the thickness of divine relationships with, and love for, all of the community of creation. Of course, he at times starkly privileges humanity over the rest of the natural order due to our rational soul, yet he still he locates humanity inside the rest of the "community of the universe" (Aquinas 1948, 1|2.91.1).

A central theme in Aquinas's work is that the grandeur of God is mirrored in God's maximal generosity that overflows in energy and grace to create and sustain a maximally diverse set of creatures who fill every possible level of being. As Aquinas states:

> Therefore, the perfection of the universe, which consists of the diversity of things, would thus be a thing of chance, which is impossible. Hence we must say that the distinction and multitude of things come from the intention of the first agent, who is God. For He brought things into being in order that his goodness might be communicated to creatures, and be represented by them; and because his goodness could not be adequately represented by one creature alone, He produced many and diverse creatures, that what was wanting to one in the representation of the divine goodness might be supplied by another. For goodness, which in God is simple and uniform, in creatures is manifold and divided, and hence the whole universe together participates the divine goodness more perfectly and represents it better than any single creature whatever (Aquinas 1948, 1.47.1).

Accordingly Aquinas offers a magnificent theological authorization for a widespread Christian commitment to preserving species and the ecosystems that sustain all species. This perspective is directly relevant as humans push climate change that alters temperature conditions and precipitation patterns of ecosystems around the globe. Such changes happening so rapidly may well make it difficult for many animal and plant species to adapt or migrate successfully.

In addition to offering a robust stress on humanity's participation with a greater community of creation, Aquinas's understanding of human action, habits, intentionality, and responsibility offers much food for thought as we try to engage the problem of the moral blindness borne from the market distortions of "externalities." Aquinas follows Aristotle closely on his understanding of the moral life. In what follows, I examine three themes in Aquinas's corpus: (1) his understanding of how circumstances can change the "species" of an act; (2) his differentiation between vincible and invincible ignorance; and (3) his stress on the role of the habits and the virtue of prudence in the moral life.

Circumstances Can Change the Nature of an Act

Aquinas has a quite sophisticated appreciation for the complexity of elements that make one type of act quite different from another. He appreciates that, for example, human law may change "on account of the changed condition of man" (Aquinas 1948, 1|2. 97.1). Indeed he holds that natural law can be changed "since many things for the benefit of human life have been added over and above the natural law" (1|2.94.6). Aquinas holds that "circumstances" matter greatly in the right understanding of the "species" of an action and in the right moral description and evaluation of that action. As he states: "And consequently that which, in one action is taken as a circumstance added to the object that specifies the action, can again be taken by the directing reason, as the principal condition of the object that determines the action's species" (1|2.18.10). This whole discussion about the significance of circumstances is basic to our adequate naming of various sorts of actions. Aquinas seems to accept the complexity of the moral life where differing circumstances can materially change the act. This is most significant in that it admits that actions across time and in different social contexts are quite often functionally different sorts of acts.

Ordinary actions like driving a car in 1919, before an understanding of greenhouse gas impacts, could appropriately be named as an efficient "act of transportation." Today driving a car needs to be named somewhat differently to help catch how we appreciate that as basic transportation it is having another unintended effect, namely as it contributes to pushing global warming trends with their increasing burden on both natural ecosystems and human communities and future generations. This leads, as Aristotle long ago suggested, that we need to attend to the "ultimate particulars" to help us make our action more virtuous and more adequately "fitting." In ecological terms today, what counts as "fitting" action is "sustainable" action. So we need to elaborate a more complex classification scheme of types of motor vehicle driving so as to differentiate between say driving a heavy vehicle, a Hummer, that gets 14-15 miles per gallon (mpg), or a lighter, hybrid vehicle that much more efficiently gets 40-50 mpg or the new Nissan Leaf, a pure electric car, whose carbon emission depends solely on the sourcing of the electricity grid that it plugs into. Across the United States, most of our electricity is generated by coal-fired plants, but that could change rather quickly as solar and wind power generation are developed.

Given the vast and threatening "new circumstance" of global climate threats—a set of "ultimate particulars" if ever there was one, we need perhaps to differentiate the "moral species" of driving these three types of vehicles. In the last two decades, we have already begun to make the distinction between "ecologically irresponsible driving" and "ecologically responsible driving" and to distinguish "sustainable zoning patterns" from zoning patterns that push suburban sprawl and motor vehicle usage.

This issue of proper naming is especially pertinent to understanding what is at stake in the basic descriptions or normative classification of acts. Years ago, for example, Congress passed legislation, the Corporate Average Fuel Economy (CAFE) standards, mandating that the major car manufactures like Ford, General Motors and others needed to have their sales of different car models average out to a certain miles per gallon median. The intent was to push car companies to increase fuel efficiency and miles per gallon across the fleets of different models produced and sold. One aim was to help America become less dependent on foreign oil. However, Detroit saw its biggest profits coming from its heaviest vehicles—SUVs and minivans. So it lobbied Congress successfully to re-label SUVs and minivans under a new normative classification as "light trucks," thereby warranting an easement from the CAFE standards restrictions. This has allowed the current American motor vehicle fleet to swell in terms of the percentage of "light trucks" that fill our highways and parking lots. This easement greatly hindered the country's movement toward on average higher miles per gallon efficiency. Indeed, words have consequences, and the right classification of products, services, and "acts" is morally and ecologically critical.

For Aquinas, "everything that is directed to an end should be proportionate to that end. But acts are made proportionate to an end by means of a certain commensurateness, which results from the due circumstances. Hence the theologian has to reconsider the circumstances.... [B]ecause the theologian considers human acts according as they are found to be good or evil, better or worse: and this diversity depends on circumstances" (1948, 1|2.7.2). Later he again emphasizes how a change in the circumstances of a human act can alter the basic type or "species" of such an act: "So it is with action. For the plenitude of its goodness does not consist wholly in its species, but also in certain additions which accrue to it by reason of certain accidents: and such are its due circumstances. Wherefore if something be wanting that is requisite as a due circumstance the action will be evil" (1|2.18.3).

Aquinas illustrates this issue of the need to carefully differentiate distinct acts one from another, by distinguishing between the proper "conjugal act" in a marriage relationship and "adultery." The "one deserves praise" while the other "blame and punishment," but as "compared to the generative power they do not differ in species." He concludes: "A circumstance is sometimes taken as the essential difference of the object, as compared to reason; and then it can specify a moral act. And it must needs be so whenever a circumstance transforms an action from good to evil; for a circumstance would not make an action evil, except through being repugnant to reason" (1|2.18.5 ad 3-4).

In Aquinas's terms, one can think of the rise of worry about climate change and its potential impact on habitat alteration and species extinction rates as part of the vast new changed set of historical circumstances that fundamentally require a new frame of understanding to rightly reclassify and reevaluate many of our ordinary human actions. "Driving my car" once sounded benign; but if my car is gasoline powered, then some of my description of my act needs to catch the chain of global impacts that my driving ties me into and to which I am contributing. In light of a range of environmental concerns, we as a general society rightly need to complexify our moral discourse about many of our basic ordinary practices to highlight our increasing attention to the widening range of impacts that flow from such day to day practices.

Vincible vs. Invincible Ignorance

Aquinas asks: "Whether ignorance is a sin?" (1948, 1|2.76.2). This is a good question for a generation facing the political and religious struggle of helping to mobilize a timely response to rising planetary threats. In a time of rapidly advancing dissemination of scientific data about the prospects of emerging threats, the issue of the culpability of ignorance or claimed ignorance gains in importance. One can certainly appreciate that in 1960, many people simply had not had a chance to hear of the emerging array of ecological threats. People are busy with the challenges of ordinary life. Yet one has waited decade after decade for our nation to act with boldness by engaging these rising ecological threats to national and global security. So also has one similarly waited decade after decade for the Christian churches to mobilize their energies by calling their communities to protect the diversity of God's creation as their top priority. And we still wait. One begins to question at what point ignorance and "business as usual" become culpable? Aquinas makes an important distinction between "vincible" and "invincible" ignorance. One of his main contributions is his dogged eagerness to make important distinctions between and among the diverse range of human actions. "Vincible ignorance" for Aquinas consists of ignorance through one's own "negligence...of what one is bound to know." Such negligence, he holds, is a sin. Whereas "invincible ignorance" is not "voluntary," it "cannot be overcome by study." Because it "is not in our power to be rid of it," it is not a sin (1|2.76.2).

With this analysis, Aquinas provides the basis for a powerful indictment against both the United States and other nations who intensively consume fossil fuel and thus disproportionately emit on a per capita basis huge levels of greenhouse gas emissions and against all religious communities, including the Catholic Church, who decade after decade fail to lift up global ecological disruption as a top ecclesial concern. By definition and also illustrated in institutional practice, if it is not held at the top or close to the very top priority, it will be little attended to. For nations and world religious bodies not to commit to planetary environmental responsibility in 1950 as the top priority is not surprising. It is "invincible ignorance." But for nations and religious communities today in 2011 to fail to place ecological responsibility as a key priority deserves Aquinas's evaluation of "vincible ignorance" and his condemnation of such negligence as sinful. In a world of people and institutions needing constant conversion the recognition of "sin" can be the helpful catalyst in an honest "examination of conscience" that can lead to renewal, new insight, and recommitment.

Habits & the Virtue of Prudence

Following Aristotle, Aquinas places great importance on habits of action and the virtues in the moral life (Nelson 1992, 69-104). Both are deeply relevant lens for examining the challenges of ecologically responsible living in the twenty-first century. Aquinas follows Aristotle's observation that "like acts cause like habits" (Aquinas 1948, 1|2.52.3). The Aristotelian and Thomist appreciation for the importance of habits in the moral life is a significant contribution. It provides a helpful resource for examining the role of "habit" in the day-to-day practices of production and consumption. The global economy is a sustained engine mobilizing such habits of shopping though relentless advertizing campaigns designed to create wants and a sense of need (Durning 1992,17-36). And Europe's decisions to raise taxes on gasoline and to promote compact and vital cities with excellent public transit options, has institutionalized in much of European society habits of apartment ownership, walking or biking as transit modes, public transit as a daily event, and shopping locally at small shops. In contrast in the United States, policies have across the decades supported superhighway growth, suburban sprawl, private motor vehicle usage, and high gas consumption. Totaled together a set of ordinary practices that have been supported across the decades in Europe has resulted in the average European using roughly half the fossil fuel imputs as the average American and thus contributing to global climate change roughly half of the way an average American does. Societal habits of production and consumption are built up and supported across the decades by governmental taxing, spending, zoning and other regulatory policies.

Indeed, as David Loy suggests, in order to appreciate the dynamism of the global economy's impact on the world's societies, we would do best to appreciate how its creation of habits of consumption are at bottom religiously valorized and ritualized, for the global market functions as a genuine world religion that offers an inner-worldly salvation via the satisfaction of a sense of lack. As he argues, it is the most successful world religion today for it is relentlessly proselytizing and converting people to its vision of the good life. Shopping malls are its new temples, its sacred space. It has sacred symbols of brands that signal group identification. It has an optimistic vision of the future, even as it viciously commodifies the world and turns eyes toward present satisfaction and away from future ecological impacts (Loy 2000, 15-28).

Reading Aquinas's *Summa Theologiae* can be a moving experience as one encounters a remarkably powerful intellect engaging both his theological tradition and the newest and most radical currents of Aristotelian science and philosophy and the commentaries of Arab philosophers and the works of Jewish sages. When one encounters his "Treatise on Prudence and Justice," one sees a medieval theologian grappling with issues that are quite pressing for anyone worried about the next century of our planet's life. A central plank of ecological ethics and policy analysis is the "prudence principle." It holds that when the stakes are very high—when the potential losses and degradation are great, we should err on the side of safety and insuring our planetary stability and our future generations' security. It is the same value assessment as any homeowner who thinks that it is smart to bear current costs to buy insurance that will mitigate future threats. Roughly it holds that we should anticipate the worst case in the future and work now and pay whatever costs there are to help prevent that case from occurring. This is sometimes called the "precautionary principle." This principle holds when a plausible risk of harm to society is established, public officials have a responsibility to take requisite measures to protect the general public. In a context of uncertainty of future risk, leaders and communities should err on the side of safety even if it means incurring costs in the present as an insurance hedge against future harm. It is moving to witness a medieval theologian reaching so directly across the centuries to embrace the hands of current ecologists struggling to save creation and to spare future generations from the full impact of massive global climate change.

For Aquinas, prudence "considers things afar off" (1948, 2|2. 47.1 ad.2). In this section he is relying heavily on Aristotle's views. Aquinas describes how prudence is "right reason applied to action" (2|2.47.9). Prudence, he tells us, "requires the memory of many things" (2|2.49.1). And he is surely right. With evidence of global climate change that we are pushing, we need to remember the decisions and missteps in policies and assumptions that have encouraged societal habits that have locked American, Canadian, and Australian societies into high-fossil fuel consumption rates. We need to remember the future—the future generations of human and non-human beings, whose well-being we hold in our hands today. Remembering runs in two directions. Prudence focuses on the need for care in very practical decisions about action. As Aquinas states: "Prudence consists in a right estimate about matters of action" (2|2.49.4). As he continues: "Consequently, future contingents, in so far as they can be directed by man to the end of human life, are the matter of prudence: and each of these things is implied in the word foresight, for it implies the notion of something distant, to which that which occurs in the present has to be directed" (2|2.49.6). For Aquinas: "Due foresight of the future belongs to prudence" (2|2.55.7 ad.2), and imprudence, which is a failure of prudence, often involves "negligence" (2|2.54.2).

Even as contemporary debates in Christian ethics have concentrated attention on the need for justice, ecologists who intently track trends of humanity's impact on global ecosystems and climate patterns stress the virtue of "prudence." Prudence as a virtue does not attract much attention in broad areas of Christian ethical concern or in everyday preaching. But the virtue of prudence is being found by ecologists to be of central importance in contemporary debates about policy choices that impact the future. Prudence has come into its own in an era of ecological challenge, precisely because today we need to frame closely our understanding of present action in light of our best estimates of its range of impacts in the future. It is not surprising that ecologists would elevate prudence to a position of primacy. As Aquinas notes, prudence looks to the future impacts—"things afar off." Prudence is sensitive to historical trends and attends closely to future risks. In this way prudence concerns itself with issues of justice owed to the future, owed to both future human communities and to future members of nonhuman living species. And while the call to "be prudent" seems at first to be a rather timid and uninspiring admonition, it is actually a rather bold and revolutionary call for long-range responsibility that flies in stark contrast to the dominant corporate and fiscal interests that push economic rationality today to be defined by the shortest of short-sighted interests that deeply discount future well-being. And if the economic sphere concentrates attention so narrowly and intensively to quarterly profit and loss statements and the rise and fall of the Dow, our political system, too, seems to pull general society's attention incessantly to the short-term focus on the immediate dramas and concerns of the next election cycle. But ecologists and concerned citizens are looking farther ahead and worrying on a broader scale.

They are engaging the vast expanse of planetary history and the rise of human dominance on Earth and are worrying about the present and future impacts of rapid population growth, of rising fossil fuel use, of expanding deforestation and habitat destruction, and rise of species extinction rates. The U. S. Catholic Bishops were surely right to center a call for prudence prominently in their 2001 pastoral letter *Global Climate Change: A Plea for Dialogue, Prudence and the Common Good* (USCCB 2001, 1-23). Aquinas's reflections on prudence and responsible care for the future and his indictment of imprudence and negligence only become more relevant each decade as helpful lenses for understanding the expanding moral challenges humanity faces in revising national policies across the globe to help institutionalize and stabilize new societal habits of ecological responsibility.

While the United States and many other nations have been shock-ingly slow to respond in timely fashion to the need to mitigate the threats posed by climate change, they have, across the last century, shown vigilance in trying to be prepared to face threats to their national security. And this expenditure of national wealth for national defense is based on taking the virtue of prudence seriously. It is clear that modern nation-states place high regard for prudence when it comes to their assessment of national security and national interest. A particularly dramatic illustration of the importance of the virtue of prudence in guiding the affairs of nations occurred in the wind-up to World War II. Hitler had been mobilizing German military strength for years as England slept. Stanley Baldwin was Prime Minister of Great Britain and he had kept British military spending down so as not to drain off economic vitality. In the House of Commons on November 12, 1936, Winston Churchill rose to give a speech con-demning the negligence, in his eyes, of the Baldwin government in its failure to mobilize in the face of a mounting German threat to Great Britain's fundamental security. Churchill drove home his condemna-tion of Baldwin's negligence by invoking a biblical image taken from Joel 2:25. Churchill spoke with mounting anger about years, precious years, lost in failures to prepare for the mounting crisis. He spoke of these years, following Joel, as the "years that the locust hath eaten." Passivity and negligence in the face of a mounting crisis constitutes a tragic lost opportunity and this negligence deserves prophetic con-demnation (Churchill 1989, 1145-128).

Churchill's dramatic attack on negligent passivity in the face of rising threat is an instructive shout across to us many decades later. Today the "years that the locust hath eaten" have wasted precious decades in which humanity—we Americans and our brothers and sisters across the many continents—could have, and should have, been enacting policies that would help reduce our fossil fuel consumption and thus our greenhouse gas emissions. In a time of mounting planetary crisis, a year is a terrible thing to waste and wasted decades are worse. Time is precious. Aquinas's whole emphasis on the virtue of prudence is that we attend to the future impacts of our actions and to the policies that we enact today and that we act responsibility in light of our awareness of future potentials.

Perhaps the great tragedy of our time is that the United States and other nations have not yet seen that our fundamental national and global security is dependent on the basic ecological sustainability of the planetary ecosystems. Notions of national security have been built up across the Nineteenth and Twentieth Centuries to take most seriously threats posed by hostile nations' military threats. But we are in new territory today. We are engaging a new world but with old maps. We follow customary concerns that entrance us about the pressing need for military preparedness to face hostile military threats, but have not yet appreciated the ecologists' warnings that rising ecological threats are of a scale to pose genuine national security theats to nations and societies. Accordingly many nations exhibit vigilance against hostile nation-state military threats, but remain clueless and unprepared for rising ecological threats to national security. Tragically most nations have not yet grasped how ecological threats can be of the same scale of importance as hostile nation threats. It is imperative that we ecologize our understanding of what counts as genuine, long-range national and global security. Only then will we be able to step up seriously to funding those policies that truly engage rising climate change threats and other ecological concerns.

CONCLUSION

Climate change is a serious concern. It threatens not just weather and temperature patterns. It threatens global habitats and the well-being of diverse animal and plant species across Earth. It also seriously threatens human communities. The Catholic community makes up roughly one-sixth of humanity. Those to whom much is given, much is required. It is imperative that the Catholic community across the globe feel the compelling planetary needs, attend to the range of concerns, examine current ecclesial and societal consumptive habits of practices, challenge "business as usual" thinking, massively scale back military budgets, and push for the redistribution of funding for "insurance policies" against global warming, habitat destruction, human population surge, and species extinction. The ancient Stoics viewed humanity as participating in a grand cosmopolis, a great universal community, and this vision helped shape the Medieval Christian understanding of the great community of creation of which humanity is but a part. In order to protect the "community of creation" today, we need to appreciate how "externalities" in our market systems, fundamentally distort our understanding of "rational" and "responsible" choice and behavior. Only national policies to internalize these "externalized" and thus "hidden" costs can move us to an market system that honestly communicates true long-range costs of various practices of production and consumption. We are now managing a planet and this is a huge and utterly new responsibility. In order to responsibly step up to this task, we need a keen eye to the true range of impacts of human actions and to the true range of costs of certain productive and consumptive choices. Loose calls for "loving nature" are insufficient. We need, as Aristotle reminds us, to attend to the details, the raw realities, the "ultimate particulars" in order to grasp the contours of our responsibilities.

Great Medieval theologians and saints such as Thomas Aquinas, Francis of Assisi, Bonaventure, and Hildegard of Bingen serve as most helpful guides through their emphasis on the graced character of the natural world. These giants serve as powerful resources for helping the global Catholic community—the one sixth of humanity—ponder what is at stake in our emerging ecological challenges. As we attempt to engage in the pressing challenges of the age confronting the threat of global climate change, we can draw on deeply helpful sources of moral reasoning offered in the heritage of Aristotle and Aquinas. Aquinas's understanding of the virtue of prudence, negligence, vincible ignorance, natural law, and the "community of the universe" all provide vital resources for thinking through the Catholic Church's responsibilities in an ecologically challenged age.

SOURCES

Aquinas, Thomas. 1948. *Summa Theologica*. Translated by the Fathers of the English Dominican Province. 5 volumes. Westminster, MD: Christian Classics.

Aristotle.1962. *Nicomachean Ethics*. Translated by Martin Ostwald. Indianapolis: Bobbs-Merrill.

Brown, Lester R. 2009. *Plan B 4.0: Mobilizing to Save Civilization*. New York and London: W.W. Norton.

Churchill, Winston. 1989. *Blood, Toil, Tears, and Sweet: The Speeches of Winston Churchill*. Edited by David Cannadine. Boston: Houghton Mifflin.

Commoner, Barry. 1971. *The Closing Circle: Nature, Man, and Technology*. New York: Bantam Books.

Curran, Charles E. 1974. *New Perspectives in Moral Theology*. Notre Dame: University of Notre Dame Press.

Daly, Herman E. and Cobb, Jr., John B. 1989. *For the Common Good: Redirecting the Economy Toward Community, the Environment, and a Sustainable Future*. Boston: Beacon Press.

Daily, Gretchen C. ed. 1997. *Nature's Services: Societal Dependence on Natural Ecosystems*. Washington, D.C.: Island Press.

Durning, Alan. 1992. *How Much is Enough? The Consumer Society and the Future of the Earth*. New York: W. W. Norton.

Friedman, Milton. 1962. *Capitalism & Freedom*. Chicago & London: University of Chicago Press.

French, William C. 1993. "The Auto and the Earthly City: Gas Taxes and Civic Renewal." *Theology & Public Policy* 5:1 (Summer 1993): 15-28.

————. 2008. "Natural Law and Ecological Responsibility: Drawing on the Thomistic Tradition." *University of St. Thomas Law Journal* 5:1 (Winter 2008): 12-36.

Gore, Al. 1992. *Earth in the Balance: Ecology and the Human Spirit*. Boston: Houghton Mifflin.

————. 2009. *Our Choice: A Plan to Solve the Climate Crisis*. Emmaus, PA: Rodale Press.

Jenkins, Willis. 2008. *Ecologies of Grace: Environmental Ethics and Christian Theology*. Oxford: Oxford University Press.

Jonas, Hans. 1984. *The Imperative of Responsibility: In Search of an Ethics for the Technological Age*. Translated by Hans Jonas in collaboration with David Herr. Chicago & London: University of Chicago Press.

Leakey, Richard, and Roger Lewin. 1995. *The Sixth Extinction: Patterns of Life and the Future of Humankind.* New York: Anchor Books.

Loy, David R. 2000. "The Religion of the Market." In *Visions of a New Earth: Religious Perspectives on Population, Consumption, and Ecology,* ed. Harold Coward and Daniel C. Maguire, 15-28. Albany: State University of New York Press.

Maathai, Wangari. 2010. *Replenishing the Earth: Spiritual Values for Healing Ourselves and the World.* New York: Doubleday.

Nelson, Daniel Mark. 1992. *The Priority of Prudence: Virtue and Natural Law in Thomas Aquinas and the Implications for Modern Ethics.* University Park, Pennsylvania: Pennsylvania State University Press.

Rubenstein, Richard E. 2003. *Aristotle's Children: How Christians, Muslims, and Jews Rediscovered Ancient Wisdom and Illuminated the Dark Ages.* Orlando: Harcourt.

Schaefer, Jame. 2009. *Theological Foundations for Environmental Ethics: Reconstructing Patristic & Medieval Concepts.* Washington, DC: Georgetown University Press.

United Nations. 1999. "The World At Six Billion." United Nations Population Division. Accessed from http://www.un.org/esa/population/publications/sixbillion/sixbilpart1.pdf on February 16, 2011.

———. 2004. Department of Economic and Social Affairs. "World Population to 2300." Accessed from http://www.un.org/esa/population/publications/longrange2/WorldPop2300final.pdf on February 16, 2011.

United States Conference of Catholic Bishops (USCCB). 2001. *Global Climate Change: A Plea for Dialogue, Prudence, and the Common Good.* Washington, DC: United States Catholic Conference.

7

CATHOLIC THEOLOGY, INTERNATIONAL LAW, & THE GLOBAL CLIMATE CRISIS

William P. George

Catholic theologians clearly have good reason to turn their attention to the global climate crisis. I assume that theology and theologically-informed magisterial teaching can both illumine various aspects of the issue and provide motivation to address it in wise and timely ways. However, responding to human-induced global climate change (GCC) is also a matter of international law. The United Nations Framework Convention on Global Climate Change (UNFGCC), the basis for ongoing negotiations and agreements (Kyoto, Copenhagen, and others), presupposes and is integral to the vast and complex arena of international law, which includes treaties on the global commons such as the oceans, customary law on various topics, widely accepted legal principles relevant to GCC and other issues. Without sacrifice of their distinct role, the more fully theologians can engage the critically important but also messy and limited system of international legal norms, instruments, organizations, and institutions, the better they will be able to address the challenges of GCC.

Given the scope, history, and complexity of both theology and international law, to say nothing of the challenge of GCC itself, this essay must be modest in its aims. One could, for instance, explore how theologians vary in their regard for international law: liberation theologians might find it biased against the Global South; strict sectarians might view it as outside their realm of concern; Niebuhrians might stress its importance but denounce its idealism; more recent voices might judge most if not all of it to be hopelessly anthropocentric; and so on. This essay does not engage in comparisons of this kind even as it welcomes a dialectical approach to both theology and law. Nor does it examine how theologically-grounded principles of Catholic social teaching might be affirmed by or brought to bear on international law, although that, too, is a critical task (George 2008, 141-145). Rather, drawing upon Thomas Aquinas and one of his intellectual heirs, Bernard Lonergan, I will attempt to sketch out two interrelated points of convergence between Catholic theology and international law, such that international law in general and climate change law in particular might more readily emerge as a viable *locus theologicus*.

The first is in terms of method, as articulated by Lonergan. Method in turn is grounded in the cognitive operations of the "subject," an important term common to international law (Damrosch, et al. 2009, 299) and to Lonergan's thought (1974), quite different initial meanings notwithstanding. Against the backdrop of the European wars of religion, successive currents of secularization, and fundamentalism of various kinds, international lawyers have long been wary of theologians' intrusions into their domain (Janis 1999; George 1999-2000). To counter the assumption that theology and international law are inherently at odds, I propose that at the level of the thinking and acting subject, theologians and international lawyers actually have much in common. As they go about their respective and diverse tasks, they arguably share a common method (George 2009).

The second meeting point is a set of virtues that links the two domains. Surely, justice is one such virtue. This includes justice between present and future generations, to which both Catholic social teaching (Agius 1998) and international law (Agius and Busuttil 1998) have been attentive. Here, however, the focus will be on charity and, more extensively, on prudence, which the U.S. Catholic Bishops herald as key to meeting the climate crisis (USSCB 2001). A focus on prudence is all the more important since international law is not only law but also politics (Henkin 1995, 4-5). Thus, attention to the concrete, the practical, the feasible—that is, the realm in which prudence must prevail if justice is to be secured—will be crucial. Included in the discussion of prudence as well are initial thoughts on "emergent probability," a world view developed by Lonergan (1988, 146-51, 234-37) but also latent in Aquinas (Byrne 1982). Grace builds on nature, according to a Thomist axiom (Aquinas 1948, 1.1.8), and both history and nature are in part matters of probabilities—for example, the greater or lesser probability that human beings can affect the outcome of the climate crisis.

Elsewhere I have argued that religiously-committed individuals and communities intent on developing a "global ethics" should not overlook international law (George 1999). The pages below return to that argument with added specificity: Catholic theologians hoping to address in meaningful and effective ways the issue of GCC should, if they are not already doing so, engage international law as fully as they can.

SUBJECT & METHOD IN THEOLOGY & INTERNATIONAL LAW

"Traditionally, international law has been seen as the law of the international community of states" (Damrosch et al. 2009, xv). While there are good reasons for retaining this state-centered understanding of international law, the list of who counts as recognized "subjects" of, and in, international law has expanded to include "international organizations, individuals, groups, multinational corporations, and other entities capable of possessing international legal personality" (Epps 2009, 3). In fact Janne Elisabeth Nigman concludes her lengthy, philosophically-sophisticated study of "international legal personality" with the assertion that "the individual is the legal personality *par excellence* in international law" (Nijman 2004, 473).

However, even in a state-centric international legal system, the individual subject and his or her multiple relations count—*de facto* if not always *de jure* in the strict sense of "legal personality." Thus, for example, Article 38 of the Statute of the International Court of Justice authoritatively lists "teachings of the most highly qualified publicists of the various nations" as subsidiary sources of international law (United Nations 1945, art. 38), and these "publicists" may also be teachers who influence generations of legal practitioners. Individuals famous for other reasons and possibly removed from international law may also contribute to its growth or direction. One thinks of Princess Diana in the case of an emergent treaty on anti-personnel mines, or Al Gore in the case under discussion in this volume. Insofar as such figures are leaders of movements, they call attention to the thoughts and actions of lesser known but equally committed individuals and groups who may indeed participate in the workings of international law, as did the politically-savvy citizens who made up the Neptune Group in the case of the 1982 Convention on the Law of the Sea (Levering and Levering 1999).

Furthermore, sometimes these subjects act for religious reasons. As legal scholar James Nafziger argues, religiously informed and motivated people play important roles in the development and maintenance of international law. They provide vision for law's growth and translate it in practice (Nafziger 1999). Religious actors may also bring about legal change through involvement in the United Nations system (Boehle 2010). Occasionally, at least, international lawyers themselves show great openness to religion and theology. Thus, a judge on the International Court of Justice can write a book on the Lord's Prayer (Weeramantry 1998), and a Scottish professor of international law can argue that a moral theologian may understand better than lawyers why international law, with its heavy positivistic tendencies, finds itself without critical grounding (Carty 2008).

The world of Catholic theology and the world of international law are not one world. Moving back and forth between the two still demands what Mark Morelli calls "horizonal diplomacy" (Morelli 1981). Nevertheless, when theologians venture into the realm of international law as a way to mediate between theological convictions and claims, on the one hand, and the realities of the climate crisis, on the other, they encounter human beings, including professionals in the field, not so unlike themselves.

Method

In their thinking and acting as authentic "subjects," theologians and international lawyers arguably share a common method for doing what they do. For Lonergan, "[a] method is a normative pattern of recurrent and related operations yielding cumulative and progressive results." Method *in theology* is rooted in "a basic pattern of operations employed in every cognitional enterprise" (Lonergan 1990, 4): "seeing, hearing, touching, smelling, tasting, inquiring, imagining, understanding, conceiving, formulating, reflecting, marshalling and weighing the evidence, judging, deliberating, evaluating, deciding, speaking, and writing" (6). These multiple and distinct operations occur on "four successive, related, but qualitatively different levels" (9) of conscious intentionality, denoted, "for brevity's sake ...by the principal occurrence at [each] level" as "experiencing, understanding, judging, and deciding" (14). Intrinsic to intentional consciousness and corresponding to these levels are certain transcendental precepts: "Be attentive, Be intelligent, Be reasonable, Be responsible" (53).

"Transcendental method," in its full-flowering, assumes self-appropriation of these recurrent and related operations through "applying the operations as intentional to the operations as conscious" (14). Self-appropriation is a very difficult achievement. Thus method-as-self-appropriation may be operative only infrequently. Still, the fundaments of this method, grounded as they are in conscious intentionality, are discernible in the workings of international law, including the international law of climate change. For like theology, international law is a vast and sophisticated set of "cumulative and progressive results" born of human attentiveness to data, of recurring and expanding insights, of well-considered judgments, and of decisions grounded in an apprehension of values—even as international law is also profoundly affected by the "flight from insight" or bias (Lonergan 1988, 214-27, 244-67). However, in order to advance the rather bold claim that theologians and international lawyers actually share a common method, it will be helpful to comment further on that method.

First, the tacit hope of *Method in Theology* is that theologians will be able to apply this method because, through self-appropriation and conversion—religious, moral, intellectual (Lonergan 1990, *passim*), they fully understand and take responsibility for their theologizing. Presumably, theologians are disposed to the kind of self-reflection needed to understand what they are doing when they are doing theology, but there is no reason to assume—or, at least, I do not assume—that international lawyers are less capable of attending to, of understanding, of making correct judgments about, and of taking responsibility for the real fundamentals of their craft, namely, the attentiveness and negligence, the insights and oversights, the sound and specious judgments of fact and value, the responsible and irresponsible decisions that contribute to or impede the "cumulative and progressive results" that are named international law. Nor do I presuppose that it is only theologians who are "converted," intellectually, morally, or even religiously.

Second, if Lonergan's account of method illumines what transpires in international law, perhaps, conversely, international law illumines method. As Lonergan insists, "[a]ny theory, description, account of [the] conscious and intentional operations [of method] is bound to be incomplete and to admit of further clarification and extensions. However, all such clarifications and extensions are to be derived from the conscious and intentional operations themselves" (19-20). So it might be that by attentively, intelligently, critically, and responsibly engaging international law, theologians will arrive at a more complete grasp of their own subjectivity and thus their own method.

Third, method is not without its context and arena of operation. Understanding method in theology requires some grasp of "the human good," "meaning," and, of course, "religion" (27-126). There can be no hope of discussing here any of these matters in depth. I can do little more than affirm Lonergan's insistence that these topics are germane to method in theology. Nevertheless, to further the argument of this essay, some modest attempt must be made to show that they also pertain to international law.

The Human Good

Under the heading of the human good, Lonergan discusses "feelings" and "skills" and "judgments of value" and "beliefs" (30-47). Certainly, these are in no short supply among lawyers and judges and members of NGOs and internet bloggers seeking or resisting legal change when it comes to such matters as GCC. As immediately relevant to our topic is the "structure" of the good, wherein multiple terms "regard (1) individuals in their potentialities and actuations, (2) cooperating groups, and (3) ends" (47). Again, it is not difficult to see how the good, so described, might be related to international law, even if that relationship cannot be detailed here. Simply consider, for instance, the extraordinarily complex structure of the human good exemplified by states' participation in the UNFGCC, or by the Intergovernmental Panel on Global Climate Change (hereafter IPCC), established "to provide the world with a clear scientific view on the current state of climate change and its potential environmental and socio-economic consequences" (IPCC 2010).

The human good also includes the "good of order" (Lonergan 1990, 49-50), not just reducing greenhouse gases for a day, but ongoing and recurrent practices by states and other international actors to slow global warming. This brings into view an aspect of the human good to which we will return: the notion of "emergent probability," wherein higher schemes of recurrence, in service of the human good, emerge or fail to emerge from lower schemes, thus leading to historical progress or decline (52-55; 1985). Effective, behavior-altering international laws may emerge to meet the climate challenge. Then again, maybe they will not.

Meaning

Just as international law can further but also hinder the human good, so, too, international law, like theology, amply proves that we live in a world "mediated by meaning," with its various "functions" and "realms" (Lonergan 1990, 76-85). For example, courts and communiqués and conferences, such as Kyoto or Copenhagen, exemplify meaning's "constitutive," "efficient," and "communicative" functions (77-78). The daily, humdrum world of international legal practice, wherein people who "know the ropes" move things along, exemplifies the "common sense" realm of meaning (81-83), just as treatises and journal articles exemplify the crucial role of "theory" (83). Whether the realm of "interiority," wherein theory and common sense are distinguished and properly related (83), is fully apparent in international law is an open question, but clearly the practice of international law is not without reflective souls (Cassese 2008, lix-lxxxi).

The "transcendent" realm of meaning (83-84), fundamental to theology, may seem to have been banished from international law long ago. As the former vice-president of the ICJ puts it, "in the twentieth century, international law …so far distanced itself from religion that the latter receive[d] scarcely a mention in the standard texts" (Weeramantry 1999). However, rumors of religion's demise may be premature. International law is rooted in a religious past, and, unless that past is to be totally denied, religion may in some fashion be tacitly retained. Indeed, questions about the role of religion and theology in international law (Janis and Evans) and international relations (Fox and Shmuel 2004) receive considerable attention today. Furthermore, if the method one finds in theology at the level of subjects' cognitive operations coincides with the method one finds in international law (again, at the level of subjects' operations, not the myriad theories about international law), then it is worth emphasizing that "the objects of theology do not lie outside the transcendental field [of method]. For that field is unrestricted, and so outside it there is nothing at all" (Lonergan 1990, 23). If the question of God is "the question that questions questioning itself" (103), if international law gives rise to question upon question (as any perplexed student can attest), then the "transcendental" question of God may be operative, if rarely explicit, in the concrete and complex "categorical" (73-74) workings of international law. But this question may rise closer to the surface. One may ask whether there is an ultimate grounding for human rights (including the rights of future generations endangered by GCC), or whether the international legal principle of "common heritage of mankind," which has been applied to the global climate (Ramakrishna 1990), is fully intelligible without a theological reference of some kind (George 1990). Method in theology and method in international law may converge, then, in a shared transcendent realm of meaning even if, to be sure, convergence in that realm is rarely overt (George 1999, 488-491).

Religion

Attention to the transcendent realm of meaning gives rise to further questions about religion—and religions—in relation to theology, to international law, and to GCC. Such questions cannot be adequately pursued here, but at least two related points may be in order. One has to do religious experiences. Self-interest of individual nations or other international actors is often viewed as international law's final word. However, Lonergan would emphasize the graced reality of "God's love flooding our hearts through the Holy Spirit that is given to us" (Rom. 5.5) (Lonergan 1990, 105) that can transform subjects' horizons. As the story of Jonah and other texts remind us, divine concern for humankind and other creatures is broader and deeper than human interest (Jonah 4.9-11). When directed toward international law, method in theology should attend to the love and faith that counters bias (Lonergan 1973, 115-118), for example, the 'group bias' of nationalism that is blind to the global common good, or the 'general bias' of common sense' (Lonergan 1988, 250-67) that precludes facing the 'big questions' about international law's true ends and its truly workable means.

Of course, not only Christians have religious experiences. If international law is to mediate between Catholic theology and the concrete issues of GCC, it will not be outside the scope of Catholic theology to engage other religious traditions independently of international law but also in regard to it. While international law undoubtedly has its roots sunk primarily in the Christian tradition, there is no reason why the connections today between international law and religion cannot be interreligious in orientation and scope, just as interreligious cooperation on climate change is now underway (Landau 2008).

Functional Specialties & Collaboration

Finally, to extend the argument that method in theology resonates with method in international law, I would note that the several "functional specialties" that Lonergan says theology entails will perhaps under different nomenclature be recognizable to scholars and practitioners of international law: "research," "interpretation," "history," "dialectics," "foundations," "doctrines," "systematics," and "communications" (1990, 125-45). I leave aside Lonergan's elaboration on these functional specialties as operative in theology. As for international law, one thinks, for instance, of the *research* or data-gathering by the IPCC; of challenges posed by the *interpretation* of environmental treaties; of the exploration of international legal *history*, with its breakthroughs and breakdowns; of the *dialectics* involved in sorting out the various, often opposing approaches to international law of GCC; of the articulation by legal theorists of the best possible *foundations* of international law; of the establishment and refinement of *doctrines*, such as the "precautionary principle" (Damrosch et al. 2009, 1499-1500, 1511-1513) so important for climate change law; or of the efforts of the United Nations to *communicate* to an ever wider public just what it is about and why international law counts.

In lieu of any such exposition, I would simply stress that if there are affinities between theology and international law in terms of these specialties, then a common method provides a basis for the collaboration that, Lonergan insists, is increasingly required today. When it comes to an issue such as GCC—and this is the central point of this essay—Catholic theologians must regard as their collaborators not only theologians specializing in functions other than their own, not only scientists, economists, and others who address various aspects of the climate crisis. Theologians must also engage the functional specialists who labor in and for international law.

CHARITY, PRUDENCE, & INTERNATIONAL LAW

I am proposing that the Thomist-Lonerganian tradition provides a basis for encouraging Catholic theologians concerned about GCC to engage international law. The focus above was Lonergan's articulation of method. Here, the major resource will be Aquinas, and my argument is that the virtues of charity and prudence provide a nexus between theology and international law. Charity begets a transcendental and increasingly expansive horizon of concern, and the *ordo caritatis* may provide a theological basis for envisioning concentric spheres of legal concern, not unlike the "spheres of justice" discussed by Michael Walzer (1983). The prudence urged by the U.S. Catholic Bishops (2001) is discernible in international law—as are its counterfeits. Thus, if Catholic theologians wish to explicate further what prudence means with regard to the global climate crisis, they will do well to examine the workings and failings of international law.

Given the constraints of space, I will emphasize only one key aspect of virtue as a nexus between theology and international law: the connections between charity/prudence and future generations. Since concern for posterity is a critical component of reflection on GCC (Page 2006), and since future generations have fallen increasingly within Catholic theology's purview (Agius 1998), Catholic theologians will do well to engage international law which has had the welfare of future generations on its agenda for some time (Agius and Busuttil 1998).

While Lonergan's account of the knowing and acting subject provides an intelligible basis for caring about future persons (George 1997), here the focus is mainly on Aquinas. As we look to the future, this return to the past may be advantageous. If Aquinas' understanding of charity and prudence entail, or at least imply, thinking about future generations, then we learn that concern for future generations has informed the Catholic conscience long before the appearance of references to future generations by popes and others in an ecologically-conscious age. Indeed, concern for posterity is rooted in patristic thought and the Christian scriptures (Agius 1998). Furthermore, by returning to Aquinas we encounter a rich account of the *virtues*, with all the issues of growth and development in moral consciousness and effectiveness that a virtue ethics implies. Such an approach, complementary to an ethics of principles (e.g., key principles of Catholic social teaching or basic principles of international environmental law such as "state responsibility") speaks to the reality of *developing* moral agents operating today in a diverse and complex world. Finally, when it comes to prudence especially, Aquinas was well aware of counterfeits, for example, the vices resembling prudence (Aquinas 1948, 2|2.55.1-8). Put theologically, a Thomistic approach to GCC and future generations by way of prudence is fully cognizant of the reality of sin.

Charity & Future Generations

For Aquinas, charity is the "form," "root," and "mother" of all the virtues" (1948, 1|2.23.8 ad 1-3). At the same time, charity cannot exist in act apart from the other virtues (1|2.65.3). Thus, if charity as love of God, self, and neighbor extends to future generations, then prudence will be required if that love is to be effective—as, of course, will justice and other virtues. Now one may argue that as love of God, charity by its very nature already extends to all whom God loves, past, present, and future. In fact, however, Aquinas explicitly cites two instances in which love of neighbor—or, more precisely, its failure—extends to persons, at least of the next generation, not yet born or even conceived. In the cases of fornication and adultery, harm to offspring, and, by extension, harm to other family members, is the primary basis for pronouncing these acts immoral and sins against charity (2|2.154; George 1992, 285-88). Moreover, unless one wants to argue that Aquinas believed that every act of intercourse results in a birth or at least a conception, the moral matter that Aquinas is concerned with is clearly a matter not of necessity but of contingency, not of certainties but of probabilities. In principle, then, for Aquinas, charity as concern for future neighbors extends to statistical matters such as GCC.

Charity for Aquinas has its order, and, while the *ordo caritatis* cannot be explored in any depth, I would at least stress that, especially when joined to the other virtues, there is in this order of charity a certain moral realism (George 1992, 289-94). If charity extends to future generations, then the order of charity renders it sensible, for instance, to be more concerned about one's immediate offspring without ignoring the need to care for those temporally and spatially more distant (Aquinas 1948, 2|2.26.8). Equally important for present purposes, cognizance of the order of charity renders more sensible the concentric circles of jurisdiction and care—sometimes interlocking, sometimes at odds—one finds in the realm of law, extending from the local to the global with various levels in between. Furthermore, Aquinas' attention to the order of charity includes this question relevant to future generations: Should we love more those who have benefitted us or those whom we benefit? Here, Aquinas gives grounds for thinking we should love more, by "closer connection," those to whom we can give good things than those who have bequeathed things to us, because in a sense they are a part of us. Our care for posterity—our "handiwork" in Aquinas's terms—says more about who *we* are than does the care

that our ancestors have shown to us (2|2.26.2). Care for or neglect of
future generations, through action or inaction regarding GCC, helps
to define who we are as intellectual, moral, and religious beings—as
"subjects," in Lonergan's terms.

Finally, whatever might have been the role of religion, and of an
anthropocentric Christianity in particular, in paving the way for the
ecological crisis and other catastrophic human blunders affecting future generations, it may also be argued that the most profound spiritual resources of religion will be required to meet ongoing challenges.
Thus, it is worth stressing that the same charity that calls for concern
for future generations both "feeds" religious devotion and "is fed by it"
(Aquinas 1948, 2|2.82.2 ad.2).

Prudence & Future Generations

The virtue of prudence, so prominently stressed by the U.S. Catholic
Bishops with regard to GCC, provides an especially apt way to develop
an ethics of future generations. However, in keeping with our topic, I
wish to show how prudence, theologically understood as participation
in God's providence (Aquinas 1948, 1|2.91.2), not only entails care
for posterity but is also intrinsic to international law.

Echoing the order of charity, prudence operates in various spheres.
Aquinas distinguishes "individual prudence," "military prudence," "do-
mestic prudence," and "regnative prudence" (2|2.50.1-4). The immedi-
ate implication with regard to future generations is that not everyone
will be expected to regard future generations in the same way. The fact
of distinct spheres and purviews is not without its tensions as evident,
for example, in the respective concerns of an unemployed parent seeking a job in an auto assembly plant and a climate scientist who warns
about more cars on the road. However, if the tensions are acknowledged, they stand a better chance of being addressed. At the very least,
in Aquinas's view, prudence does not proceed as though distinct social
roles and responsibilities did not exist. Clearly, such distinctions reso-
nate with the multi-layered world of international law, starting with
the sometimes highly problematic relationship between international
and national law (Damrosch, et al. 2009, 652-754).

However, by looking at the various "quasi-integral" parts of prudence
we reach the heart of the relationship between prudence and future
generations (Aquinas 1948, 2|2.48). These parts are "memory," "under-
standing," "docility," "shrewdness," "reason," "foresight," "circumspection,"

and "caution." No one can consider this aspect of Aquinas's understanding of prudence without confronting the multi-faceted nature of prudence. Prudence cannot be reduced to any of its parts, and "perfect" acts of prudence require the proper functioning of all. Let me discuss each of these parts—a bit out of turn. Problems of extrapolating from the *Summa Theologiae* to, for example, the expiration of the Kyoto Protocol notwithstanding, prudence as Aquinas describes the virtue is discernible in the workings of international law.

Foresight (Aquinas 1948, 2|2.49.6): "Foresight" is, in a sense, first among the quasi-integral parts since, as Aquinas explains, prudence (*providentia*) gets its name from foresight. Foresight also provides the most direct link between this virtue and future neighbors who will be affected by GCC, since foresight is that part of prudence by which moral agents recognize possible future realities as contingent on human action in the present. Furthermore, questions of contingencies inevitably include questions of probabilities with the assumption that probabilities may be shifted. Thus, for example, the probabilities regarding a rising ocean temperature due to GCC might be shifted through a truly effective and enforceable international climate regime. Prudence-as-foresight, then, may be reconceived in terms of "emergent probability," which refers to the way in which higher integrating schemes of recurrence may emerge according to schedules of probability from lower schemes (Lonergan 1988, 146-51, 234-37).

Much more would need to be said to link prudence, emergent probability, GCC, and future generations, but I would suggest at the very least a connection between the lower and higher schemes of recurrence and the various spheres of prudence. Prudence operative at individual and local levels, for example in the work of NGOs, can give rise to prudent policies and practices at more expansive, possibly global levels. However, emergent laws may be defective or even sabotaged by biases of various kinds, such as nationalism or narrow-interest politics (George 1996, 157-160). One might ask and emergent probability might help explain why the 2009 Copenhagen Conference did not seem to deliver on its promise, just as emergent probability might explain why the United Nations secretary general recommends a "small steps" response to GCC (MacFarquhar 2010). For along with the vector of progress that runs through history, including the history of international law, there is also a vector of decline (Lonergan 1985;

1990, 52-55) to which theologians and international lawyers, in their common subjectivity, can and should carefully attend.

However, foresight is not the whole of prudence; it must work in concert with the other parts. Thus, while foresight most directly relates present moral agents to persons and situations not yet in existence, the quality of that relationship will be affected by the exercise or failure to exercise prudence in its other quasi-integral parts. So we must shift our focus there.

Memory (Aquinas 1948, 2|2.49.1): Here, Aquinas means learning from the past. If prudence is derailed by failures of memory, foresight will also suffer. Put more positively, those who want to lay stress on future persons will not be distracted if they improve their moral memories—and their international legal memories as well. This would include learning from past actions that have brought the present generation harm or great benefit. It would also mean recalling international law's own heritage, including its deep roots in a theologically-informed understanding of natural law that, it was assumed, bound the entire human community (Carty 2008; George 1999-2000).

The role of memory in international law is complex. One thinks of the difficult question of how best to establish, through a review of past state behavior and intent, what counts as customary international law (Damrosch, et al., 2009, 55-121). Or, one notes the raft of Latin phrases that punctuate texts and court cases and thus testify to international law's classical and even theologically-rich heritage, even as that heritage is suppressed (George, 1999-2000). None of this can be explored here, but, it is germane to our discussion to note that international legal memory need not be crudely anthropocentric. For example, in his extraordinary speech to the U.N. General Assembly in 1967 on the need for a new ocean regime, Arvid Pardo. a diplomat and scholar from the tiny country of Malta, recalled that "the dark oceans were the womb of life: from the protecting oceans, life emerged. We still bear in our bodies—in our blood, in the salty bitterness of our tears—the marks of this remote past" (Pardo 1967). To develop regard for future generations is a moral challenge, and meeting that challenge will involve both learning from mistakes (why did past negotiations fail?) and drawing on the wisdom of the recent and the very remote past.

Understanding (Aquinas 1948, 2|2.49.2): By understanding, Aquinas means a grasp of first principles, such as "Good is to be done, and evil avoided." Although I cannot argue the point here at length, it seems feasible to interpret these principles as heuristic guides that are affirmed precisely in the convergence of moral decision-making in its various facets, much as the principle of non-contradiction will not on its own give the answer to specific questions but will be affirmed in correct judgments about particular matters. In any event, one function of prudence is to draw guidance from various principles, and "understanding" is prudence rightly informed by those principles. I suggest, therefore, that prudence might embrace or at least consider within that orientation the kinds of principles that are most fundamental in Catholic social thought (the dignity of the human person, the common good, preferential option for the poor, solidarity, etc.), along with the often parallel principles found in international law, such as concern for the global commons (George 1990). Not only can prudence embrace these principles; the very nature of prudence as foresight bends them towards the future and future generations: the dignity of future persons, as well as the living; the common good of all, including those not yet born; a preferential option for future generations due to their vulnerability at the hands of our choices; an appeal for help from higher governing bodies (subsidiarity) when the good not just of the present generation but also of future generations cannot be secured or assured by local action alone—as is the case with GCC.

Docility (Aquinas 1948, 2|2.49.3): For Aquinas, docility is an eagerness to be taught. This highlights, for example, the importance of science for the development of international law. However, dialogue with experts is not only important. It is also very difficult, and the openings for bias appear to be legion. The "group bias" (Lonergan 1988, 247-250) of nations or political parties or the "general bias of common sense" (250-67) that would settle every problem with a sound bite can render people deaf to input from thoughtful people. Again, the question of the "subject" and "subjects," converted or unconverted, is never far away. Two additional observations are in order. First, while prudence requires the counsel of experts, there are warrants for including among those "experts" individuals and groups often placed on the margins or, in this case, those most likely to be affected by GCG. If foresight regards future contingents, if some of those contingents are future persons and their environment, if charity extends to those future persons as future neighbors, then perhaps in some way we can "listen" to the voice of their authority. It has been proposed that future generations have a guardian at the United Nations to speak on their behalf (Stone 1998), so one might ask who that guardian is to hear so the guardian might speak. Does "docility" as understood by Aquinas require that we "listen" to future generations? The future orientation of prudence would so suggest. Finally, under the heading of docility, I would place the ongoing task, falling to lawyers as well as theologians, of listening to other religious traditions, and what they have to say about the future, future generations, and GCC.

Shrewdness (Aquinas 1948, 2|2.49.4): For Aquinas, shrewdness complements docility. If docility requires slowing down to listen, shrewdness requires moving with alacrity towards decision. The lack of shrewdness is apparent whenever the response to a glaring problem such as GCC is "we need to do another study," or some other stall tactic, when, in fact (and establishing this fact requires serious judgment), the data is sufficiently clear, the call for action is real, and what is needed is all too apparent but also costly to me or us rather than to somebody else. In this case, that "somebody else" may be future generations. What shrewdness calls for is both boldness and creativity in finding the means best to address current moral challenges (Lonergan 1985).

Reason (Aquinas 1948, 2|2.49.5): According to Aquinas, reason is the opposite of what I have elsewhere labeled "angelism"—the naïve assumption that humans can resolve immensely difficult issues through a single bright idea, rather than through an accumulation of insights, a process of learning, a painstaking effort to think things through (George 2000). Prudence is hospitable to future persons and the contingencies that affect them. But if "reason" is not given its due, regard for future generations will degenerate into the quick fix, the ideological mantra, a new international law that is neither reasonable nor perhaps even new.

Circumspection (Aquinas 1948, 2|2.49.7): If foresight regards contingent matters that can be directed to the end of human life, circumspection regards the circumstances surrounding the means chosen to attain the desired end. When considering complex issues such as GCC, the challenge of circumspection is daunting. However, it should not be overwhelming. In answer to the objection that circumstances are infinite, Aquinas responds that "though the number of possible circumstances is infinite, the number of actual circumstances" is not. Prudence as circumspection "fixes" the circumstances, including, in the case of GCC, those disclosed by science or dictated by the practice of law, so that the moral agent(s) may act accordingly. To complain end-lessly about complex circumstances may be simple moral avoidance. As complex as they might be, the circumstances surrounding GCC are not unknowable in every respect.

Caution (Aquinas 1948, 2|2.49.8): Equating prudence with caution is not unusual. This is not the case with Aquinas; caution is but one of eight quasi-integral parts. Still, caution is integral to prudence, and it is certainly relevant to considerations of future generations. While for Aquinas prudence should not be equated with fear, since fear is a bad counselor (1|2.44.2), awareness of the very real prospect of ir-reparable harm to future others should no doubt feature in prudential judgments. Caution has another focus: reluctance to proceed down a road that would leave the decision-maker less virtuous. Thus, pres-ent agents should exercise caution lest their failure to regard future persons in their contingency constrict their own moral horizon and sphere of moral concern.

Imprudence and vices resembling prudence: Some may regard Cath-olic social thought as too optimistic, placing too little emphasis on the power of sin (Curran 2002, 129). And, there is no shortage of people quick to question overly optimistic claims of international law (Goldsmith and Posner 2005). Thus, an emphasis on what Aquinas says about the vices juxtaposed to prudence is all the more important (1948, 2|2.53-55). It should be clear that, given its multiple demands, acting prudently is exceptionally difficult. So much can go wrong: bad memory, failure to listen to others, faulty reasoning, bad science, the multiple biases discussed by Lonergan, and so on. In every case, future neighbors stand to lose. The vices opposed to prudence including "im-prudence" (2|2.53) and "negligence" (54) render explicit the possibil-ity of moral failure that threatens proper regard for posterity. More subtle, perhaps, but still harmful to future generations (George 1992, 198-199), are "vices resembling prudence" that include "prudence of the flesh," "craftiness," "guile," "fraud," and disordered "solicitude" about temporal things or about the future (55.1-8).

These vices deserve more attention than can be given them here, but to focus on vice alongside virtue at least brings some balance to the theological engagement of international law. It also brings to the fore Aquinas's convictions about how little humans can accomplish apart from grace (1|2.109.2). To focus on moral failure with regard to prudence is to emphasize that justice, including justice between and among both nations and generations, may not always be realized or at least realized fully. And this returns us to a topic taken up earlier, namely the theological virtues, particularly the virtue of charity, as *theological*. It is significant, for example, that prudence is infused along with charity. Grace perfects nature, including the natural capacities to remember and to reason and to look ahead, but first of all grace heals (1|2.109.4). Although I have focused on the virtue of prudence, this emphasis should not detract from the grace of charity as love of future neighbors. What "channels of grace" might broaden moral horizons and behavior to take future neighbors into account? Where in the Church but also within the sometimes meandering, sometimes rapid course of international law might those channels run? This, I insist, is a question to which theologians concerned about GCC should be especially attuned.

CONCLUSION

At the outset I noted the modest aims of this essay, and it may be well to close with remarks on limitations in order to move beyond them. While my purpose has been to encourage theologians to engage international law more fully, these pages do not reach the level of engagement required. For one thing, much more would need to be said about the specifics of international law regarding GCC. However, even if more were said on that topic, the kind of engagement I am counseling here involves international law not on a single issue—although this may be the starting point—but in its many aspects, covered, for example, in introductory but often lengthy texts and casebooks. If theologians are serious about engaging international law on the issue of GCC, they will soon be asking just how treaties are made, about the very problematic relationship of international law to national law, about the relationship between law and politics, about international law's historical and philosophical underpinnings, about the relationship of environmental law to economic law, and many other things. After all, as the editors of a prominent casebook warn, "international law, is not a 'course'; it is a curriculum" (Damrosch et al. 2009, xvi).

Secondly, while I have focused on method and virtue, especially prudence, as two points of convergence between theology and international law, I have done little to show that method and virtue themselves converge. Indeed, I believe they do, but I can only suggest how this is so. For one thing, as the eight functional specialities of method discussed by Lonergan invite collaboration, so the eight parts of prudence, operative in international law, may call for collaboration, too. No one can be in equal measure historian (memory) and scientist-consultant (docility), bold spirit (shrewdness) and necessarily cautious soul (caution). However, no matter one's role in the communal project of prudence, the attentiveness, intelligence, critical-mindedness, and responsibility of method will be required.

This leads to a third and final point about limitations and also about opportunities. Bringing theology and international law together is only in part a theoretical matter, discussed in an essay such as this. The real unfinished task is ongoing engagement in the concrete, as theologians and lawyers, virtuous and methodical, pool their insights, refine their judgments, and share with myriad other actors responsibility for our world and its future. The climate crisis lends more than a little urgency to this inter-disciplinary but also inter- subjective endeavor.

SOURCES

Agius, Emmanuel. 1998. "The Earth Belongs to All Generations." In *Caring for Future Generations: Jewish, Christian, and Islamic Perpsectives*, ed. Emmanuel Agius and Lionel Chircop, 103-22. Westport, CT: Praeger.

Agius, Emmanuel and Salvino Busuttil, eds. 1998. *Future Generations and International Law*. The Law and Sustainable Development Series. London: Earthscan Publications Ltd.

Anand, Ruchi. 2004. *International Environmental Justice: A North-South Dimension*. Burlington, VT: Ashgate Publishing Company.

Aquinas, Thomas. 1948. *Summa Theologica*. Translated by Fathers of the English Dominican Province. 5 vols. Westminster, MD: Christian Classics.

Boehle, Joseph. 2010. "The UN System and Religious Actors in the Context of Global Change." *Cross Currents* 60/3: 383-401.

Cameron, James, Will Wade-Gary, and Juli Abouchar. 1998. *Future Generations and International Law*, ed. Emmanuel Agius and Salvino Busuttil, 93-113. London: Earthscan Publications Ltd.

Carty, Anthony. 2008. "The Moral Theologian, Oliver O'Donovan and International Law." *Political Theology* 9/3: 339-362.

Cassese, Antonio. 2008. *The Human Dimension of International Law*. Selected Papers. Oxford: Oxford University Press.

Curran, Charles. 2002. *Catholic Social Teaching: A Historical, Theological, and Ethical Analysis*. Washington, DC: Georgetown University Press.

Epps, Valerie. 2009. *International Law*, 4th ed. Durham, SC: Carolina Academic Press.

Damrosch, Lori Fisler, Louis Henkin, Sean D. Murphy and Hans Smit, eds. 2009. *International Law: Cases and Materials*. 5th ed. American Casebook Series. St. Paul: West Publishing Company.

Fox, Jonathan and Sandler, Shmuel. 2004. *Bringing Religion in to International Relations*. Culture and Religion in International Relations. New York: Palgrave MacMillan.

George, William P. 1990. Envisioning Global Community: The Theological Character of the Common Heritage Concept in the Law of the Sea. Ph.D. dissertation, University of Chicago.

———. 1992. "Regarding Future Neighbours: Thomas Aquinas and Concern for Posterity." *The Heythrop Journal* 33/3 (July): 283-206.

———. 1996. "International Regimes, Religious Ethics, and Emergent Probability." *1996 Annual of the Society of Christian Ethics*: 145-70.

———. 1997. "Anticipating Posterity: A Lonerganian Approach to Contingent Future Persons." In *Contingent Future Persons: Philosophical and Theological Challenges*, Theology and Medicine Series, ed. Jan C. Heller and Nick Fotion, 191-208. Boston: Kluwer Academic Publishers.

———. 1999. "Looking for a Global Ethic? Try International Law." In *Religion and International Law*, ed. Mark W. Janis and Carolyn Evans, 483-504. Boston: Martinus Nijhoff Publishers.

———. 1999-2000. "Overcoming Textbook Bias: Grotius, Theology, and International Law." *The Journal of Law and Religion* 14/2: 605-631.

———. 2000. "'Angelism' and Its Devilish Effects on Education." *Chicago Studies* 39 (Summer): 194-210.

———. 2008. "Why Catholic Universities Should Engage International Law." *Journal of Catholic Higher Education* 27/1 (Winter): 137-157.

———. 2009. "International Law as Horizon," paper delivered at the 36th Annual Lonergan Workshop, Boston College, June 21-26.

Goldsmith, Jack L. and Eric A. Posner. 2005. *The Limits of International Law*. New York: Oxford University Press.

Henkin, Louis. 1995. *International Law: Politics and Values*. Developments in International Law, vol. 18. Dordrecht: Martin Nijhoff Publishers.

Intergovernmental Panel on Climate Change (IPCC). 2010. Accessed from http://www.ipcc.ch/ on October 20.

Janis, Mark W. 1999. "Religion and the Literature of International Law: Some Standard Texts." In *Religion and International Law*, ed. Mark W. Janis and Carolyn Evans, 121-44. Boston: Martinus Nijhoff Publishers.

Landau, Christopher. 2008. "Faith Leaders Urge Climate Curbs." BBC News Mobile, November 28. Accessed from http://news.bbc.co.uk/2/hi/europe/7753784.stm on November 3, 2010.

Levering , Ralph B. and Miriam L. Levering. 1999. *Citizen Action for Global Change: The Neptune Group and Law of the Sea*. Syracuse: Syracuse University Press.

Lonergan, Bernard J. F. 1974. "The Subject." In *A Second Collection: Papers by Bernard J.F. Lonergan*, ed. William F.J. Ryan, S.J. and Bernard J. Tyrell, S.J., 69-86. Toronto: University of Toronto Press.

———. 1985. "Healing and Creating in History." In *A Third Collection*, ed. Frederick E. Crowe, 100-09. New York: Paulist Press.

———. 1988. *Collected Works of Bernard Lonergan*, vol. 3, 5th ed. rev. and aug., *Insight: A Study of Human Understanding*, Toronto: University of Toronto Press. Original edition, London: Longmans, Green & Co., 1957.

———. 1990. Reprint. *Method in Theology*. Toronto: University of Toronto Press, 1990. Original edition, New York: Herder & Herder, 1973.

MacFaquhar, Neil. 2010. "U.N. Chief Recommends Small Steps on Climate." *New York Times*, August 9. Accessed from http://www.nytimes.com/2010/08/10/science/earth/10nations.html on November 1.

Morelli, Mark. 1981. "Horizonal Diplomacy." In *Creativity and Method: Essays in Honor of Bernard Lonergan*, ed. Matthew L. Lamb, 459-74. Milwaukee: Marquette University Press.

Nafziger, James A.R. 1999. "The Functions of Religion in the International Legal System." In *Religion and International Law*, ed. Mark W. Janis and Carolyn Evans, 155-76. Boston: Martinus Nijhoff Publishers.

Nijman, Janne Elisabeth. 2004. *The Concept of International Legal Personality: An Inquiry in to the History and Theory of International Law*. The Hague: T.M.C. Asser.

Page, Edward A. 2006. *Climate Change, Justice and Future Generations*. Cheltenham, England: Edward Elgar Publishing.

Pardo, Arvid. 1967. UN Assembly, Twenty-second Session, First Committee, 1515 meeting, November 1, 1967. Accessed from http://www.un.org/Depts/los/convention_agreements/texts/pardo_ga1967.pdf on November 3, 2010.

Ramakrishna, Kiliparti. 1990. "North-South Issues, Common Heritage of Mankind and Global Climate Change." *Journal of International Studies* 19/3: 429-445.

Soltau, Friedrich. 2009. *Fairness in International Climate Change Law and Policy*. New York: Cambridge University Press.

Stone, Christopher. 1998. "Safeguarding Future Generations." In *Future Generations and International Law*, ed. Emmauel Agius and Salvino Busuttil, 65-79. London: Earthscan Publications Ltd.

United Nations. 1945. Statute of the International Court of Justice. Accessed from http://www.icj-cij.org/documents/index.php?p1=4&p2=2&p3=0 on November 1, 2010.

Weeramantry, C.G. 1998. *The Lord's Prayer: Bridge to a Better World*. Liguouri, MO: Ligouri Publications.

————. 1999. "Preface." In *Religion and International Law*, ed. Mark W. Janis and Carolyn Evans, ix-xii. Boston: Martinus Nijhoff Publishers.

United States Conference of Catholic Bishops (USCCB). 2010. *Global Climate Change: A Plea for Dialogue, Prudence, and the Common Good*. Washington, DC: NCCB/USCC.

Walzer, Michael. 1983. *Spheres of Justice*. New York: Basic Books.

PERSPECTIVES INSPIRED BY
20TH-CENTURY THEOLOGIANS

8

A TEILHARDIAN VALUE-SUPPORTIVE
WORLD VIEW

Robert Faricy, S.J.

Any set of fixes or solutions to the global warming problem depends on a generally accepted world view as the basis for valuing and acting accordingly. Unless rooted in a faith in which the world is understood as dependent upon a higher being who creates the world of which our species is a part and gifts us with the capacity to know that we are a part of the world and to act responsibly toward it, fixes and solutions will always be bandaids. Technical solutions and fixes are never remedies per se to a problem as immense and problematic as changes that humans are forcing on the global climate.

Examples of an adequate world view from a Christian faith perspective can be found in the great Eastern Fathers of the Church, St. Thomas Aquinas, Pierre Teilhard de Chardin, and Ignatius Loyola. They considered everything—all things and all things together—as gifted by God who is present in and acting for the good of all including the good of each person. From their faith perspectives, God can be found in all things—in the universe, in nature, in global warming, in our problems, and even in their partial solutions.

Teilhard de Chardin's spirituality is a Christian ethic of love built on a theology that integrates biblical doctrine, especially the Christology of St. Paul, within a contemporary evolutionary perspective. Through this spirituality, the person-nature rapport is understood as positive, not as an opposition, and unitizing. When he uses the term "nature," human nature is included.

The Teilhardian theology of creation in Christ was adopted by the Bishops of the Second Vatican Council in *Gaudium et Spes*. Jesus risen is the future and active focal point of all evolution, including social evolution. A personal relationship with Jesus crucified and risen motivates the Christian to co-create, in the structure of the cross, with Jesus Christ. Not everyone who engages in this personal relationship with Jesus is Catholic or even Christian.

I present Pierre Teilhard de Chardin's theology of creation as a way of showing a world view that is centered on, held together in, and moving convergently toward the risen Jesus Christ as its future focal point at his second coming. This is a faith-based view of the world that gives meaning to efforts aimed at restraining and neutralizing human-driven climate change.

TEILHARD'S THEOLOGY OF CREATION

With regard to a Christian theology of creation, Teilhard's purpose, as always, was to construct a theory that has maximum coherence, meaningfulness, and make as much sense as possible in the light of all the data available. What is more, Teilhard always aimed for a *theory*, a coherent understanding of creation that will provide a maximum of interest and motivation for human effort. For Teilhard, the criterion of truth for any understanding or theory of creation was precisely the extent to which it provides a coherent and meaningful vision of creation and somehow concurrently activates us to respond to the problems of our time. Any understanding of creation that would undermine human effort must be re-examined.

Teilhard's objection to the Reformation theology of creation that has been dominant since the Middle Ages is that it seems to discourage human effort. The problem of any theology of creation, and the problem that medieval scholastic theology tried to solve, is this: it would seem that God can only be thought of pantheistically as if He constituted all being. From a pantheistic perspective, either the world would be only a mirage or it would be a part, an aspect, or a phase of God. To solve the problem, scholastic metaphysics developed the notion of "participated being," an inferior or secondary form of being gratuitously drawn out of nothing by a special act of God's transcendent causality—the *creatio ex nihilo* (creation from nothing). The ontological distinction between the Creator and the creature is, of course, absolutely necessary from a Christian perspective, and Teilhard found no fault with scholasticism on this point. On the contrary, he strongly reaffirmed the distinction. However, from the point of view of human action and effort, Teilhard found difficulty with the scholastic theology of creation. An understanding of creation that insists on the complete self-sufficiency of God and consequently on the utter contingency and even arbitrariness of his creation risks making the Christian lose all taste for and interest in the world. This risks conveying a strong depreciation of God's creation and tends to discourage the necessary human effort to work and to contribute to the world's progress. A theology of creation that stresses God's goodness and the goodness and lovableness of his creation is all right as far as it goes, but it does not go far enough. If that theology seems to detract from the value of human effort to make progress in the world and to make that effort despised, then the theology is positively dangerous. Thus, in reaction to the theology of creation that he learned as a seminarian, in reaction to a scholasticism perhaps more rigid and "closed" than the teaching of many Thomist theologians today, Teilhard tried to rethink the idea of creation in terms of his own system of thought.

Teilhard's ideas were tentative, and his theological reflections consisted of theological hypotheses that were quite open to further development and refinement or, for that matter, rejection. This is true of nothing so much as of his theology of creation.

THE THEORY OF "CREATIVE UNION"

Teilhard referred to his theology of creation as the theory of "creative union" (*l'union créatrice*). From his perspective, creative union is not a metaphysical doctrine. It is much better described as a sort of empirical and pragmatic explanation of the universe. One must keep in mind that Teilhard's theology of creation is not metaphysical, particularly since many authors insist on thinking that it is (e.g., Barbour 1969; Brunner 1959; Russell 1962).

Teilhard did not consider creation strictly in terms of being, as would be the case if his approach was traditionally metaphysical. Rather, he described being in terms of "union." For Teilhard, being in its active sense meant to unite oneself or to unite others; in its passive sense, being meant to be united or unified by another.

As we shall see, "to create" means "to unite," in Teilhard's writing, and "to be created" means "to be united." Creation is not an instantaneous act. It is a process or synthesizing action. This will become clearer as we proceed through the four steps of Teilhard's theory of creative union.

In a first step, Teilhard assumed the existence of a divine and self-sufficient First Being. The second step is the recognition according to the revelation of God as Trinity that the existence of this First Being, this divine Center, consists in the act of opposing and uniting himself in a Trinitarian manner. It is in the third and fourth steps that Teilhard described creative union. In the third step, God, in the very act by which He opposes and unifies himself in his unique existence, *ipso facto* causes another type of opposition to arise, not within himself but at his antipodes. There is a self-subsisting unity at the pole of being and, as a necessary consequence, there is a multiplicity all around at the periphery. This unity is a pure multiplicity, a "creatable void" that is nothing but which nevertheless, by its passive potency for arrangement and union, is a possibility and an appeal for being. Teilhard considered this void a correlative to God's existence and pictured it as an infinite multiplicity with the possibility of being united. In the fourth step he described creation as union, and this presents a problem. He believed that there is necessarily an infinite multiplicity antecedent to any creative act. He did not deny creation from nothing; he affirmed it. However, he pictured the void, the "nothing," as an infinite multiplicity.

In classical philosophy or theology, creation tends to be presented as an almost arbitrary gesture of the First Cause, executed by a causality that is analogous to efficient causality and executed according to a mechanism that is completely indeterminate—truly an "act of God" in the pejorative sense of the expression. Conversely, in a metaphysics of union the creative act takes on a significance and a structure that are well defined, even though the self-sufficiency and self-determination of the Absolute Being remain inviolate. Creation results as a reflection of God, although this reflection is outside God's self instead of within himself. Creation is "pleromization," as St. Paul might have said—a giving of reality to participated being by arrangement and totalization to appear as a sort of replica or symmetry of Trinitization. This pleromization becomes expressible in the terms that we used to define being.

To create means to unite, Teilhard explained, bringing together previously disunited elements. Creation itself is a process, the expression of which is evolution. Evolution is the expression in space and time of creation, now taking place in the human sphere through various kinds of progress, interrupted by problems, war, natural disasters, and other events. It is the outward form of God's creation taking place partly through humankind's efforts.

Evolution has a direction, according to Teilhard. It follows an axis of increasing complexity-consciousness. According to the law of complexity-consciousness, the degree of consciousness of a material entity varies according to the entity's structural complexity. This law is valid universally, even in cases that cannot be verified by observation. To say that evolution proceeds along an axis of increasing complexity-consciousness means that the universe is evolving not only in the direction of greater spirituality or consciousness. At the same time and correlatively, the universe is evolving in the direction of greater unity. Evolution is the expression of God's continuous creation by which He more and more unifies the world. As Teilhard stressed, God creates by uniting. Creation is a process that has never ceased. Its act is a great continuous movement spread out over the totality of time. It is ongoing. Incessantly but imperceptibly the world emerges more and more from nothingness. The operation that creation gives rise to and that it forms is infinitely refracted in creatures in which the work of creation is materialized and accumulated.

Of course, this continous creative act of God is not split up; it is one indivisible act. However, the term of this ongoing creative act is the entire universe in all its extension and all its duration. Creation began with an infinite multiplicity of elements, and, by a gradual process of unification of elements, creation is approaching a point of maximum unity. This point, of course, is the Parousia, the point of maximum union of the universe with Christ to which I now turn.

THE PRIMACY OF CHRIST IN
CONTINUOUS CREATION

Far from a merely rational speculation, Teilhard's theory of creative union depends on revelation and is, therefore, intrinsically theological. His theology of creation does not prescind from the fact of the Incarnation; it is constructed with the Incarnation in mind. More exactly, it is built around a keystone, and that keystone is Christ. It is developed from the belief that Christ is the Head of creation. The Incarnation is the renewal and the restoral of all the forces and the powers of the universe. Christ is the instrument, the Center, the Term of all creation. By him everything is created, sanctified, and vivified. Nor is Teilhard's theology of creative union a theology of some possible world, an abstract reconstruction of the metaphysical mechanics of what God's creative act is "in itself." Teilhard was concerned with *this* world, the world that God continues to create and is created in Christ. This is the world in which God has concretely involved himself through the Incarnation.

Teilhard not only reaffirmed the primacy of Christ over all creation; he also dispelled the notion that the creation of this world is arbitrary. He stressed the mutual complementarity of the Creator and his creation and the mutual completion of God and the world that makes Christianity live. God is entirely self-sufficient, yet creation brings to him something necessary. The object of God's creative act can be understood as a mysterious product of completion and fulfillment for himself.

In the concrete and for God as well as for us, creation is creation in Christ. God should not be thought of as personally independent of the present world that He has created in Christ. In virtue of the Incarnation, God cannot—at least in the present order and from now on—do without the Many in which he has immersed himself. And, on the other hand, the reality of "God plus the Many" in Jesus Christ seems, in Christian practice and in Pauline spirituality, to represent a perfection which, no matter how qualified it is as extrinsic to God, carries with it a real completion in the balance of universal Being.

To repeat Teilhard's belief, the present world with its supernatural order is not something that God can just do without. It is not an utterly contingent world and its creation is not an arbitrary act. The world is not superfluous to God. God is not indifferent to the world that He has created. God is personally involved in the world.

As already mentioned, Teilhard understood God's continuous creation as a process of unification, of building up the universe in the direction of increasing unity directed to Christ. The term of this continuous creation in Christ is the Pleroma, the final state of the world, the consummation of all things in Christ. God's continuous creation is directed to the quantitative repletion and the qualitative consummation of all things, the mysterious Pleroma in which the substantial One and the created many fuse without confusion into a whole which, without adding anything essential to God, will nevertheless be a sort of triumph and generalization of Being. The final result of God's creation will be the Pleroma. Teilhard's idea of the Pleroma is traditional—not an absorption of creatures into God that involves a loss of identity for the creatures, but an absorption of creatures that is

a maximum union with God. Since union differentiates the individu-
alities of the elements united, the Pleroma will be the state of maximum union without any confusion of identities while remaining in a
state of minimum separation and distinction between God and God's
creatures.

The Pleroma, the mysterious synthesis of Creator and the created,
the great fulfillment of the universe in God both quantitatively and
qualitatively, finds its physical principle, its expression, and its stability
in the figure of Christ-Omega, the Universal Christ. Creative union is
the gradual formation of the Pleroma; this "pleromization" is directed
toward Christ-Omega, toward Jesus risen in his place as the Omega of
the world's evolution (Teilhard 1963, 26-27).

For Teilhard, creation is creation in Christ. God should not be
thought of as personally independent of the present world created in
Christ. On the one hand, in virtue of the Incarnation, God cannot do
without the Many in which he has immersed himself, at least in the
present order to its end. On the other hand, the reality of "God plus
the Many" in Jesus Christ seems, in Christian practice and in Pauline
spirituality, to represent a perfection which, no matter how qualified it
is as extrinsic to God, carries with it a real completion in the balance
of universal Being. God is not indifferent to the world that He has
created. God is personally involved in it (see Mooney 1966, 174-76).

Thus, Teilhard understands God's continuous creation as a process
of unification, of building up the universe in the direction of increasing unity. The entire process of creation is directed to Christ. The term
used is "Pleroma," the final state of the world, the consummation of all
things in Christ. God's continuous creation is directed to "the quantitative repletion and the qualitative consummation of all things... the
mysterious Pleroma in which the substantial One and the created
many fuse without confusion into a whole which, without adding anything essential to God, will nevertheless be a sort of triumph and generalization of Being" (Teilhard 1960, 122). The final result of God's
creation will be the Pleroma. Teilhard's idea of the Pleroma is the traditional one: not an absorption of creatures into God that involves a
loss of identity for the creatures, but an absorption of creatures that is
a maximum union with God. Since "union differentiates" the individu-
alities of the elements united, the Pleroma in the state of minimum
separation between God and his creatures will be in the state of maximum union without any confusion of identities. Creative union is a

continuous process whose expression is the converging evolution of the world and whose term is the fullness of the Pleroma.

The Pleroma, "the mysterious synthesis of the Uncreated and the created, the great fulfillment-both quantitative and qualitative-of the universe in God, ...finds its physical principle, its expression, and its stability in the figure of Christ-Omega, the Universal Christ" (Teilhard 1963, 26-7). Creative union is the gradual formation of the Pleroma; this "pleromization" is directed toward Christ-Omega, toward Jesus, the Center toward whom all moves. Creation, for Teilhard, is accomplished not "from behind" by a causality analogous to efficient causality, but "from up ahead" by the unifying influence of Christ. "All energies hold together, are welded deep down into a single whole, and what the humanity of Our Lord does is to take them up again and re-weld them in a transcendent and personal unity" (Teilhard 1965, 93). Christ gives himself to us through a world which will reach completion even on a natural level by reason of its relationship to him.

THE UNIVERSAL PRESENCE &
INFLUENCE OF CHRIST

A central theme of *The Divine Milieu* is the universal presence and influence of Christ which is developed explicitly near the end of the book in terms of the divine milieu and the Universal Christ (Teilhard 1960, 121-32). Teilhard identifies Christ's omnipresence and the divine milieu. The action by which God maintains us in the field of his presence is a unitive transformation. The consummation of all things in Jesus Christ risen is the mysterious Pleroma in which the substantial One and the created many fuse without confusion into a whole which, without adding anything to God, will nevertheless be a sort of triumph and generalization of being.

What is the active center, the living link, the organizing soul of the Pleroma? St. Paul proclaims it to be He in whom everything is reunited, in whom all things are consummated, and through whom the entire created edifice receives its existence—the risen and coming Jesus Christ. The divine omnipresence translates itself within our universe by the network of the organizing forces of the total Christ. God exerts pressure, in us and upon us, through the intermediary of all the powers of heaven, earth, and hell, in the act of forming and consummating Jesus Christ who saves and animates the world. Since in the course of this operation Jesus himself does not act as a dead or passive point of convergence. Instead, Jesus acts as an active center of radiation for the energies that lead the universe back to God through his humanity so the layers of divine action finally come to us impregnated with his organic energies. The Universal Christ is Christ in his cosmic role as Omega, the end, the consummation, and the Consummator.

CONCLUSION

Teilhard aimed to reformulate the Christian theology of creation in terms of a genesis, a "becoming" of the universe in Christ. The word he finally coined after years of reflection is "Christogenesis." It may appear to be an awkward word, but it is one that sums up the evolutive structure of the universe as Teilhard saw it—a dynamic movement directed toward the final unity of all things in Christ in the fullness of the Pleroma.

Teilhard's view of the universe centered on Jesus risen and moving toward him as its focal point and future fulfillment is not a widely shared view in the United States nor elsewhere. Nevertheless, something like it is necessary to address the climate crisis. God in Jesus has the whole world in his hands. We have no need to fear. Everything is in God's hands. The martyrs testify to this. We do not even need to fear death. However, we do need to see the world as moving to its future focal center—Jesus risen. And we need to take responsibility to help that movement by trying to restrain evils present and future, including human activities that are forcing changes in the global climate. Other religious or philosophical views may serve the same function.

Unless a basic view of the world is common in a civilization, the motivation will not be there to check evils like human-forced climate change. The problem seems too far away, not local, not pressing, not immediate. The response is to leave the climate crisis to the experts. Yet the problem will not go away as it continues to have immediate and growing influence on our lives and projects to influence the lives of future persons, species, systems, and the biosphere. An understanding of the world that Teilhard presents can help us make a positive response through which to address this difficult problem with the sense of responsibility with which we are gifted by God and expected to exercise.

SOURCES

Barbour, Ian G. 1969. "Teilhard's Process Metaphysics." *The Journal of Religion* 49.2 (April): 136-59

Brunner, "Pierre Teilhard de Chardin," *Stimmen de Zeit* 165 (1959): 210-22.

Mooney, Christopher. 1966. *Teilhard de Chardin and the Mystery of Christ.* New York: Harper & Row.

Russell, J. 1962. "The Principle of Finality in the Philosophy of Aristotle and Teilhard de Chardin." *Heythrop Journal* 3: 347-57.

Teilhard de Chardin. 1963. "La parole attendue." In *Cahiers Pierre Teilhard de Chardin.* Vol. 4, 22-29. Paris: Desclee.

———. 1965. *The Making of a Mind.* Translated by Rene Hague. New York: Harper & Row.

———. 1960. *The Divine Milieu.* Translated by B. Wall, A. Dru, N. Lindsay, D. MacKinnon, et al. New York: Harper & Row.

9

TEILHARD & THE LIMITS TO GROWTH

THE EVOLUTIONARY DYNAMIC TOWARD "ULTRA-HUMANITY"

Richard W. Kropf

In the face of growing evidence that this planet is undergoing a warming trend that heralds long-term climate change, the human species faces an unprecedented challenge. Either we successfully adapt to these changes or run the risk of a major setback in the future of humanity.

This essay analyzes and addresses this challenge from the viewpoint of two reference points. One is the evolutionary perspectives developed by Pierre Teilhard de Chardin, S.J., the Jesuit priest-scientist whose long-suppressed writings on evolution and the future of humanity captivated many in the world immediately after the publication of his masterwork, *Le Phénomène humain*, following his death in 1955. The other reference point is the 2004 update of the highly controversial study entitled *The Limits to Growth*, which was originally written by four researchers at the Massachusetts Institute of Technology in 1972. First commissioned by the influential Club of Rome, a consortium or "think tank" comprised mostly of European sociologists, economists, and industrialists, the computer-simulated projections in *The Limits to Growth*, indicating what could happen to our planet and its population if steps to limit growth were not taken, raised storms of protest, denial, and occasionally outright invective. The "scenarios" generated by these authors are fleshed out with the aid of subsequent sources that deal directly with climate change.

Over all, the object of this essay is to show how the limits to growth projected by the Club of Rome/MIT study and exacerbated by recent indications of global warming associated with large-scale climate change largely confirm Teilhard's analysis of human evolutionary development. These growth limiting indicators also force us to confront the critical choices or "grand options" that he saw facing our species. First, however, we must focus on the general pattern or dynamics of evolution as Teilhard understood it, both on a biological as well as human scale.

THE EVOLUTIONARY DYNAMIC: CONVERGENCE & THE LAW OF COMPLEXITY-CONSCIOUSNESS

From Teilhard's perspective, biological evolution proceeds in a series of three steps or movements: First, the physical convergence or joining of disparate elements to form new entities; next, the complexification of these entities; and, finally, the propensity of these complexified entities to exhibit the phenomenon of sensitivity, awareness, or even reflective consciousness. The importance of this concept of complexity and its dynamics in the advance of evolution is emphasized in studies such as Eric J. Chaisson's *Cosmic Evolution: the Rise of Complexity in Nature* (2001). Although Chaisson takes some pains to disassociate himself from those who, like Teilhard, see some teleological force or principle at work in the evolutionary process, Chaisson's views of the process itself strikingly resemble Teilhard's.

For the purposes of this essay, Teilhard's notion of biological *convergence* is especially important. However, his use of the term "convergence" differed markedly from its current use in evolutionary biology. When evolutionary scientists speak of "convergence" today, they are referring generally to the phenomenon of similar types of organisms evolving in a parallel fashion from different evolutionary lines of development. For example, marsupial "tigers" or other species of animals found in Australia are similar to other species in Southeast Asia both in general form and function, but marsupials did not have the same biological ancestry as these other species.

Teilhard's use of convergence was more accurate than used by evolutionary biologists, though his use was not initially exclusive or uniform. For example, in *The Human Phenomenon* (1959, 60),[1] the first draft of which was completed in 1936, Teilhard used the term "grouping" to describe the converging aspect of evolutionary development. A year or so later, in his essay on "Human Energy" (1969, 144), he

---

1   Title of the 2000 edition; however, the quotations with page numbers are
    taken from the more widely disseminated edition of 1959.

introduced this more general understanding of the term when referring to "any unification by *convergence* that operates in the field of our
experience: a grouping of cells in a living body, a grouping of individuals and functions in a social organism, a grouping of souls under the
influence of a great love."

His notion of convergence was also much more graphic. In several
of his essays, he spoke of how the human species is, like any other
species, confined to a limited area (e.g., an island or a continent) and
has to either live in co-operation with others of their species as their
numbers increase or eventually suffer the consequences of conflict. In
at least one of his personal "Journals" or notebooks (Teilhard 1948,
173), he illustrated this understanding in the form of a drawing of a
globe representing our planet where lines spreading upward from the
bottom (South Pole) diverge until they reach the mid-mark (Equator)
and then begin to converge as they rise toward the top (North Pole).
The first he labeled "the phase of Expansion," followed by the second
label, "the phase of Contraction." In this case, the overall application of
this comparison appears to have been the growth or evolution of human awareness, because he inscribed the words "space limited" at the
top of the drawing (above the North Pole). Three days later, he noted
the need to distinguish between the growth of *science* (the discovery
of new elements, new animals, etc.) and the growth of *conscience* (after which he added, in parenthesis, underlined twice, "*Welt*") that may
have meant a new or expanded *weltanschauung* or worldview (ibid.
174). Conscience, or in this case, human consciousness, already had
a special meaning for Teilhard as a concise way of describing *reflective*
awareness or thought—"the power acquired by consciousness to turn
in upon itself, to take possession of itself as an object endowed with its
own particular consistence and value: no longer merely only to know,
but to know that one knows" (Teilhard 1959, 165).

From Teilhard's perspective, whether or not this increased power
of thought would lead to greater unanimity among humans or to
eventual chaos was questionable. He recognized several major choices
facing humanity. First outlined in a 1939 essay titled "Le Grand Option" (1964) were the choices of Pessimism vs. Optimism that pitted
withdrawal against engagement in the evolutionary process. The next
choice was between Plurality vs. Unity, the latter implying many social, psychological, and even political ramifications that he foresaw. In
"Les Directions et Conditions de l'Avenir" that he prepared in 1947,

he began to worry not so much about the human species blowing itself up, a pandemic disease wiping out all human life, or running out of basic mineral resources or fuels. His concern was running out of enough food to properly feed the world's exploding population that he predicted would reach 2.5 billion within the next quarter century. While both of these essays were translated and republished in *The Future of Man* (1964, 37-60 and 232-34), Teilhard's estimate of future population growth proved to be understated. In fact, according to the United Nations in 1955 (UN Department of Economic and Social Affairs 2009), the world's population had already reached 2.76 billion, and, by 1970 (two years short of the quarter century of Teilhard's 2.5 billion prediction), the population was 3.69 billion.

Whether or not this rapidly expanding number of people would be able to live amiably and in peace with one another greatly concerned Teilhard. Instead of referring to the "convergence" of humanity, he began to categorize the movement as "external compression" and the instinctive reaction of "repulsion" that this movement might trigger (Teilhard 1964, 235). In January of 1950, the choice he saw was the one he held in 1939—between political totalitarianism or some new breakthrough into a new state of human "unanimisation" about which he wrote in "On the Probable Coming of an 'Ultra-Humanity'" (ibid. 270-80).

Before venturing further into what Teilhard meant by "ultra humanity," a closer look at growth trends is warranted. They had begun to concern Teilhard, leading him to speculate about new possibilities for the future of human evolution.

GROWTH: PROJECTIONS, PREDICTIONS, & REALITY

In their introduction to the 30-Year Update, three of the original authors of the *Limits to Growth* have gone to some length to emphasize that while their computer-generated projections were never meant to be predictions, they did contain warnings as to what might happen if the then-current growth patterns (based on 1970 data) were allowed to continue. The book occasioned many reviews, ranging from high praise to outrage (Meadows 1992, preface notes 4-6; Meadows 2002, preface 17).

However, the unfazed authors did continue, and they produced a second book, *Beyond the Limits*, which constituted a 20-year update based on additional data acquired by 1990 (Meadows 1992, Preface

14). This update confirmed the accuracy of *most* of the earlier projections and explained the apparent miscalculations regarding petroleum supply following the OPEC embargo and the spectacular results of increased efforts to find new reserves elsewhere. As explained in the second book (Meadows 1992, 182-84), these miscalculations and efforts resulted eventually in a plunge in oil prices that seemed to completely undercut the authors' 1972 warning that the world might find itself in the predicament of having exhausted all its readily available petroleum reserves by the end of the 20th century. Most significantly, the second book (Meadows 1992, chaps. 2, 8) put new emphasis on the concept of *overshoot*—what happens when a crisis is foreseen but corrective action is not taken soon enough. This second book (Meadows 1992, 47-54) presented growing evidence that the overshoot phenomenon had already begun to occur in several respects, while the third book (Meadows 2002, pref. 12, 14) was able to confirm this trend. For example, despite the "green revolution," the world's grain production was in decline from its peak in the mid-1980s and the world's marine fish stocks were in serious decline, even as the population has boomed (Meadows 2002, 57-66, 229-34).

Finally, in keeping with the world's growing concern over the environment, the third book adopted the concept of the human "ecological footprint" when compared with Earth's carrying capacity—the planet's ability to renew itself in the face of increased human demands on its resources. Measured in this way, the latest edition of *Limits to Growth* claimed that 1.2 Earths would be required to satisfy human demands at the present time (Meadows 2002, 15; n. 7, 295-6). However, the latest estimate issued in 2009 by the Footprint Network that was recommended by the *Limits to Growth* authors concluded that 1.4 Earths are needed to supply humanity's current demands. Today we are failing to achieve the goal of a sustainable planet, which everyone seems to agree is the only way humanity can survive in the long run. In other words, all indicators point to the basic logic underlying all three of the "limits" books—unlimited or infinite growth cannot be sustained in a finite or limited world.

The Mathematics & Psychology of Overshoot

To better understand the reasoning behind these projections, one must also appreciate the difference between linear (mathematical) and geometrical (exponential) growth. The first type of growth occurs primarily mostly when we are dealing with yearly fees or dues, over-due book fines, or naturally renewing processes (e.g., the rate at which polluted water can be recycled and purified through ground filtration or, alternatively through evaporation and returned to Earth as precipitation). However, when we are dealing with biological reproduction (or with compounded bank or credit card interest), another matter looms. When couples have more than two children (or 2.1 to allow for infant and childhood mortality), population inevitably grows. If infant mortality was entirely eliminated and the average life-span reached 65 years world-wide, the additional .1 would double the human population within 924 years!

The phenomenon of overshoot plays an especially crucial part when estimating population growth. While the mathematics may be clear enough, the actual transition from one standard to another takes time—often several generations to take effect. the human population will grow to about 9.15 billion by mid-century. These experts hope that the world's population will subsequently stabilize, mostly because industrialization has generally resulted in couples' choosing to have smaller families. As indicated in Europe, industrialization and the pursuit of material prosperity have played a major role in a greatly diminished birth rate. However, if industrialization is delayed, perhaps due to shortages in energy or raw materials, or is curtailed by pollution and other environmental problems, what happens remains an open question.

All scenarios represented on the graphs in the 2004 update of *Limits to Growth* project a world population peak of somewhere between 7 to 8 billion by 2040. Only scenario #10 that reflects heeding the warnings of the first *Limits to Growth* projected a world population peak of slightly over 6 billion. Nevertheless, according to the authors of the updated *Limits to Growth*, a sustainable world still might be achieved, but only at the cost (at least to the affluent nations) of cutting back our standard of living. The authors also cautioned that threats like world-wide pandemics, major (especially nuclear) wars, and other possible catastrophes might persist.

However, the above estimates need to be compared to the UN estimates according to which the current world population has already reached 6.9 billion and will reach 8.8 billion by 2040 and 9.15 billion by 2050 (UN Department of Economic and Social Affairs, 2009) when allowing for the "medium variant."[2] When asked about these UN projections, one of the *Limits to Growth* authors, Dennis Meadows, expressed his opinion to this writer that the UN population projection of over 9 billion by 2050 may never be reached. Instead, he believes that a major crash of the world's ecological system and a major increase in the world's population has become increasingly questionable but can be avoided before mid-21st century. What part is climate change likely to play in these speculations?

Climate Change & Global Warming

While the gradual warming of the planet might be viewed by some few remaining optimists who anticipate new opportunities for increased agricultural development as a boon, the fact is that the disruptions in the more normal weather patterns are already causing major problems for human populations in many parts of the world. Extended droughts in some places, sudden floods in others, rising sea levels and related storm damage, as well as earthquakes when coupled with tsunamis as occurred in Southeast Asia in 2005 or with denuded mountain sides as in Haiti in 2010 immensely compound the problems already facing the future of our species.

Many who resist acknowledging the threats that climate change poses or who, even while they admit these changes, nevertheless deny any significant human causality, appeal to natural cycles, ranging from more or less short term weather phenomena such as El Niño and the Northern Pacific Oscillation (e.g., Spencer 2010). Still others point to periodic factors such as solar activity peaks, solar precession, the tilt of Earth's axis, and the eccentricity of Earth's orbit around the sun, which normally varies by about 3.5% each year but can become as great as 23% about every 95-100 thousand years (University of California-San Diego 2002).

Of these periodic phenomena, the eccentricity of Earth's orbit coupled with the "solar forcing" that seems to result from the occasional coincidence of a number these other phenomena seems to have had a measurable effect on Earth's long term climate change. The maximum eccentricity, thought to be caused by the gravitational pull of a particular arrangement of Jupiter and the other planets in their various orbits, appears to have occurred in synchronicity with the major ice ages that

---

2 The "Low Variant" would account for a population of 7.96 billion, the "high variant" 10.46 billion, and the "constant fertility variant" 11.03 billion.

have taken place over the past million years. There have been about ten of these, with some minor ones interspersed. They come about rather slowly, taking many tens of centuries for the ice to accumulate, yet are followed by a warming period that occurs relatively quickly. The last ice age, which reached its peak between 20,000-30,000 years ago, finally lost its grip on the northern hemisphere (except for Greenland) only about 10,000 years ago (NOAA 2010).

Currently, scientists predict that, despite occasional anomalies, Earth will not begin to experience the onset of a new ice age until at least about 30,000 years from now. Meanwhile, Earth's surface has been gradually warming. In the century prior to 1995, the average surface temperature of Earth has risen from .2 to .8 degrees Centigrade (Silver 2008, 10) depending on the location, with the greatest differences occurring near the poles and the least differences in equatorial regions. The average ocean surface temperature has risen by .8 degrees C over the past century and a half (ibid. 35). Besides the argument over whether or not human agency is a significant cause of global warming, the only other real argument seems to be over how quickly all this is happening. Have the oceanic and the atmospheric temperatures accelerated more quickly than might be expected? In terms of the last century and a half, the answer seems to be definitely yes, with the temperatures rising especially since the 1970s.

Explaining why the global temperature has risen, the majority of climate scientists blame the so-called "greenhouse gases" (carbon dioxide, methane, and a few others—the latter mostly from the chlorofluorocarbon group). Almost all of these gases, with the partial exception of methane (and, of course, the exhalation of carbon dioxide by oxygen-breathing organisms) are the product of human activity of one sort or another, particularly the burning of fossil fuels. According to the prevailing science, when these gases are released into the atmosphere, they block the heat in the form of infrared light that radiates from Earth. Hence, while sunlight in visible rays adds heat to the surface of the Earth, these gases, by blocking the escape of that accumulated heat back into space, are causing Earth to overheat or at least to heat up more quickly than would be the case from the gradual warming following the last ice age. However, to complicate matters further, some aerosol pollutants, such as the sulfur dioxide spewed by many industrial operations (as well as by volcanoes and some other natural phenomena) may actually have a cooling effect, mitigating, at least to some effect, the current warming trend.

What will this mean for the human species if global warming continues unabated? Between 2.2 to 1.5 million years ago, there was only *Homo habilis*, who, although using primitive tools, probably had not discovered how to make fire. The likely descendent, *Homo erectus*, apparently did and took this primitive technology out of Africa into Eurasia, managing to survive several of the ice ages at least until about 250,000 years ago.[3] *Homo sapiens*, who seems to have originated during a relatively dry period in Africa while the northern hemisphere was in the grip of the Illinoian Ice Age about 200,000 years ago, seems to have come close to becoming extinct (or at least was reduced to greatly isolated remnants of population) during the intermediate cooling period that seems to have followed the gigantic eruption of the Tubo volcano in Indonesia about 70,000 years ago. Meanwhile, *Homo heidelburgensis* or at least its offshoot, the Neanderthals, who had moved up into the Middle East and Europe long before that, survived the onset of the Wisconsonian Ice Age that began some 35,000 years ago, only to slowly disappear as the ice began to retreat 10,000 years or so later (Smithsonian Museum of Natural History 2008).

What none of these human types survived, except *Homo sapiens*, was the warming that followed. We are not sure exactly why. There has been considerable debate especially regarding the disappearance of the Neantherthals toward the beginning of the end of the last ice age. While some scientists think the Neanderthals killed off all the big game that existed south of the retreating ice sheets and were simply too intellectually incapable or too stubborn to try anything else to keep alive, there seems to be growing consensus that the primary factor was that they lost out as a distinct group in the competition for the available resources as anatomically modern humans gradually migrated up into Europe from Africa (Banks et al. 2008). In any case, it appears that a combination of intelligence, adaptability, and willingness to change is the key to human survival. The question now being asked is: having not only survived but thrived to the point of having successfully populated nearly all of the habitable places on this planet,

---

3   Perhaps the diminutive remains found recently on the Indonesian island of Flores (*Home floresiensis*) indicate that a subspecies of Homo erectus may have survived until about 12,000 years ago, thus surpassing the survival of the last remaining Neanderthals.

will we be intelligent and willing enough to adapt to the changes that will be demanded of us? Or will the technology—derived primarily from our ability to use fire—also bring about the destruction of the ability of Earth to sustain us?

The Challenge of Sustainability

Climate change is only one of the challenges facing the human species. In fact, it is only a part of one of the five major topics or areas of concern addressed in the original *Limits to Growth*. The first to be considered was population, followed by the availability of natural resources (both renewable and non-renewable), pollution, food production, and industrial output. Studied within the general topic of pollution and Earth's ability to dispose of or recycle waste, climate change prompted by excessive heat affects food production positively in some areas and negatively in others as well as the availability and reliability of water resources. For example, although wrongly reported to totally disappear by 2035, the accelerated melting of the Himalayan glaciers will greatly reduce the availability of a major source of water needed for agriculture by over a billion of the inhabitants of Southeast Asia (Columbia Water Center, Columbia University). Meanwhile, catastrophic flooding (such as seen during the 2010 monsoon season in Pakistan) as well as rising sea levels will displace many millions of persons who have lived and worked in the fertile agricultural river deltas lying close to sea level.

However, none of these challenges, even if effectively met or addressed, will ensure humanity's future well-being unless we also face the economic and other social pressures generated by the human desire for higher standards of living. Several times in recent years, as evidenced most recently at the world climate change summit held in Copenhagen in December 2009, the issue of wealth and prosperity sharply divides the nations. Those who are comparatively rich (especially in Europe and North America) have been pleading with developing nations (primarily in Asia and the southern hemisphere) to undertake the same measures that industrialized nations know must be adopted, while those that are developing industrially are demanding that they be allowed the chance to expand their development to the levels that have caused climatic problems.

However, if over-development of the industrialized nations is already a major problem, we can be even more certain that a similar over-development of other nations would spell global disaster. For example, if China, with approximately one-fifth of the world's population, were to reach the standard of living current in the United States where one-twentieth of the world's population has been consuming about a fifth of the world's resources and producing about a fifth of its pollution, China would consume—at least theoretically—*all* of the world's resources! Certainly, China's leadership must know that consumption at this level is impossible or would spell disaster for our planet. In fact, despite China's aggressive multiplication of coal-fired electrical power generating plants (a new one currently coming on-line most every week), Chinese leaders seem to be planning far ahead of the rest of the world in terms of future non-carbon-dioxide producing sources of electricity, such as nuclear, wind, and solar power. India, which is presently in the throes of industrial development, is growing at the current rate of 8.8%, thus approaching China's 10.3% (Trading Economics, 2010), while its population, despite repeated campaigns to lower the birth-rate, is estimated to be growing by over 42,000 persons per day, has already reached over 1.1 billion (MedIndia, 2010), and is expected to exceed China's rate of population growth by mid-century.

Clearly, none of this growth will be sustainable for long. Even apart from the expected leveling off of the world's population at nine billion, the increase in industrialization and standards of living that may motivate a halt or slowing of population growth will have to be balanced, as computer models indicate, by a limit to human aspirations to even more prosperity and comfort. In other words, what has to be faced is the implication that upset many of the critics of the 1972 *Limits to Growth*: to insure peace and stability in the world (in other words, to ensure social or distributive justice), the now-rich societies will have to voluntarily cut back on their consumption and waste in order to provide the opportunity for a more decent and moderately adequate standard of living for the rest of Earth's inhabitants.

Of course, this will sound to many like some kind of blueprint for world socialism, though the authors deny that this is necessarily a foregone conclusion. Instead, they have proposed that enlightened thinking and planning can move humans toward voluntary measures that will lead to a more stable future. Thus, the last two chapters of the 2004 update of *Limits to Growth* are devoted to the topic of *achieving sustainability* and the *tools* that will be required to bring this about. Among these are a set of seven guidelines, the first three of which address the gathering, management, and communication of information and the next three with the management of resources, both renewable and nonrenewable. The seventh on the list, however, is the most critical: "Slowing down and eventually stopping the exponential growth of population and physical capital (e.g., industrial output)" (Meadows 2004, 260). The remarks of these authors regarding this requirement are especially noteworthy: "There are limits to the extent that the first six items on this list can be pursued. Therefore this last item is the most essential. It involves institutional and philosophical change and social innovation. It requires defining levels of population and industrial output that are desirable and sustainable. It calls for defining goals around the idea of development rather than growth. It asks, simply but profoundly, for a *larger and more truly satisfying vision of the purpose of human existence* that mere physical expansion and accumulation" (Meadows 2004, 260, my emphasis).

TEILHARD'S VISION OF AN 'ULTRA-HUMANITY'

Where are we to find this vision that will satisfy the purpose of human existence? One source is Teilhard de Chardin's vision of an "Ultra-Humanity." As nebulous as this term may seem, Teilhard repeatedly used this language to emphasize the need for humanity to enter into a new phase of its own evolution. As he saw it, the *first* phase happened long ago when our ancestors first emerged into the state of reflective consciousness. As he described this phase in his 1951 essay "From the Pre-Human to the Ultra-Human: The Phases of a Living Planet": "Closely related though he is to the other major primates, among which he is only one of the family, Man is psychically distinguished from all other animals by the entirely new fact that *he not only knows, but knows that he knows*" (Teilhard 1964, 292-93; my emphasis).

The *second* phase of human evolution, according to Teilhard, involved not just this reflective consciousness. This phase also involved "*co-consciousness*," a term he coined to convey a kind of collective awareness brought about by the convergence of human beings. From his perspective, the earlier period of expansion is now being replaced by compression or intensification of human communication and thought that constitutes the "*Noosphere*" over the surface of Earth. Finally, he envisioned fast approaching on the horizon, a *third* even more critical phase facing humanity. This third phase, which he first described in *The Human Phenomenon* as "a completely new mode of phylogenesis" and a kind of "Mega-synthesis" (1959, 243-44) was further elaborated in terms of a "planetisation," a status he described in his 1945 essay "A Great Event Foreshadowed: The Planetisation of Mankind" (Teilhard 1964, 124-39). How is this "Planetisation" to take place? As he explained in *The Human Phenomenon*: "[T]here are two ways, through two stages, in which we can picture the form mankind will assume tomorrow—either (and this is simpler) as a common power and act of knowing and doing, or (and this goes much deeper) as an organic superaggregation of souls. In short, science or unanimity" (Teilhard 1959, 248).

The scientific route to knowledge at the time of Teilhard's analysis had already progressed from a simple exploration of nature (*"knowledge for its own sake"*) into a "conquest of matter put to the service of the mind" for *"knowledge for power,"* *"increased power for increased action,"* and *"increased action for increased being"* (ibid. 249, Teilhard's emphasis). He continued: "However far science pushes its discovery of the essential fire and however capable it becomes someday of remodeling and perfecting the human element, it will always find itself in the end facing the same problem—how to give to each and every element its final value by grouping them in the unity of an organized whole" (Teilhard 1959, 250).

In *The Human Phenomenon*, Teilhard encouraged going beyond science to achieve a kind of "mega-synthesis" or "unanimity" (1959, 243-45, 248, 251, 268). He described these terms in two essays written in 1950, the first of which described his belief "On the Probable Coming of an 'Ultra-Humanity'" (Teilhard 1964, 270-80) and the second began with the title in the form of a question: "How May We Conceive and Hope that Human Unanimisation Will Be Realized on Earth?" (ibid. 281-88). For the most part, both of these 1950 essays revisit a concern that Teilhard had already expressed in the last chapter of the final section of *The Human Phenomenon* titled "The Ultimate Earth" (1959, 273-90) where he presented two possibilities: "We can entertain to almost contradictory suppositions about the physical and psychical state our planet will be in as it approaches maturation. According to the first hypothesis which expresses the hopes towards which we ought to in any case to turn our efforts as to an ideal, evil on the earth at its final stage will be reduced to a minimum. Disease and hunger will be conquered by science and we will no longer need to fear them in any acute form. And, conquered by the sense of the earth and human sense, hatred and internecine struggles will have disappeared… Some sort of unanimity will reign over the entire mass of the noosphere. The final convergence will take place *in peace*" (Teilhard 1959, 288). Or positing another possibility: "Obeying the law from which nothing in the past is ever been exempt, evil may go on growing alongside good, and it too may attain its paroxysm at the end in some specifically new form. There are no summits without abysses" (Teilhard 1959, 288).

Which will it be? From the conclusion of the last chapter of *The Human Phenomenon*, its Epilogue (concluded in 1940) on "The Christian Phenomenon" (291-99), and the Appendix (added in 1948), and "Some Remarks on the Place and Part of Evil in a World in Evolution" (311-313), Teilhard's thinking on the subject, reinforced by the horrors of the Second World War and the growing threat of yet a third, evidenced a growing sense of foreboding. Indeed, the first hypothesis—the vision of total peace and the elimination of all evil and suffering—may never exist in this world. Instead, in one of his last essays on this topic aptly titled "The End of the Species" that was first published in *Psyché* in February 1953 (Teilhard 1964, 298-303), he questioned whether or not the human species, after a long period of development that seemed to afford "a scientific justification of faith in progress … is now confronted by an accumulation of scientific evidence pointing to the reverse—the species doomed to extinction" (ibid. 299).

This pessimism was not new for Teilhard. In one of his personal notes dating back to 1948, he wrote that humanity faced an insurmountable "wall," and he described the typical human reaction to this wall as either the *extroversion* of "escape" or else the *introverted pessimism* of Sartre's "existentialism" (Teilhard 1945-55, 15. 3, 170). Similarly, in a 1953 essay, in the face of what he again termed as a "presentment of a blank wall ahead," he noted that writers and teachers had begun to appeal either to an *infinity* of *Time* for the human species insofar as most species up to that time survived for approximately a million years or else to appeal to *Space* into which humans can escape to insure the future of humanity (Teilhard 1970, 398). To the contrary, he pointed out, strategies of escape would be at best only postponements of the inevitable. The only solution presented in an essay written in 1952 titled "The End of the Species" (Teilhard 1964, 298-303) is not in "an improvement of living conditions," as desirable as that might be: "[I]t is not *well being* but a hunger for *more-being* which, of psychological necessity, can alone preserve the thinking earth from the *taedium vitae* ...it is upon its point (or superstructure) of spiritual concentration, and not upon its basis (or infra-structure) of material arrangement, that the equilibrium of Mankind biologically depends" (Teilhard 1964, 303).

Finally, in a meditative essay "The Death Barrier and Co-Reflection" begun on January 1, 1955, barely four months before his death (Teilhard 1970, 395-404), he again spoke about the necessity of our somehow surmounting that wall, that "death-barrier" that faces us, both as individuals and as a species. In this essay we can find some revealing insights. But we can also find another dilemma: "Today, ninety-nine percent of men, perhaps, still fancy that they can breathe freely this side of an unbreakable death barrier—provided it is thought to be sufficiently far away. Tomorrow (and of this I am certain because, like so many other people, I am already experiencing it) mankind would be possessed by a sort of panic claustrophobia simply at the idea that it might find itself hermetically sealed inside a closed universe"[4] (Teilhard 1970, 403). All this is because deep within us, perhaps without our suspecting it, reflective beings have been oriented in our very substance towards a super-life to which there can be no end. Furthermore, we cannot have a 'mass-perception' of this primordial polarization until, all around us, co-reflection has attained a certain critical value (ibid.).

Yet once this point of co-reflection has been reached, a major problem occurs, a problem that he confronted in an appendix added five days later: "From the preceding analysis it follows that when biological evolution has reached its *reflective* stage ('self-evolution') it can continue to function only in so far as man comes to realize that there is some *prima facie* evidence that the death-barrier *can* be broken" (Teilhard 1970, 403). However, he pointed out that this evidence is only "quasinegative," or at least as he saw it—as evidenced in the writings of the existentialist philosophers of his day (especially Sartre and Camus) or what Viktor Frankl was soon to diagnose a few years later as a variety of "noögenic neuroses" afflicting many persons in the modern world—among them the rise of boredom that the philosopher Schopenhauer predicted would increasingly afflict the future of humankind (Frankl

---

4   "Closed" not in the cosmological sense (as contrasted to an "open," i.e., infinitely expanding, universe) but in the psychological as well as biological sense of the universe providing no future possibilities for human life.

1959, 160-64; 1964, 122-23, 183-97). In other words, in the face of all these difficulties, what is really at stake is not merely a matter of physics, biology, or even psychology. In fact, as Teilhard saw it, the problem goes even beyond the speculative or "the 'philosophical' problem of immortality" to "the question, in appearance completely 'theological,' of a revelation" (Teilhard 1970, 404), without which we run the danger of "the terrestrial noosphere being unable to complete its evolution" (ibid. 405) in other words, the long slow process of human evolution running into a final dead-end.

At this point, I can imagine that a number of readers will demure, asking what theology or religion can possibly add to our concerns about this planet's future. From Teilhard's evolutionary perspective, however, the danger that he is most worried about is that humanity, in losing its faith in God, is also in danger of losing what he called in the title of one of his post-war essays its "Zest for Living" (1970, 229-43). This particular essay was written in November 1950 as an outline for a lecture he was to give the next month to a broadly ecumenical or inter-faith group that called itself the *Congrès Universel des Croyants*. In it he admitted that in the past, due to their unawareness of the extent and destiny of the universe, "the various creeds still commonly accepted have been primarily concerned to provide every man with an *individual* line of escape" (ibid. 240) and for this reason they failed to "allow any room to a global and controlled transformation of the whole of life and thought in their entirety" (ibid.). This, Teilhard insisted, should no longer be. What we need instead is "[n]o longer simply a religion of individual and of heaven, but a religion of mankind and of the earth—that is what we are looking for at this moment, as the oxygen without which we cannot breathe" (ibid.).

Teilhard's call for a revision of our understanding of religion drew mixed reactions. In France, his privately circulated essays had caused quite a stir among intellectuals, even before their systematic publication began after his death. Predictably, these explorations into new and more evolutionary ways of approaching transcendence and ultimate meaning drew sharp criticism from those who considered themselves bastions of orthodoxy and "defenders of the faith." For example, Jacques Maritain charged that Teilhard is guilty of a "reversal of the Christian perspective," "a new Gnosticism," and of "kneeling before the world" (1968, 116-17).

Likewise, what had been the general impression of Teilhard's optimism about the future of the world and humanity had begun to draw criticism and doubt. As far back as 1927, long before the posthumous 1956 publication of *Le Phénomène humain*, Teilhard had concluded his treatise on evolutionary spirituality, *Le Milieu divin*, a treatise on the spiritual life that was also intended to be his defense of his orthodoxy, with a strong appeal to the Christian concept of the *parousia*— the belief that at the end of time, Christ will return, not simply as Judge, but also as the restorer of creation, to bring about, in the words of Isaiah (65:17) repeated in 2 Peter (3:13) and again in Revelation (21:1), "a new heavens and a new earth." However, Teilhard remained privately convinced, despite his growing worries, that this final state would not come about until humankind had completed the course of its evolution. Or, as he wrote several years earlier in his essay "Mon Universe": "It is then, we may be sure, that the Parousia will be realized in a creation that has been taken to the climax of its capacity for union" (1965, 84). In other words, "a new heavens and a new earth" would occur only after humanity has not only reached but has exercized its full capacity for evil as well as good or again, for good or for ill, tried everything.

These reflections led to an interesting debate in which Teilhard participated on "Theology and Cosmology" shortly after his return to Paris when World War II ended. In this exchange, the Dominican theologian, Pere D. Dubarle, objected to Teilhard's optimistic view that God permits evil and suffering so a greater good might emerge. Debarle pointed out that a comet could destroy Earth, wiping out all human progress. Teilhard replied: "That will never happen," implying that God would not permit it. When Dubarle objected, saying that was an answer not of a philosopher but of a theologian, Teilhard responded: "Our life should be only a life of believing [*de croyant*]" (Cuénot 1965, 258)

Of course, Teilhard could have been wrong. Scientists generally acknowledge today that the impact of an asteroid brought about the end of the age of dinosaurs about 65 million years ago. And the impression of what such an astronomical accident could mean for us was especially re-enforced by the disturbance caused by the impact of comet Shoemaker-Levy on Jupiter in 1994, leaving a mark on that distant planet's atmosphere that was much larger than Earth. However, Teilhard was more worried about a burgeoning population running short of food and prematurely reaching the avoidable limits to its growth before its psychological and spiritual potentialities had fully matured. Thus, in an unusually extensive entry into his personal notebook on August 11, 1948, Teilhard discussed the implications of Lemaître's "hypothesis" (later nick-named the "Big Bang"), its "explosive force," and, even more, its entropic implications for our understanding of time and space (*Journal* 15 (3), 141). This in turn seems to have led to a note added to *The Human Phenomenon* (1959, 47-48) as well as to his reflections in the final chapter, where he wrote about "the event that comes that comes nearer every day that passes: the end of all life on our globe, the death of the planet, the ultimate phase of the phenomenon of man" (Teilhard 1959, 273).

CONCLUSION

Had Teilhard lived another two decades, long enough to have seen the publication of the first 1973 edition of the *Limits to Growth* as well as the sharp reactions it prompted, I am reasonably sure that he would have shared the authors' concerns, as well as of those at the Club of Rome who had commissioned that study. Likewise, had he lived still longer, he probably would have shared the sober *déjà vu* reminder of their 1992 update, *Beyond the Limits*. What might his reaction have been?

I suspect that while Teilhard might be sad to see that humanity has so badly mistreated our planet, especially when we could have done otherwise, but that he would not entirely have been surprised nor completely lost hope. As we have already seen, evolution was not an exclusively biological phenomenon from Teilhard's perspective. Evolution was an all-encompassing paradigm and force operating on all levels of reality, from cosmic materiality to the most sublime spirituality. Evolution is, as he once termed it, "Le Milieu divin," within which the *Pleroma* is realized. He defined this latter term, borrowed from the later Pauline epistles, in his 1940 essay "The Awaited Word" as "the mysterious synthesis of the uncreated and the created—the grand completion (both quantitatitve and qualitative) of the universe in God" (Teilhard 1973, 97).

Viewed from this perspective, while the premature foreshortening of human history on Earth may be seen as regrettable, even sinful, it does not violate the fundamental movement of evolutionary *convergence*—the bringing together of diverse elements, organisms, and even the currents of human thought. For Teilhard, this convergence was not only predictable. It was inevitable, especially as human population growth has expanded.

Evolution also has gradually led to the near exhaustion of what is now termed the "carrying capacity" of our planet. Certainly, it is one of the great ironies of our own human evolutionary history that the single most important invention, the discovery of how to ignite and use fire, has in turn led to the greatest human-caused danger to our survival. Yet, for Teilhard, this was all seen as preliminary to the final transformation of the material energies of Universe into a final product—an "Ultra-humanity," the final product of the evolution of matter into spirit or what we might call a "universe of souls."

Thus, I conclude as someone long familiar with Teilhard's thought, that while as a human he would regret the mistakes leading to the crisis to which humanity has come, as a scientist he would also realize (apart from any human agency) the inevitability of it all. Accordingly, from this perspective, I would guess that, for Teilhard, arguments over the causes of global warming and what we might do about it would not come down to a serious debate over whether or not (and to what extent in the face of recurring ice ages) human life can be prolonged on this planet before its and its parent Sun's eventual demise. Instead, I suspect that for Teilhard, it would be a matter of concern, as it already was for him, that the world's burgeoning population, entering what he seemed to believe was its final phase of evolution on this planet, forcing us to confront what amounts, at least in terms of its physical existence, "The End of the [in this case *Our*] Species" (Teilhard 1964, 294-303).

Finally, as a believer who was always in touch with the latest scientific developments, I think that Teilhard had finally reached the point, at least toward the end of his life, where his deep Christian faith had been forced to expand to truly cosmic dimensions. In doing so, he came to the realization that neither Earth nor humanity may be the only instance of a planet harboring intelligent life. Indeed, as he admitted in his 1953 essay "A Sequel to the Problem of Human Origins: The Plurality of Inhabited Worlds" there could be millions of such worlds (Teilhard 1969, 229-36). Although he did not reflect on what this would mean in terms of human ecological consciousness and responsibility, we cannot avoid concluding that, faced in these terms, the fate of the human species for the foreseeable future has been largely left up to us. Although Teilhard may have believed that the end of humankind would not come until it had exhausted all its potentials, we also have to face the fact that perhaps we are already close to having done so, and that Divine Providence is not going to save us from our own folly. Instead, at least as Teilhard saw it, God may have a much bigger picture in mind.

SOURCES

Banks WE, d'Errico F, Peterson AT, Kageyama M, Sima A, et al. 2008. Neanderthal Extinction by Competative Exclusion, *Public Library of Science.* Accessed from PloS ONE 3(12): e 3972. doi:10.1371/journal. pone.0003972 on September 20, 2010.

Bhattacharjee, Meghna. 2009. "Himalayan Glaciers Melting," Columbia Water Center. Accessed from http://www.water.columbia.edu/HIMALA-YAN%20GLACIERS%20MELTING.pdf on September 21, 2010.

Chaisson, Eric J. 2001. *Cosmic Evolution: the Rise of Complexity in Nature.* Cambridge: Harvard University Press.

Cuénot, Claude. 1958. *Teilhard de Chardin: les grandes étapes de son évolution.* New York: Helicon.

Frankl, Viktor. 1959. *Man's Search For Meaning: An Introduction to Logotherapy.* New York: Pocket Books.

———. 1967. *Psychotherapy and Existentialism: Selected Papers on Logotherapy.* New York: Simon & Schuster.

Global Footprint Network. 2009. "What Is Earth Overshoot Day?" Accessed from http://www.footprintnetwork.org/en/index.php/GFN/page/earth_overshoot_day.htm on September 25, 2010.

Maritain, Jacques. 1968. *The Peasant of the Garonne.* New York: Holt, Rinehart, Winston.

Meadows, Donella H., Dennis L. Meadows, Jørgen Randers, and William W. Behrens III. 1972. *The Limits to Growth: A Report for the Club of Rome's Project on the Predicament of Mankind.* Washington, DC: Potomac.

———. 1992. *Beyond the Limits: Confronting Global Collapse, Envisioning a Sustainable Future.* Post Mills, VT: Chelsea Green.

———. 2004. *Limits to Growth: The 30-Year Update.* White River Junction, Vt: Chelsea Green.

MedIndia. 2010. Accessed from http://www.medindia.net/patients/calcula-tors/pop_clock.asp on September 21.

NOAA, National Climatic Data Center. 2010. "Paleoclimatology." April 14. Accessed from http://www.ncdc.noaa.gov/paleo/ctl/clihis100k.html on September 25.

Silver, Jerry. 2008. Global Warming and Climate Change Demystified. New York: McGraw Hill.

Smithsonian Institution, Smithsonian Museum of Natural History. 2009. "What Does It Mean To Be Human?" Accessed from http://www.humanorigins.si.edu on September 20, 2010.

Spencer, Roy. 2010. "Global Warming, Natural or Manmade? Accessed from http://www.drroyspencer.com/global-warming-natural-manmade/ on September 21.

Teilhard de Chardin, Pierre. 1945-1955. Journals. Archives of the Paris Province, Society of Jesus.

————. 1955. *Le Phénomène humain*. Paris: Editions du Seuil.

————. 1959. *The Phenomenon of Man*. New York: Harper & Row.

————. 2000. *The Human Phenomenon*. Translated by Sarah Appleton. Eastbourne: Sussex Academic.

————. 1957. *Le Milieu divin*. Paris: Editions du Seuil.

————. 1960. *The Divine Milieu: An Essay on the Interior Life*. Translated by Bernard Wall. New York: Harper & Row.

————. 1959. *L'Avenir de l'Homme*. Paris: Editions du Seuil.

————. 1964. *The Future of Man*. Translated by Norman Denny. New York: Harper & Row.

————. 1963. *L'Activation de l'Energie*. Paris: Editions du Seuil.

————. 1970. *The Activation of Energy*. Translated by René Hague. New York: Harcourt, Brace, Jovanovich.

————. 1969. *Comment je crois*. Paris: Editions du Seuil.

————. 1971. *Christianity and Evolution*. Translated by René Hague. New York: Harcourt, Brace, Jovanovich.

————. 1973. *Les Directions de l'Avenir*. Pais: Editions du Seuil.

————. 1975. *Toward the Future*. Translated by René Hague. New York: Harcourt, Brace, Jovanovich.

Trading Economics. 2010. Accessed from http://www.tradingeconomics.com/Economics/GDP-Growth.aspx?Symbol=INR & CNY on September 21.

United Nations, Department of Economic and Social Affairs. 2009. "World Population Prospects: 2008 Revision." March 11. Accessed from http:// esa.un.org/unpp/ on March 17, 2010.

University of California, San Diego. 2002. "Climate Change, Past and Future: Ice Age Climate Cycles." Accessed from http://earthguide.ucsd.edu/ virtualmuseum/climatechange2/03_1.shtml on September 22, 2010.

CLIMATE CHANGE & THE THEOLOGY OF KARL RAHNER

A HERMENEUTICAL DIALOGUE

Denis Edwards

I
n many different parts of the world, theologians have been engaged in responding to ecological issues, some specifically focusing on climate change. My intention here is the modest one
of attempting to clarify the kind of dialogue involved in this
emergent ecological theology. In particular, I will ask about the
hermeneutical principles involved in bringing a scientifically informed
ecological consciousness into creative dialogue with the Christian tradition. In order to pursue this aim, I think it is helpful to be specific
and concrete, and so I will take up one example of a major ecological
issue that confronts us, climate change, and engage with one specific
interpretation of the Christian tradition, the theology of Karl Rahner.

One of the positions I will advocate is that there are two sides to a
real conversation, so there ought to be two sides in a theology that engages with ecology. There are examples of ecological hermeneutics that
are one-sided where, for example, ecological insights are seen as challenging aspects of the Christian tradition, but this tradition appears to
have nothing to add to the conversation. On the other hand, there are
Christian theologies that ignore or marginalize ecological concerns.
Both perspectives will need to be engaged critically in an authentic
ecological theology. An ecological consciousness has the possibility of
not only raising crucial questions for Christian theology, but also of
opening up fruitful new perspectives. An ecological theology will not
only raise critical questions about assumptions within the ecological
movement, but may also insight into the meaning and the value of

God's creation. A creative dialogue between these perspectives may lead to a critically aware ecological theology that has something to say in the context of global climate change.

Global climate change involves many disciplines besides science and theology, including those of economics and politics. For the sake of clarity my focus will be restricted to the interaction between a scientifically informed ecological consciousness and theology. Obviously this can only offer a partial view of the whole. But I hope it can clarify some aspects of the dialogue that goes on in the work of ecological theology. As a small step in this direction, I will attempt to articulate hermeneutical principles that emerge from a scientifically informed consciousness of global climate change on the one hand and from the theology of Karl Rahner on the other.

A SCIENTIFICALLY INFORMED
CONSCIOUSNESS OF CLIMATE CHANGE

What does a scientifically informed consciousness bring to the task of constructing a Christian ecological theology that can offer a response to the issue of global climate change? My proposal is that a scientific consciousness can be thought of as bringing to the table one general challenge to theology and five more specific hermeneutical principles.

A General Challenge

An ecological consciousness confronts theology with a stark challenge: *Global climate change, understood in relation to all its associated ecological and human issues, is the most urgent issue facing the community of life on our planet in the twenty-first century and therefore it is also an issue that Christian theology must address.* In our time we see the loss of Arctic sea-ice, the northern permafrost thaw, the retreat of glaciers that are the sources of great rivers and the thermal expansion of the ocean. It seems clear that large areas of our planet will become far hotter while others will be subject to more violent storms and floods. While some species will be able to migrate to new habitats, many others will become extinct as their habitats are lost. Human beings will be displaced and many will be forced to become refugees. Great food growing areas will become subject to extreme drought, flooding or inundation by the seas. Water will become very scarce in many parts of the planet. There is a real danger of wars over sources of water and over land for food, as well as over the sources of energy. The continuing loss of biodiversity challenges us at every level. All the other major issues facing the human community, from the inequality between rich and poor in our world to the threat of nuclear war, are inextricably interconnected with the issue of global climate change.

There can be little doubt that global climate change, in the context of the other issues with which it interconnects, is a great challenge we face in this century. Responding to it will require all the creative energy, good will and cooperation that the human community can muster. When the human community faces this kind of challenge, it is involves the deepest dimensions of the human. Religious faith, including Christian faith, is challenged to ask itself two critical questions: (1) Do its formulations and practices contribute to the problems facing the community? (2) In what ways might its traditions and liturgies become a source of energy, life and motivation for a genuine ecological commitment in this new time? A scientific and ecological consciousness puts to theology a challenge to rethink its doctrines and practices in the light of the global crisis facing our planet.

Five Interpretative Principles

An ecological consciousness can also be seen as offering specific contributions to the discussion. I will propose that it can offer at least five further guiding ideas or interpretative principles to the building of a genuine ecological theology in the context of global climate change.

> 1. *Global climate change needs to be addressed from within a scientific worldview, which includes an understanding of the observable universe as expanding and evolving, and of life on Earth as evolving by means of natural selection.*

An ecological consciousness is necessarily informed by the biological sciences, all of which are shaped by the theory of evolution. Biologists have a common vision of life as beginning about 3.8 billion years ago with the simple bacterial cells, and emerging through the evolution of complex cells into multicellular creatures, and then into all forms of life, including dinosaurs, flying reptiles, mammals, birds and flowering plants. About four million years ago, chimpanzee-like apes gave rise to various hominid species (the *Australopithecines*), then, with a large increase in brain size, to various species of *Homo*, including *Homo erectus*, and about 200,000 years ago, to modern humans. Everything is interconnected in the history of life and everything depends upon evolution, which is driven at least in large part by natural selection.

Life on Earth is unthinkable without the stars, something now recognized in the establishment of departments of Astrobiology at NASA and various universities. It was only through the work of Einstein and Hubble in the early first decades of the twentieth century that we came to know that our universe is not static but expanding. We can now trace it back to a time when it was extremely small, hot and dense 13.7 billion years ago. Biological life is deeply connected to the story of the universe. We are a carbon-based life-form and all the carbon atoms that make us up, along with the nitrogen and oxygen, are produced by the process of nucleosynthesis in stars. An ecological consciousness is informed by the story of the universe offered by scientific cosmology and of life offered by evolutionary biology.

What questions or challenges does this principle bring to the dialogue with theology?

Obviously it challenges a theology that is content to simply repeat what has been said in the past. The cosmological worldview taken for granted by the Bible, Augustine or Aquinas cannot be taken for granted today. The theological insights contained in the ancient classical texts may be as relevant as ever, but they need to be rethought from within a new worldview. We know wonderful things about the dynamic nature of God's creation and its evolutionary emergence that were not available to Paul, Augustine or Aquinas. An ecological theology will need to be a theology articulated within a consciousness of the 13.7 billion year history of the observable universe and the 3.8 billion year history of life on Earth. This is not only a challenge, but an opportunity for theology to speak about God's action in creation and salvation in ways that take seriously the new dynamic understanding of the world available to us. An ecological theology will be one that takes seriously the story of the universe and of life given to us by the sciences.

> 2. *The regular shifts in the global climate, the current scientific consensus on human-induced climate change, and the provisional and revisable nature of this science need to be taken into account.*

In a scientific approach to climate change, our present experience of climate change is situated in a larger context of regular shifts in the global climate, caused by variations in green house gasses, plate tectonics, volcanism, solar radiation and the Earth's orbit. Since the 1970s, scientists have linked long-term climate change to three variations in the orbit of the Earth around the Sun (the Milankovitch theory). These variations occur in predictable cycles. One cycle, caused by a wobble in the axis of the Earth's rotation, called precession, occurs every 22,000 years. The others, caused by the tilt in the Earth's rotational axis and by the shape of its orbit, occur every 41,000 and 100,000 years. These three cycles alter the distribution of solar energy reaching the Earth. Over the last three million years, these variations have produced a series of ice ages followed by warmer interglacial periods. The last ice age was about 20,000 years ago and the present interglacial period (the Holocene) is well advanced.

The Intergovernmental Panel on Climate Change, in its Fourth Assessment Report (2007) concludes that global warming has accelerated over recent decades and there is new and stronger evidence that it is to be attributed to an increase in greenhouse gasses caused by human activities. By the end of the 21st century average world temperatures are likely to be between 1.4 and 5.8 degrees higher than in 1990. The IPCC predictions are tentative and revisable. Its conclusions are based on a current consensus in peer-reviewed science. Many of the challenges and attacks directed against the IPCC that appear in the media are not based on peer-reviewed climate science. When theology engages with climate science, I think it needs to engage with what appears to be the best information available from peer-reviewed science. It must also recognize that science is revisable in the light of further evidence. In discussions on climate, theology engages not with what is certain or absolute, but what is provisional.

> 3. *The costs of evolution are intrinsic to the process of the emergence of life on Earth. Suffering is built into the natural world.*

The earthquakes and tsunamis that have killed hundreds of thousands of people in recent years are caused by the meeting of tectonic plates. It is this dynamic system of tectonic plates that allows for the emergence of mountain ranges, rivers, rain forests and fertile plains, providing habitats that allow life to evolve in new ways. The evolution of life, with its abundance and beauty, is accompanied by terrible costs to human beings and to other species. The costs are built into the universe and its laws. Evolutionary emergence involves not only cooperation but also competition for resources. The pattern of evolution depends on death, and the continuing cycle of generation. The diversity of life, with all its beauty and goodness, arises by way of increasing complexity through emergent processes that involve tragic loss. The costs are evident in the history of life with its predation, death and extinctions. We know, as no generation has known before us, that these costs are intrinsic to the processes that give rise to life on Earth in all its wonderful diversity.

This principle challenges a tendency to see the ecological as always harmonious, peaceful and in balance. What is natural can also be ugly and it can be painful. This principle challenges all romantic notions of the natural world. It challenges Christian theology, not to abandon its conviction of the goodness and beauty of creation, but to see it for what it really is in all its complexity. It also puts suffering at the centre of a Christian theology of creation. Suffering, death, and the extinction of species that have occurred throughout evolutionary history cannot be seen, in the way that an earlier theology might seem them, as the outcome of human sin, but are part of the way God has created the diversity of life. This new insight into the suffering built into creation sets and important agenda for twenty-first century theology. It needs to return to the age-old issue of suffering. It needs to respond to the costs built into evolutionary emergence by God and to do this in the light of its deepest insights into the meaning of Christ and his promise.

> 4. *Other living creatures are to be valued not only because of their usefulness to human beings but because they possess intrinsic value.*

One of the fundamental claims of ecological ethics is that the value of life-forms is not determined simply by their value for humans. Holmes Rolston III, a founding contributor to the new field of environmental ethics, has argued for the objective intrinsic value of the natural world by means of an analysis that is based closely on the biological sciences. At the level of each organism, he finds a life being defended and valued. Because the genetic set for an organism is conserved primarily at the species level, Rolston sees the species as well as the organism as being the object of moral consideration. At the level of the ecosystem, and the biosphere of the planet, Rolston finds not only intrinsic value but "systemic" value, above all in the natural world's capacity to evolve diverse and complex life forms. He finds value in living things from the tiniest organism to the biotic community of Earth, and insists that this value is objectively there in the natural world and that it makes moral claims on human beings (Rolston 1999, 38-53).

What this principle challenges is uncritical anthropocentrism, whether it appears in the biblical texts, in the Christian theological tradition or in the life of the churches. Uncritical anthropocentrism is the view that other creatures exist only to serve human beings. The principle of intrinsic value does not, in my view rule, out a strong view of the unique dignity of the human person. It supports the idea other creatures also have their own unique dignity and value. This principle is a fruitful one for Christian theology. It is congruent with, and can be brought into creative dialogue with, fundamental biblical positions: that the Creator holds all things in existence, finds all of creation good and enables it to flourish in all its fertility and abundance (Gen 1:20-31); that God loves and cares for each creatures (Wis.11:24; Luke 12:6); that the diversity of creatures is the self-expression of God (Psalm 104: 24); that, in God's eschatological future, each creature, in the heavens, on earth, under the earth and in the sea will sing praise to the one seated on the throne and to the Lamb (Rev 5:13).

> 5. *All forms of life on Earth are interconnected in one web of life, and are interrelated with the land, the seas, the rivers and the atmosphere.*

This is perhaps the most pervasive principle at work in an ecological consciousness. The Earth is a place of abundant and exuberant life. The five to ten million species that inhabit our planet have emerged over the last 600,000 years. They have a common heritage that goes back to the origins of bacterial life more 3.8 billion years ago. They have evolved in relationship to each other, interconnected in delicate ecological systems. They are interdependent not only with each other, but also with the Earth's atmosphere, its seas, rivers and lakes and the land itself. Human actions, such as ruthless fishing practices, the dumping of industrial and urban waste, the destruction of river systems and uncontrolled land-clearing destroy both known and unknown species and make Earth a more sterile and dangerous place.

This principle is a critical principle that can guide human action into an ecological future. At a deeper level, it points to the idea that our very identity is that we exist only in relation to others, to other human beings, to other species, and to all that supports life on our planet. When we act as if we can disconnect ourselves from the rest of biological life, we ignore our true nature as interdependent in the community of life on our planet. While this challenges the individualism found in some interpretations of the Christian tradition, it can be a stimulus for conversation with the deepest traditions of Christian faith, which finds an inter-relational view of reality grounded in an inter-relational view of God as a God of mutual relations. In a fully ecological theology, the ecological relationships that characterize life on Earth can be understood as springing from the relational life of the Trinity.

FROM THE PERSPECTIVE OF RAHNER'S INTERPRETATION OF THE CHRISTIAN TRADITION

So far I have proposed that from the perspective of an ecological consciousness, a general challenge is offered to Christian theology along with five interpretative principles that can contribute to the building of an ecological theology. Turning now to Rahner's thought, I will suggest one fundamental theological stance and five further interpretative principles that can contribute to this same project.

A Fundamental Theological Stance

I proposed above that the issue of global climate change is a fundamental issue facing the human community in the twenty-first century and, as such, is an urgent issue for Christian theology. The fundamental response offered in the theology of Karl Rahner is that it is precisely in the engagement with what confronts us in our daily and communal lives, such as global climate change, that we encounter the mystery of God. Rahner holds that when we go out of ourselves to our world, in knowledge, love and commitment, we find ourselves also open to the infinite mystery that transcends all objects of immediate experience. This experience of radical openness to mystery is always mediated by our knowledge and love of other persons and entities in our world. It is in our engagement with our world, with persons, with issues, with great challenges like climate change, that we are brought to a sense of the incomprehensible mystery in which all of this dwells. Rahner insists that because of revelation, we can accept the world as the place of grace, the place of the Holy Spirit (Rahner 1978, 51-71).

When we engage with the world, when we are drawn to an awareness of our planetary community of life, there is an openness to holy mystery in our knowledge of our fragile planet. We know our limited and finite planet only against a horizon that is boundless. When we are drawn to loving commitment to the Earth and its creatures, our love opens our beyond itself to a love that is without limits. Rahner insists that the boundlessness of the human heart and mind that occurs in our engagement with our world is the place of God's grace. God has freely chosen to give God's self to us, and we encounter this God in our knowledge of, love for, and commitment to the Earth and its creatures. To adapt Rahner's words from his famous article on the unity of love of God and love of neighbor: We can only love the God that we do not see by loving the planetary community that we do see. And to truly be committed in love for the community of life on our planet is already to love, at least implicitly, the living God (Rahner 1969b, 231-49).

We encounter the living God in negative as well as positive experiences of our engagement with our world. Involvement with the issue of global climate change will lead, at least at times, to the experience of failure and frustration, and living with ambiguity, unable to control the outcome of political decisions. Rahner would say that when we resist cynicism and despair, and continue our commitment, continuing to hope and to act, then this is the place of God and God's liberating grace (Rahner 1983, 189-210). And also when we are taken out of ourselves by the exuberant life of a rain forest, or by the song of a single bird, or by the experience of human solidarity, this too can be a place for encountering God's Spirit. To see this as the silent presence of God, to give ourselves to this holy one, is to be involved with what Rahner calls the mysticism of everyday life (Rahner 1983, 203). In terms of this discussion, it might be said what Rahner's theology offers to the discussion of climate change, at the most fundamental level, is the idea that there is a genuine mysticism of commitment to the Earth and all its creatures.

Five Further Interpretative Ideas from Rahner's Theology

What I have been describing above is sometimes called Rahner's transcendental method. It refers to the way he consistently seeks to see our experiences in this world as the place where we encounter a gracious and mysterious God. But, as scholars like Francis Schussler Fiorenza point out, Rahner's work is not limited to any one methodology (Fiorenza 2005, 65-82). He uses a variety of approaches. In particular he consistently sees the Word made flesh in Jesus of Nazareth as the norm for his theology. Everything, including the experience of grace in everyday life, is understood from the perspective of the Christ-event.

There is a christological centre to Rahner's theology. Based on his view of God's self-communication in Christ, five further interpretative principles can be found in Rahner's theology.

> 1. *The creation of the universe and the Incarnation can be seen as distinct dimensions of the one act of divine self-bestowal in love.*

According to Rahner, the central insight of Christianity is that God gives God's self to us in the Word made flesh and in the Spirit poured out in grace. God is revealed to us as a God who bestows God's self. God gives God's self to creatures. Based on what is revealed in Christ, Rahner sees this self-bestowal as defining every aspect of God's action in creation, redemption and final fulfillment. The story of the universe, and everything that science can tell us about its evolution, is part of a larger story, the story of divine self-bestowal. He sees the creation of the universe as an element in the radical decision of God to give God's self in love to that which in not divine (Rahner 1974b, 219). When God wills to bestow God's self in love, creation comes to be as the addressee of this self-bestowal. Rahner insists that the story of salvation is the real ground of the history of nature, and not simply something that unfolds against the background of nature (Rahner 1975, 1442). The history of the evolution of the universe, and of life on our planet, exists *within* this larger vision of the divine purpose.

In this theology, the incarnation is not something that comes about simply as a remedy for sin. It is not simply a corrective for a creation that has gone wrong. It is not thought of an add-on, or as an afterthought, to creation. With the Franciscan school of theology, exemplified in Duns Scotus (1266-1308), Rahner holds that God freely chooses, from the beginning, to create a world in which the Word would be made flesh and the Spirit poured out. Harvey Egan finds that the briefest possible summary of Rahner's theological enterprise in found in "his creative appropriation of Scotus's view that God creates in order to communicate *self* and that creation exists in order to be the recipient of Gods free gift of self" (Egan 2005, 16). Once sin exists, the incarnation is the great expression of divine forgiveness. But even more radically, the incarnation expresses the meaning and purpose of creation, divine self-bestowal. Creation and incarnation and united in the one act of God: they are "two moments and two phases of the *one* process of God's self-giving and self-expression, although it is an intrinsically differentiated process" (Rahner 1978, 197).

This principle challenges two opposed theological positions. On the way hand, it challenges a theological tradition that centers the whole redemptive act of God in Christ on human sin, and atonement for human sin. On the other hand, it also challenges some of the proponents of creation spirituality, who tend to reject or abandon the theology of redemption as dangerously anthropocentric. This principle proposes a theology in which creation and redemption are held together as inseparable dimensions of God's one act of self-giving love. This, in turn can provide the foundation for a theology of redemption that involves the whole creation.

> 2. *God's act of redemption involves the deification of human beings and with them the whole creation.*

Rahner sees the incarnation and creation itself from the perspective of the resurrection. He contrasts the Western juridical notion of redemption focused on the death of Jesus with the theology of the East, where the resurrection plays a fundamental role in the theology of salvation: "The redemption was felt to be a real ontological process which began in the incarnation and ends not so much in the forgiveness of sins as in the divinization of the world and first demonstrates its victorious might, not so much in the expiation of sin on the cross as in the resurrection of Christ" (Rahner 1974a, 126). There are three themes in this Eastern theology that fundamental for Rahner's thought: salvation as ontological rather than juridical, the idea of the divinization of the world, and the resurrection of Christ as the beginning of this divinizing transfiguration.

In the death of Jesus a piece of this world is handed over freely into God, in complete obedience and love, and is fully taken up into God. In the resurrection, God irrevocably adopts creaturely reality as God's own reality. Because of the unity of the world that springs from the Creator, this is an event for the whole world. What occurs in Jesus, as part of the physical, biological and human world, is *ontologically* and not simply juridically, "the embryonically final beginning of the glorification and divinization of the whole of reality" (Rahner 1974a, 129). The resurrection is "the beginning of the transformation of the world as an ontologically interconnected occurrence" (Rahner 1975, 1142). The final destiny of the world is decided and already begun. The risen Christ is the "pledge and beginning of the perfect fulfillment of the world." He is the "representative of the new cosmos" (Rahner 1975, 1142).

This means that we Christians are really "the most sublime of materialists." Christians hold that matter will last forever, and be glorified forever in Christ. It will, however, undergo a radical transformation, "the depths of which we can only sense with fear and trembling in that process which we experience as our death" (Rahner 1971, 183). The transfiguration of the world has begun in the risen Christ and is "ripening and developing to that point where it will become manifest" (Rahner 1971, 184). This position challenges any theology of the redemption that limits the meaning of Christ to human beings, because it sees the whole creation as involved in God's work of redemption and deification. At the same time, however, because it gives a radically central place to the incarnation and the resurrection, it sees God as forever united to humanity in Christ, and challenges claims, sometimes put forward from an ecological perspective, that undermine the dignity of the human being.

What it offers is a theology of redemptive deification that involves the whole creation, clearly fundamental for a Christian ecological theology. It opens up the possibility of pursuing questions not considered by Rahner: To what extent might we think of the incarnation as a deep incarnation in the sense that God embraces and transforms not only human flesh, but all flesh, the whole of our evolutionary history, and the interconnected web of life? How do we think of the final salvation not only of matter, but also of nonhuman biological creatures, both species and individuals?

3. The transcendent God's relationship with creation is characterized by radical immanence.

Rahner is a theologian who is convinced that we stand before a God who is an abiding, absolute and incomprehensible mystery to us, a mystery that theology can approach only by means of analogy. In his final theological reflection Rahner writes: "theologians are worthy of the title only when they do not seek to reassure themselves that they are providing clear and lucid discourse, but rather when they are experiencing and witnessing, with both terror and bliss to the analogical back and forth between affirmation and negation before the abyss of God's incomprehensibility" (Rahner 2005, 301). In stressing the divine transcendence, Rahner is not unique, but stands in the great theological tradition of the Gregory of Nyssa, Augustine and Aquinas. It is precisely and only because God is radically transcendent that God can be, as Augustine says, more interior to me than I am to myself. God's radical immanence to creatures is the fruit of divine transcendence.

Rahner goes further when, in a late article he asks himself: what is most specific to the Christian view of God? His answer is the idea that God bestows God's very self to creation (Rahner 1988b, 185–95). God creates creatures that are able to receive God's life as their own fulfillment: "Only when this statement is made, when, within a concept of God that makes a radical distinction between God and the world, God himself is still the very core of the world's reality and world is truly the fate of God himself, only then is the concept of God attained that is truly Christian" (Rahner 1988b, 191). Rahner makes the staggering claim that *God's self-giving is the very core of the world's reality and the world is truly the fate of God*. This claim is based on the conviction that in creation, incarnation and in its culmination in resurrection, God commits God's self to this world, to this universe and its creatures, and does this eternally.

Rahner sees God's action on creation not according to the model of efficient causality, but that of formal causality: God creates through self-bestowal. God gives God's self to us and we are made a new creation. God, then, does not create simply by producing something different from God's self, as a carpenter makes a table. Rather God creates by communicating God's own divine reality and making it a constitutive element in the fulfillment of the creature. God remains God but radically determines our being, and we are transformed from within (Rahner 1978, 120-3). We become a "new creation" (2nd Corinthians 5:17) and are made "participants of the divine nature" (2nd Peter 1:4). In some works, Rahner calls this *quasi*-formal causality, with the *quasi* indicating the uniqueness of this kind of formal causality, in which both divine transcendence and creaturely integrity are fully maintained (Rahner 1978, 120-3; and 1988a, 35-6). This relationship that Rahner sees as characteristic of grace, he sees as an appropriate analogy for the relationship that God has with the whole universe and all its creatures, since what is true of grace is always valid "in an analogous way for the relationship between God's absolute being and being which originates from him" (Rahner 1988a, 36). God is creatively present to every entity and process of the universe, in such a way that "the reality of God himself is imparted to the world as its supreme specification" (Rahner 1974a, 225). Self-bestowal is not only the goal of creation, but that which moves creation from within to the goal. This self-bestowal of the transcendent God is "the most immanent factor in the creature" (Rahner 1973, 281).

Because of its stress on the absolute transcendence of God Rahner's theology questions tendencies in ecological and scientific thought which seem to make the universe into the Creator, or which would make evolution itself into something divine, or which suggest that the creativity of nature is "God enough." It insists on the transcendent God who creates through secondary causes, which include the universe itself, evolution and the creativity of nature. With the idea that this God gives God's self to the universe of creatures in such a way that this self-giving is the very core of the world's reality and the world is the fate of God, this theology vision questions the widespread idea that transcendence is opposed to immanence, and that there is some kind of need to move from a theology of transcendence to one of immanence. What Rahner's theology offers is a vision of God who is present in self-giving love to every aspect of creation, to every galaxy and to each insect in a rain forest, as the source of its existence, its creativity and its final fulfillment.

4. God acts in a noninterventionist way through creaturely causes.

Thomas Aquinas saw God as working through creatures that are themselves truly causal, which he calls secondary causes. God is the primary cause who is always creatively and providentially at work in all created causes. It is by God's power that every other power acts. While God enables creaturely causes to exist and to have effect, Aquinas insists that secondary causes are genuinely causal in their own right. It is through these secondary causes which have their own proper independence that God cares for creation: "Divine Providence works through intermediaries. For God governs the lower though the higher, not from any impotence on his part, but from the abundance of his goodness imparting to creatures *the dignity of causing*" (Aquinas, *Summa theologiae* 1.22.3). God respects the dignity of secondary causes, and bestows on them their own integrity. According to Aquinas God works consistently through secondary causes, except for the case of miracle where God acts without secondary causes.

Rahner develops the tradition of Aquinas. He proposes that the great events of the history of salvation as well as special divine acts in our lives might be understood in noninterventionist terms. They are instances where God freely brings to particular, concrete and historical expression that one act by which God is immanent to creation from the beginning: "God in his free grace, from the very beginning and always and everywhere, has communicated himself to his creation as its innermost energy and works in the world from the inside out" (Rahner and Weger 1980, 78-79). God is never simply one object amongst others in our world, but it "embedded" in this world from its origin (Rahner 1978, 87). A special act of God is an "objectification" of God's one self-bestowing action. In such a divine act, a created reality expresses and mediates the immanent action of God. Because a created entity reality really does give expression to divine action, it is appropriate to speak of *objective*, special divine action (Rahner 1978, 88). God's actions in creation, in the history of salvation and in our own lives can be understood as God acting through secondary causes. And these secondary causes have their own integrity and proper autonomy.

A fundamental principle of the God-world relationship, grounded in the tradition of Aquinas, and one that Rahner often repeats, is expressed in the axiom: Radical dependence on God and the genuine autonomy of the creature are directly and not inversely related (e.g., Rahner 1978, 78-9). The closer creatures are to God, the more they can be truly themselves. Creaturely integrity and autonomy are not diminished by this relationship of ongoing creation, but are enabled to flourish.

This view of divine action challenges not only creationism, but also recent attempts at the more sophisticated theory of "intelligent design." If one holds, with Aquinas and Rahner, that in creating God acts consistently through secondary causes, then God is not an alternative to what science can discover. Gaps in science are to be filled by science, not by invoking God. God is not an alternative to natural selection, but works through natural selection, through randomness and lawfulness, through all the processes of nature, which it is the role of science, not theology, to discover. This view of divine action can also respond to the scientific critics of religion who suppose that, if science can explain, for example, the origin of life on Earth, then this will do away with the idea of a Creator. What this theology supports is a healthy and creative relationship between science and theology. A theology that sees God as acting in and through secondary causes does not compete with science, but delights in the discoveries of science, seeing them as pointing to the way God creates. It sees God acting in and through the whole of creation, enabling the universe and all the entities that make it up to exist, to interact creatively, and to evolve.

5. *God creates by giving to creation itself the capacity for self-transcendence.*

The concept of divine self-bestowal describes the divine act from the side of God. Rahner proposes a second principle that describes this same act from the perspective of its impact on creatures: *creation has the capacity for self-transcendence.* Self-transcendence means that an entity is enabled to go beyond what it is to become something new. This idea is explored in Rahner's anthropology and evolutionary Christology, but it functions in many aspects of his work (Rahner 1965, 98–101).[1] He considers the transitions to the *new* in the history of the universe, particularly when matter becomes life, and when life becomes self-conscious spirit. He argues that the traditional view, that God confers existence and the capacity to act on all things, needs to be developed to take account of an evolutionary universe. Rahner proposes an evolutionary dynamism that is truly intrinsic to creation, but which occurs through the creative power of the immanent God. The idea of *self*-transcendence indicates that at the empirical level of science, the emergence of the new is completely open to explanation at the scientific level.

In this view, the Creator not only enables things to exist and act, but also enables them to become something radically new, as when life first appears in a lifeless universe. The immanent presence and "pressure" of the divine being enables creation to become more than it is in itself. Rahner proposes a large pattern of evolutionary self-transcendence, one that brings out the inner connection between evolution and christology. The material universe transcends itself in the emergence of life, and life transcends itself in the human. In human beings, the universe becomes open to self-consciousness and freedom, and to a fully personal response to God's self-bestowal in grace. The Christ-event is the radical self-transcendence of the created universe into God. As a creature Jesus in his life and death lives the radical response of love to God's self-bestowal. In his humanity he is, like us, part of the evolutionary history of life on Earth, and a product of long history of the

---

1   See also Rahner 1969a, 157-92; 1978, 178-203.

universe. Unlike us, he is wholly open to God. If the Christ-event is considered from below, it can be seen as the self-transcendence of the evolving universe into God. Jesus, in his life, death and resurrection, is the culmination of the process of evolutionary emergence, one that has not yet reached its final fulfillment. If considered from above, Jesus Christ can be seen as God's irreversible self-bestowal to creation. In this one person, we find the event of salvation: God's irreversible self-communication to creatures and full creaturely acceptance of this self-bestowal (Rahner 1978, 193).

This view of divine action seeks to situate Christology within an evolutionary view of the world. It puts a question to Christian theologies that ignore the scientific revolutions associated with the discovery of the expansion of the universe and the evolutionary history of life on Earth. It is a response, at least in part, to evolutionary thinkers who attack Christianity for what they see as opposition to evolution. What it offers is the beginning of an engagement between the evolutionary scientific worldview and the heart of Christian faith. Such an engagement will not only be essential for all those who inhabit this worldview and want to be committed to faith in Jesus Christ, but it will also be the necessary foundation for an ecological theology.

CONCLUSION

A scientifically informed ecological consciousness brings to the dialogue first of all a sense of global climate change as an urgent issue that theology needs to address. It brings five further hermeneutical principles: the worldview of big bang cosmology and evolutionary biology; the science of climate change and its provisional nature; the costs of evolution; the intrinsic value of nonhuman creation; and the interconnectedness of all things.

Karl Rahner's theology brings a theology of grace which enables us to see engagement with the issue of global climate change as the place of God. It, too, offers five further hermeneutical principles that can contribute to an ecological theology: creation and redemption as distinct dimensions of God's one act of self-bestowal; redemption as the deification of human beings and the whole creation; God's relationship to creation characterized by radical transcendence and radical immanence; God's action as noninterventionist action through secondary causes; God's creation as enabling creaturely emergence through self-transcendence.

As Michael Petty wrote, Rahner's theology is "profoundly ecological in that he brings to each theological problem and issue a fundamental vision which sees all reality, God, the world, and human being, as interrelated" (Petty 1996, 39). At the same time the ecological consciousness I have described challenges Rahner' theology at several points and invites further developments. In particular it raises questions about biology. While Rahner certainly takes matter seriously, he seldom discusses animals or the biological world in general. And he does not take us to the issue of nonhuman suffering. The costs of evolution, put before us by contemporary science, call for a renewed ecologically aware theology of divine action.

In this paper I have not attempted an ecological theology that responds to climate change, but simply to tease out some of the hermeneutic issues involved. Such a theology needs to be based, at least in part, on a creative dialogue between a scientifically informed ecological consciousness and a consciousness informed and shaped by the Christian tradition. This dialogue will need to be two-sided, so that interpretative ideas that spring from both sides can shape the theological outcome. What I think is suggested by this exercise is the need to hear from both sides of the dialogue that constitutes ecological theology, the desirability of allowing critical questions from one side raise challenges for the other, and the possibility that insights from both sides might contribute to a new and in part unforeseeable outcome, a truly ecological theology that responds to the great issues of our time.

SOURCES

Harvey D. Egan. 2005. "Theology and Spirituality." In *The Cambridge Companion to Karl Rahner*, edited by Declan Marmion and Mary E. Hines, 13-28. Cambridge: Cambridge University Press.

Fiorenza, Francis Schussler. 2005. "Method in Theology." In *The Cambridge Companion to Karl Rahner*, edited by Declan Marmion and Mary E. Hines, 65-82. Cambridge: Cambridge University Press.

Petty, Michael W. 1996. *A Faith that Loves the Earth: The Ecological Theology of Karl Rahner*. Lanham: University Press of America.

Rahner, Karl. 1965. *Hominisation: The Evolutionary Origin of Man as a Theological Problem*. Translated by W. T. O'Hara. London: Burns and Oates.

———. 1969a. "Christology within an Evolutionary View of the World." In *Theological Investigations* 5, trans. Karl-H. Kruger, 157-92. Baltimore: Helicon Press.

————. 1969b. "Reflections on the Unity of Love of Neighbour and the Love of God." In *Theological Investigations* 6, trans. Karl-H. and Boniface Kruger, 231-49. New York: Seabury.

————. 1971. "The Festival of the Future of the World." In *Theological Investigations* 7, trans. David Bourke, 181-85. London: Darton, Longman and Todd.

————. 1973. "Immanent and Transcendent Consummation of the World." In *Theological Investigations* 10, trans. David Bourke, 273-89. London: Darton, Longmann & Todd.

————. 1974b. "Christology in the Setting of Modern Man's Understanding of himself and of his World." In *Theological Investigations* 11, trans. David Bourke, 215-39. New York: Seabury Press.

————. 1974 a. "Dogmatic Questions on Easter." In *Theological Investigations* 4, trans. Kevin Smyth, 121-33. London: Darton, Longman and Todd.

————. 1975. "Resurrection: D. Theology." In *Encyclopedia of Theology: A Concise Sacramentum Mundi*, edited by Karl Rahner, 1442. London: Burns and Oates.

————. 1978. *Foundations of Christian Faith: An Introduction to the Idea of Christianity*. Translated by William Dych. New York: Seabury.

————. 1983. "Experience of the Holy Spirit." In *Theological Investigations* 18, trans. Edward Quinn, 189-210. New York: Crossroad.

————. 1988a. "Natural Science and Reasonable Faith." In *Theological Investigations* 21, trans. Hugh M. Riley, 16-55. New York: Crossroad.

————. 1988b. "The Specific Character of the Christian Concept of God." In *Theological Investigations* 21, trans. Hugh M. Riley, 185-95. New York: Crossroad.

————. 2005. "Experiences of a Catholic Theologian." In Declan Marmion and Mary E. Hines, *The Cambridge Companion to Karl Rahner*, 297-310. Cambridge: Cambridge University Press.

Rahner, Karl, and Karl-Heinz Weger. 1980. *Our Christian Faith*. London: Burns and Oats.

Rolston, Holmes III. 1999. *Genesis, Genes and God; Values and Their Origins in Natural and Human History*. Cambridge: Cambridge University Press.

CHANGING OUR MINDS

BERNARD LONERGAN & CLIMATE CHANGE

Richard M. Liddy

The story is told that Bernard Lonergan, one of the 20th century's outstanding theologians, was once asked to dine with the board of the Woodrow Wilson Institute in Washington, DC. After some polite chatter, someone asked, "Father Lonergan, if you were to found a research institute, what would its question be - its focus?" Without missing a beat, Lonergan replied, "Why do people change their minds?" Indeed, in all of Lonergan's many writings on theology, philosophy, methodology and economics, one could say that his fundamental interest was in tracking "why people change their minds."

In *Insight: A Study of Human Understanding*, his major philosophical work, Lonergan's interest was, first, in analyzing the change of mind involved in doing science; and secondly, in drawing out the correct implications of such scientific performance. Lonergan's conclusion was that an adequate analysis of science, far from justifying a positivist or empiricist vision of scientific understanding, was the most important contemporary evidence for a critical realism and the philosophical basis for a dignified view of the human person. Such critical realism is the most adequate basis for integrating all the sciences within a coherent view of the universe as well as raising the question of God and grounding moral and religious concerns. Of course, the key moment in appropriating such a critical realism was a change of mind regarding mind itself. As Lonergan cautioned: "Thoroughly understand what it is to understand, and not only will you understand the broad lines of all there is to be understood but also you will possess a fixed base, an invariant pattern, opening upon all further developments of understanding" (1992, 22).

This understanding of understanding, an event Lonergan elsewhere called "intellectual conversion," was a breakthrough from inadequate images of knowing, objectivity, and reality to a full appreciation of the world of theory within an adequate knowledge of our own minds (1996a, 238-40). It was a radical breakthrough from the "already out there now" image of reality so prevalent even among scientists and philosophers, to a full appreciation of the world of science and theory, and to the real world to be known—not through imagination, but through refined understanding and tested judgment (Liddy 2006).

Thus, although Lonergan largely began his analyses from the modern empirical sciences, and in all his writing fully appreciated the remarkable discoveries of the sciences, he never naively acquiesced in "what scientists say" about science. His critical realism, rooted not just in experience, but in penetrating understanding and tested judgment, was more developed and nuanced than that. In essence, he had a much wider view of humanity and history than "what science says." Furthermore, his *Method in Theology* and other writings provide a set of heuristic categories for analyzing humanity's history.

Lonergan did not address the issue of climate change directly, but his work does present a heuristic framework for analyzing the different sets of questions that arise in relation to this pressing issue. Some of these questions are empirical: What in fact does the destruction of the rain forests have to do with climate change? What about the depletion of the ozone layer? What has this to do with aerosol emissions and with modern processes of industrialization? All of these empirical questions are related to another set of questions about the particular configuration of human economic activity: What in fact is our understanding of the economic system that we are daily involved in? And, by what scale of values do we evaluate our present lifestyles and economic activities and how are we to evaluate future possibilities in the light of what we know both from climate change science and from other sources? What notion do we have of "who we are," the *humanum*, in relation to these questions? Do we have a notion? As he once wrote in relation to large-scale economic crises, they are not due to "the reason on which simple-minded moralists insist. They blame greed. But the prime cause is ignorance" (Lonergan 1992, 82).

Therefore, these sets of questions are related to each other. Some sets pertain to empirical facts about the natural world, including the destruction and dissolution of natural structures. Another set has to do with human economic choices and their effect on underlying natural processes. Beyond these sets of questions are additional ones about the very "light" within which we ask and answer them. Here philosophical questions arise: In light of what scale of values should we be addressing these issues? And ethical and normative questions: What ought we do? Additional questions are religious and theological: How are religious people to think about all these questions in the light of faith?

Lonergan was perhaps the most unabashedly "scientific" of modern theologians, yet he was basically a theologian. He did not believe the world went along the smooth rails of scientific progress. He believed there was the surd, the irrational, the unreasonable. Surveying human history, he would not have been surprised at clear evidence of anthropogenic climate change that is endangering large numbers of people and, at the same time, massive human resistance to either considering the evidence, listening to those who have considered the evidence, or, having considered the evidence, doing anything about it. He knew about the surd of sin—the basic lovelessness that threatens humanity, the original sin of the inability to achieve sustained development (Lonergan 1992, 750-70). At the same time, Lonergan the theologian emphasized "redemption." He was convinced that in order to understand human history in all its concreteness, one had to take into account three "vectors" of history: progress, decline, and redemption (1996b, 271-72). If not, one is only analyzing an abstraction. From this three-fold perspective, we will consider the changes of mind relevant to the issue of human-caused climate change.

To fully spell out progress, the first vector of human history, Lonergan gave a detailed analysis of the development of the modern sciences as a primary example of the self-transcending character of human consciousness as it moves from attentiveness to experience to refined understanding to critical judgment. He analyzes this process in the development of mathematics and the natural sciences in *Insight*. His analysis of the modern sciences resulted in his overall view of world process as emergent probability, a topic on which we will focus in our first section.

However, Lonergan also highlighted the need for intellectual conversion in order to understand science correctly and how the sciences are linked together in an overall view of human history. There are the biases against intelligence and its smooth functioning, and there are also concrete and massive moral failures. His *Insight, Method in Theology*, and other writings provide ample analyses of the historical vector of decline.

Finally, in his theological works, beginning with his doctoral dissertation, *Grace and Freedom in Aquinas*, and in his discourses on the Trinity and the Incarnate Word, there are ample analyses of the dynamics of redemption and conversion, the change of mind and heart brought about by the grace of Christ. Lonergan's contributions were, therefore, on the following three basic levels on which this essay will touch in relation to the issue of human-forced climate change:

1. *An analysis of contemporary scientific methods and the basic worldview of emergent probability that surfaces from this analysis.* Closely related to this world-view is an analysis of human consciousness as ethical, as deriving the "ought" of moral obligation from the "is," the facts of the situation. If it is true that human activities result in carbon emissions, deforestation, and other adverse effects that force changes in the global climate that are harmful to the natural environment and, thereby, to human survival in the world, the "ought" of appropriate public policy emerges;

2. *A more fundamental analysis of the human biases against recognition of the "is" revealed by intelligence and the "ought" revealed by ethical questioning.* This constitutes an analysis of the biases and aberrations affecting human intelligence and activity in the world; and

3. *An illustration of answering the question about God, about the absolute intelligence at the basis of human intelligence as well as the absolute good presupposed to ethical analysis that arises in human consciousness* (Lonergan, 1996a, 101-03). One way to formulate this question would be: What is God doing about the absence of intelligibility in the world, specifically the absurdities revealed in the climate change debate? In response to this absence or a decline in intelligence, the Christian assertion of the historical redemption in Christ and the gift of the Holy Spirit becomes relevant. How does transforming grace, a "falling in love with God," heal both ill will and distorted reasoning? Does grace involve a "leap to reason" and a willingness to do what people previously did not want to do—for example, adequately address the contemporary issue of human-forced climate change?

PROGRESS & EMERGENT PROBABILITY

One can begin by asking how many sciences go into analyzing humanly induced climate change: physics, chemistry, biology, psychology, sociology, economics, political science, and, most likely, others. As mentioned above, some of these are natural sciences about the empirical, ascertainable world, and some are human sciences—sciences that in one way or another deal with human understanding, judgment, and decision-making. Analysis prompts the following question: How can all of these sciences be linked together in such a way in order to reach an adequate understanding of the issues involved in climate changes that are human-induced?

An article in the journal *Science* lists nine crucial issues involved in climate change science: climate change itself; the rate of biodiversity loss (terrestrial and marine); interference with the nitrogen and phosphorous cycles; stratospheric ozone depletion; ocean acidification; global fresh water use; change in land use; chemical pollution; and, atmospheric aerosol loading (Rockström et al. 2009). The authors seek to identify acceptable boundaries which, if transgressed, threaten human and other biological life. Although the authors describe the planetary boundaries in terms of individual quantities and separate processes, the boundaries are tightly coupled. If one boundary is transgressed, other boundaries are also under serious risk. "For the first time, we are trying to quantify the safe limits outside of which the Earth system cannot continue to function in a stable, Holocene-like state" (474). How did the authors arrive at this conclusion? Lonergan would ask: What areas of scientific inquiry and what methods did the authors pursue in arriving at their conclusions? The answer:

> This approach rests on three branches of scientific enquiry. The first addresses the scale of human action in relation to the capacity of Earth to sustain it. This is a significant feature of the ecological economics research agenda, drawing on the knowledge of the essential role of the life-support properties of the environment for human wellbeing and the biophysical constraints for the growth of the economy. The second is the work on understanding essential Earth processes including human actions, brought together in the fields of global change research and sustainability science. The third field of enquiry is research into resilience and its links to complex dynamics and self-regulation of living systems, emphasizing thresholds and shifts between states (Rockström et al. 2009, 474-75).

One thing to note about these various methods is that some of them are natural sciences regarding Earth's processes and some of them are human sciences regarding the relation between human action and the life-support properties of the environment. Some sciences touch on both. For example, "ecological economics" would relate both natural and human sciences in a common goal of understanding. At this point, one can ask: What is the notion of "human" operative in this science? Is it a reductionistic notion whereby the human is reduced to the status of biological or mechanical? (Lonergan 1996b, 143). Can "the human," including the ability to do magnificent science, be conceived in another way? A fuller way? A way that is more comprehensive?

Finally, Rockström and his colleagues ask what level of probability should attach to their conclusions:

> Although we present evidence that these boundaries have been overstepped, there remain many gaps in our knowledge. We have tentatively quantified seven boundaries, but some of the figures are merely our first best guesses. Furthermore, because many of the boundaries are linked, exceeding one will have implications for others in ways that we do not as yet completely understand. There is also significant uncertainly over how long it takes to cause dangerous environmental change or to trigger feedbacks that drastically reduce the ability of the Earth system, or important subsystems, to return to safe levels (2009, 475).

This is the kind of question Lonergan would ask, so we shift now to Lonergan's over-arching philosophical viewpoint that involved an understanding of how all these scientific inquiries might be related. In *Insight* he begins his analysis of scientific consciousness with the analysis of mathematics: for example, as exemplified in the distinction between understanding a circle as "something round" in a descriptive or common sense way, and, on the other hand, understanding a circle in a theoretical way as "a locus of co-planar points equidistant from the center." Description relates things to ourselves and our own senses while explanation relates things to each other within a universal viewpoint: for example, the various chemical elements within the periodic table. Explanatory understanding demands the creation of a theoretical language to mark such a differentiation of consciousness.

This clear distinction between description and explanation is central to all of Lonergan's analysis of scientific method. In the natural sciences explanatory understanding is made possible by mathematics, but as one proceeds toward the human sciences, mathematics takes second place in relation to understanding human understanding and free decision-making (1992, 488).

Lonergan's analyses of empirical physical science involves an understanding of classical scientific laws as correlations verified in data. The correlation of the time squared with the distance travelled—"the law of falling bodies"—is a correlation that, "all things being equal," has been verified for the last four centuries. It is quite distinct from Aristotle's descriptive understanding of bodies falling according to some force within them—a *vis materiae insita* (1992, 47). Galileo made the transition from such imaginative understanding by abstracting from such visual images and limiting himself to understanding correlations actually attainable in the case of falling bodies (57-59).

However, a major point for Lonergan can be found in the phrase "all things being equal." For all things are not always equal. What has become abundantly clear in the history of science is that there is another method at work in scientific understanding, and that is statistical method. Classical laws hold under proper conditions and "other things being equal." Statistical method deals with the probabilities of how often other things are in fact equal (1992, 70-92).

This combination of classical and statistical methods yields an understanding of the universe as emergent probability. For example, Lonergan remarks that the focus of Darwin's evolutionary account was the gradual accumulation of small "sensible qualities," that is to say, observable and describable phenotypic characters (1992, 290). By way of contrast, the focus of his own account is what he calls "schemes of recurrence." The small variations of classical Darwinism do not merely pile up; rather, they must be understood in their intelligible relationships to the internal and external functioning of the organism and its environment: "[T]he concrete living of any plant or animal may be regarded as a set of ... recurrent operations.... Within such schemes [of recurrence] the plant or animal is only a component. The whole schematic circle of events does not occur [solely] within the living thing, but goes beyond it into the environment" (Lonergan 1992, 156).

By "scheme of recurrence" Lonergan means a series of events or operations that are intelligibly linked together by the natural laws of physics, chemistry, and biology. Schemes of recurrence can be represented by the series of conditionals: if A occurs, B occurs; if B occurs, C occurs; if C occurs, A will recur (1992, 141). The intelligible connection between the occurrence of A and B, between B and C, etc. is determined by some law of physics, chemistry, or biology. Simple examples of schemes of recurrence include the hydrogen-helium fusion cycles in the interiors of stars, the Krebs cycles in cells that continually regenerate energetic ATP from depleted ADP, and the mutual regeneration of atmospheric CO_2 and O_2 by animals and plants (Byrne 2003, 8).

Lonergan's analysis of classical and statistical scientific methods and their inter-relationships leads to his insistence on the "conditionality" of the laws of science. "To put the matter bluntly," Byrne asserts, "the laws of science in and of themselves determine nothing. It is only the laws plus specified conditions that determine concrete events" (2003, 9).

In other words, if all of the other appropriate conditions happen to be already fulfilled, then the occurrence of A will result in the occurrence of B and if B occurs, C will occur; if C occurs, A will recur. Lonergan goes on to note that schemes of recurrence are usually far more complex, involving intricate sub-loops and alternative pathways. Lonergan argues that science seeks a correct understanding of how events are intelligibly connected within schemes of recurrence. When new schemes begin to function, new intelligibilities emerge, and it is the task of science to correctly understand those newly emergent intelligibilities.

According to Lonergan's explanation of emergence, specific kinds of schemes of recurrence form the prior conditions for other subsequent schemes (1992, 145). For example, a biological species "is an intelligible solution to a problem of living in a given environment," "later species are solutions that ... rise upon previous solutions," and "a solution is the sort of thing that human insight hits upon" (1992, 290). Simply put, the intelligibility of strictly natural emergent probability is compatible with the termination and even extinction of particular schemes of recurrence.

Lonergan's way of situating the "conflict" of development within the larger context of evolutionary emergent probability raises a significant issue. On the one hand, emergence depends on the continued functioning of prior recurrent schemes that constitute ecosystems. If these are violently destroyed, both they and the subsequent emergent forms that depend upon them are lost. On the other hand, development cannot proceed without the transformation of prior schemes by later ones. In nature, emergent probability "respects" its underlying conditions, and yet it does not leave prior schemes untouched. Human development can profit by learning to respect this delicate interplay (Byrne 2003, 11).

Lonergan sets this understanding of development within his world view of emergent probability. If the point of metaphysics is to be able to talk in a general way about the relationships among the disciplines and the dimensions of the world they explore, the question of the finality of the universe surfaces. If the universe is intelligible, if it responds to our questioning as science presupposes, where is this development heading? If human questioning heads toward answers, where is the universe heading? "By finality we refer to a theorem of the same generality as the notion of being. This theorem affirms a parallelism between the dynamism of the mind and the dynamism of proportionate being. It affirms that the objective universe is not at rest, not static, not fixed in the present, but in process, in tension, fluid. As it regards present reality in its dynamic aspect, so it affirms this dynamism to be open" (Lonergan 1992, 470).

Lonergan works out various characteristics of finality as the dynamism of the real, but two characteristics relevant to the climate change debate are important. One characteristic of the directed dynamism of the universe is that it is realistic. Here again Lonergan's initial distinction between description of things related to ourselves and explanation of things related to one another is important:

> Men are apt to judge the universe by anthropomorphic standards. They look for the efficiency of their machines, the economy of their use of materials and power, the security of their comprehensive plans, the absence of disease and death, of violence and pain, of abuse and repression that reflects the desires and the aspirations of their hearts. But human utopias are paper schemes. They postulate in the universe more perfect materials than those with which it builds. They suppose that the building can be some extrinsic activity apart from the universe itself. They forget that they themselves and all their great achievements and all their still greater hopes and dreams are but byproducts of the universe in its proper expansion in accord with its proper intelligibility (Lonergan 1992, 473-474).

Furthermore, this finality of the universe is universal:

> It is no less the sadness of failure than the joy of success. It is to be discerned no less in false starts and in breakdowns than in stability

and progress. It is as much the meaning of aberration and corruption and decline as of sanity and honesty and development. For finality is an immanent intelligibility operating through the effective probability of possibility. Effective probability makes no pretense to provide an aseptic universe of chrome and plastic. Its trials will far outnumber its successes, but the trials are no less part of the program than the successes. Again, in human affairs, finality does not undertake to run the world along the lines of a kindergarten; it does undertake to enlighten men by allowing their actions to have their consequences, that by this cumulative heaping of evidence men may learn; and if one tribe or culture, one nation or civilization, does not learn, finality will not stoop to coaxing and pleading; it lets things take their course, that eventually tribes and nations, cultures and civilizations may reach the degree of intelligent and rational consciousness necessary to carry forward the task of finality in transcending limitations (Lonergan 1992, 474).

Does universal finality mean that the universe is ultimately cold and unfriendly to humanity, merely "nature red in tooth and claw" as held by some authors (Byrne 2010, 4-6; Dawkins 2006, 2)? Besides many inadequate common sense misunderstandings, might this ultimate assessment of the universe also be an extra-scientific opinion? We will return to this question in the last section of this essay that focuses on theology.

Emerging Schemes of Human Cooperation

Lonergan goes on to show the relevance of emergent probability in the realm of human affairs and human decision making. He points out that the human world is permeated by schemes of recurrence. As in the fields of physics, chemistry, and biology, there are schemes of recurrence in the fields of human events and relationships. The advent of humanity does not abrogate the rule of emergent probability. Human actions are recurrent. Their recurrence is regular, but their functioning is conditioned, not inevitable (1992, 234-35).

Human social and economic schemes operate, not with the blind laws of natural selection, Lonergan insists, but through the conscious, self-correcting activity of human inquiry and insight. Human schemes consist of intelligible patterns of relationships that "condition the fulfillment of each man's desires by his contributions to the desires of others" (1992, 239). For example, every commercial enterprise is a scheme of recurrence of human actions that involves recurrent transactions among suppliers, workers, buyers and recurrent patterns of payments that condition their functioning. Every family involves recurrent schemes that not only continually take sustenance from the economy and regularly disposes of waste products, but also regularly develops or declines according to its intelligent or unintelligent adaptation to the social and historical opportunities open to the family (Byrne 2003, 11-12).

What radically distinguishes human schemes of recurrence from natural ones is that their emergence and survival depend upon acts of human intelligence and choice. Human "practical intelligence devises arrangements for human living" (Lonergan 1992, 239). These arrangements are largely patterns of cooperation that depend upon understanding "what one can expect" of the other person (1992, 248). Human practical intelligence or "common sense" is the accumulation of innumerable such insights that make possible the participation in human economic, social and political institutions.

Human insights not only maintain the schemes of human living; they also constantly transform and bring about new schemes. Prior schemes "set problems calling for" insights into inventions, insights into how to organize and distribute the fruits of production in ever better ways, insights into how to reach group agreements and decisions ever more fairly and effectively (Lonergan 1992, 233-34). The ongoing development and emergence of human schemes follows what Lonergan calls a "self-correcting" process that consists of (1) existing human schemes that give rise to (2) questions about how to do things better that give rise to (3) insights for improvements that give rise to (4) actions that modify the schemes that give rise to (5) further questions and insights.

The processes of environmental science, advocacy, and the long painful revisions of public policy illustrate self-correcting human schemes of recurrence: "The creative task is to find answers. It is a matter of insight, not of one insight but of many, not of isolated insights but of insights that coalesce, that complement and correct one another, that influence policies and programs, that reveal their short-comings in their concrete results, that give rise to further correcting insights" (Lonergan 1985, 193).[1]

---

1   See also Lonergan 1985, 186 on the importance of feedback processes.

The Role of Belief

Scientists believe other scientists. Lonergan often emphasized the large role belief plays, not only in human life in general, but in science as well:

> Human knowledge results from a vast collaboration of many peoples over uncounted millennia. The necessary condition of that collaboration is belief. What any of us knows, only slightly results from personal experience, personal discovery, personally conducted verification; for the most part it results from believing. But the eighteenth-century Enlightenment was not content to attack religious belief. It prided itself on its philosophers. It set up a rationalist individualism that asked people to prove their assumptions or else regard them as arbitrary. In effect it was out to destroy not only the religious tradition but all tradition. Such rationalist individualism in the twentieth century seems to have infected our educationalists. Students are encouraged to find things out for themselves, to develop originality, to be creative, to criticize, but it does not seem that they are instructed in the enormous role of belief in the acquisition and the expansion of knowledge. Many do not seem to be aware that what they know of science is not immanently generated but for the most part simply belief (Lonergan 1996b, 185-186).[2]

Most of what forms the furniture of the scientific mind consists in believing what other scientists have concluded. For example, scientists do not begin their investigations by doubting the legitimacy of the various measuring instruments that they use in conducting their own experiments. If on occasion they are led to re-check a measurement or a means of calculation, the majority of their work consists in believing the conclusions of other scientists and working from there.

Nevertheless, control of belief through empirical testing characterizes the scientific process. Not only is there direct confirmation through the repetition of experiments; there is also indirect confirmation through the long and difficult process of peer review, judgment by others, and the gradual acceptance of conclusions. This process involves a great deal of believing others in the process of arriving at probably warranted assertions. This is a long social and communitarian process. We are within that process presently with regard to climate change science and concomitant public policy decisions. In other words, there is within the practices of the human community a major

---

2   See also Lonergan 1992, 725-40 on the structure of belief.

place for reasonably arrived at belief ultimately controlled by an appeal to empirical tests and the process of communal review. This is because the scientific judgments generally arrived at are not certain; they are only probable. As Lonergan would so often put it, scientific conclusions are "the best available opinion of the time" (1985, 43; 1996a, 315).

How, then, can we make reasonable and responsible decisions? How can we make public policy based only on the best available opinion at the time? John Henry Newman wrote about "a convergence of probabilities" asymptotically approaching certitude to justify action. In the history of his own intellectual development he writes about reaching the conclusions: "[P]robabilities which did not reach to logical certitude, might suffice for a mental certitude; that the certitude thus brought about might equal in measure and strength the certitude which was created by the strictest scientific demonstration; and that to possess such certitude might in given cases and to given individuals be a plain duty, though not to others in other circumstances" (Newman 1967, 31).

One piece of evidence by itself might not warrant a judgment of value, but there can be a convergence of evidence that allows arriving at an objective and genuinely probable judgment that justifies a decision. Thus, from the depletion of the forests, the rise in water temperature, ozone concentration, and other evidence, there emerges a moral necessity for responsible decision-making and appropriate public policy regarding climate change.

Climate change science has been ongoing for many years, and there is a history of this decision-making process. Climate science has prompted important groups of scientists to advocate public policy changes with regard to human cultural and social practices. The cultural changes include the prioritization of human values. Societal changes include the exercise of political preferences, economic expenditures, the uses of technology, and other actions. Obviously, this whole process has been contested. Climate change deniers have declared that the evidence has been fudged, skewered on behalf of vested interests, for example, governmentally or internationally funded agencies whose very existence depends on the articulation of the problem in a particular way. These critics have succeeded in sowing doubt in the public mind resulting in a lowering of public support for policies aimed at addressing humanly induced climate change issues

("Scientists Becoming Active in Defending Climate Data" 2010). In response, a group of 250 scientists published a letter ("Open Letter" 2010) claiming that the basic science behind the assertion of human-induced climate change is valid. Some have pointed out that the vested interests behind the assertions of climate change deniers (Brin 2010, 13-17).

That is the contentious situation today, and Bernard Lonergan's psychological, philosophical and theological analyses can be very helpful in clarifying the issues. They provide a template for understanding how human development and education is constituted by two dynamics. One is a development from above constituted by the influence of people and groups we trust: our parents and teachers, even the scientific "experts" and communities of experts. There is also a development from below constituted by our own experience, understanding, judgment, and decision by which we come to personally appropriate and critically control our own beliefs. Ideally, these two developments proceed concurrently, and the communities that have a beneficial influence on us can bring us to use our own heads and come to our own measured judgments. Nevertheless, bias, prejudice, and skewered judgments can occur, and we look at Lonergan's analysis of these factors in our next section.

DECLINE: BIAS AND SELF-DESTRUCTION

Lonergan notes in *Insight* that fully intelligent and ethical choices "cannot consistently" undertake initiatives that destroy their underlying conditions, including natural ecological conditions (1992, 629). Bias and self-destruction can become key issues. If the human mind is made for 'being," for the answers to our questions, it is also obvious that "learning makes a bloody entrance." For example, besides psychological blind-spots, there is the individual bias of selfishness: "[P]rior to the criteria of truth invented by philosophers, there is the dynamic criterion of the further question immanent in intelligence itself. The egoist's uneasy conscience is his awareness of his sin against the light. Operative in him, there is the Eros of the mind, the desire and drive to understand; he knows its value, for he gives it free rein where his own interests are concerned; yet he also repudiates its mastery, for he will not grant serious consideration to its further relevant questions" (247).

Prominent also in the effort to arrive at decisions is the group bias of prejudice and tribalism that skewers the operations of intelligence in the social and political arena. Bias can be seen in phrases such as "my country right or wrong, but my country" and "my party right or wrong, but my party." These biases may be supported by normal intersubjective feelings, while individual bias has to overcome these feelings. Group-feelings certainly characterize progressives' bonding together to change the status quo and conservatives' wanting to protect vested interests. As Lonergan states: "Just as the individual egoist puts further questions up to a point, but desists before reaching conclusions incompatible with his egoism, so also the group is prone to have a blind spot for the insights that reveal its well-being to be excessive or its usefulness at an end" (1992, 248).

Group bias is responsible for the historical conflicts between the privileged and the oppressed, the haves and the have-nots, that distort social progress. "The sins of group bias may be secret and almost unconscious. But what originally was a neglected possibility, in time becomes a grotesquely distorted reality" (Lonergan 1992, 250). Through group conflict the have-nots sometime replace the haves in the halls of power in what Lonergan calls "the shorter cycle" of human history. Eventually, however, the dialectic, the back-and forth between vested interests, brings to light the fact that all groups are subject to a bias against intelligence as such. This is what Lonergan calls the "general bias" responsible for the longer cycle of human decline. Such a general bias opposes the priority of intelligence as such: the practical conviction that we can get somewhere by understanding, that understanding is what we are made for. Ordinary human common sense does not know this. "Common sense knows, but it does not know what it knows nor how it knows nor how to correct and complement its own inadequacies. Only the blind and destructive blows inevitable in even a partial breakdown of social order can impress on practical common sense that there are limits to its competence and that, if it would master the new situation, it must first consent to learn" (ibid., 241).

The general bias against the priority of intelligence manifests itself in common sense rejections of intellectual and scientific contributions to solving human problems. Inversely, general bias can manifest itself in a dedication to purely technological solutions to human problems.

Applied science and consequent inventions have given us our vast industrial, commercial, financial, adminstrative, educational, military complex. Technicians are the people with the task of figuring out the most efficient use of currently available hardware. The more successful they are, the greater is the domain that they organize, and the less the domain under the control of old-style decision-makers, of managers, directors, mayors, governors, presidents. Again, the more brilliant they are, the less is it possible to explain to the uninitiated why things are done the way in which they are done. Finally, the more thorough the application of the principle of efficiency, the more must men adapt themselves to its dictates in all their labor hours and in all the goods and services they purchase from the technological establishment. Yet we must bear in mind that anything less than the most efficient procedures threatens the survival of the mass of mankind (Lonergan 1996b, 186).

The general bias against the priority of intelligence also surfaces in inadequate philosophies and symbolically expressed world-views. Among examples are materialist, empiricist, reductionist, and relativist accounts of the sciences that either implicitly or explicitly negate the priority of human intelligence. Other examples are symbolic world-views expressed in popular culture that would negate the possibility of attaining to convictions on the nature of the human person. As Lonergan noted when writing on modern distortions of human knowledge: "A second distortion occurs in man's apprehension of man. Positivists, naturalists, behaviorists insist that human sciences have to be conducted on the same lines as the natural sciences. But the resultant apprehension of man, if not mechanistic, is theriomorphic. Nor is this view of man as a machine or as an animal confined to some rarefied academic realm. It is applied. The applications reach out into all departments of thought and into all walks of life. They have the common feature of omitting advertence to human dignity and respect for human morality" (ibid.).

Is this kind of general bias present in the climate-change controversy? Because the questions might not seem "immediately practical," might they not be particularly important to ask nonetheless? Are we truly seeking to understand the valid points that "the other side"— our adversaries—are seeking to make? This kind of real encounter with the other challenges us to the core: "The needed higher viewpoint is the discovery, the logical expansion and the recognition of the principle that intelligence contains its own immanent norms and that those norms are equipped with sanctions that man does not have to invent or impose" (Lonergan 1992, 259).

Unlike natural ecologies, innovations in human social and economic arrangements are all too frequently implemented without the fullness of intelligent self-correction (Byrne 2003, 13-15). Real self-correction can occur only when the full complement of further pertinent questions and problems are taken into account and answered with creative solutions. According to Lonergan, biased courses of action that evade intelligent self-correction initiate the downward spirals of decline, degradation, and destruction not only in the natural world, but also in the cultural environment itself. "Decline disrupts a culture with conflicting ideologies. It inflicts on individuals the social, economic, and psychological pressures that for human frailty amount to determinism. It multiplies and heaps up the abuses and absurdities that breed resentment, hatred, anger, violence" (Lonergan 1996, 117).

Certainly, the massive industrialization characterizing modernity was not accompanied by sufficient reflection on industrialization's perverse effects on nature and humanity within nature. And, when perverse consequences became clear to many, others refused to listen. In the face of significantly more scientific evidence, is there still an unwillingness to face the facts—an unwillingness to listen? An unwillingness to grow in understanding and/or an unwillingness to act in the light of what is with high probability the case? Is this kind of moral failure attributable to a basic lack of love? A basic lovelessness? Lonergan is in fundamental agreement with St. Augustine's characterization of evil as nothing but the removal of good until finally no good remains (Byrne 2003, 14).

Opposite religious love is a hostility toward and hatred of nature embedded in the seminal works of some founders of modernity like Machiavelli and Bacon. There is also misanthropic hatred to be found in certain strains of environmental activism (Byrne 2003, 15).

How can this iciness be broken? For Lonergan the theologian, there is need for healing and creating love.

REDEMPTION: HEALING & CREATING

For Lonergan, redemption consists in a falling in love with God that opens the eyes of the heart to God and to what God wants to do in the world. Redemption also gives the strength to collaborate with others in the transformation of the world. As a Christian theologian, he affirmed that the reversal of sin and its devastating social consequences is accomplished by grace, that is, a falling in love with God that changes our hearts of stone into hearts of flesh and, besides elevating us to friendship with God, helps us to do what previously we did not at all want to do. According to Byrne, "religious love is love of God, and to love God unconditionally is to love everything God loves—all of natural and human creation. Grace is religious love, and it sets about undoing hatred. Loving grace makes possible healing, discerning, and creatively intelligent responses to degenerating situations" (Byrne 2003, 15).

Reductionistic forms of Darwinism tend to eliminate theological considerations of grace as superfluous and even as dangerous distractions from urgent mundane matters. Contrarily, Lonergan integrates God's grace into the evolving character of the world and shows the relevance of grace to the challenges of human existence. The Christian doctrine of the Word of God finds insertion into the understanding of the world marked by science and the human world challenged by meaninglessness and disintegration.

Early in his life, Lonergan found a great deal of inspiration in Saint Paul's letter to the Ephesians where he writes of the reconciliation of all things in Christ: "Such is the richness of the grace which he has showered on us in all wisdom and insight. He has let us know the mystery of his purpose, the hidden plan he so kindly made in Christ from the beginning to act upon when the times had run their course to the end: that he would bring everything together under Christ, as head, everything in the heavens and everything on earth" (Ephesians 1:8-10).

This seemed to be the ultimate thrust of Lonergan's *Insight*: to link all the natural and human sciences into a vision of the human person and humanity needing to be redeemed by an incarnate savior who can truly "change our minds" and our hearts. Certainly Lonergan would be against any "interventionist" view of God's relationship with the world (Edwards 2006). At the same time, in line with Aquinas' understanding, Lonergan held that God controls each event in the universe by divine action on the whole. As early as his 1940 dissertation, Lonergan's insistence on the importance of moving beyond "picture thinking" to explanatory thinking influenced his conception of the divine-human relationship: "The first fallacy lies in a misconception of time.

To a temporal being our four-dimensional universe has three sections: past, present and future. To an eternal "now" this division is meaningless. On this point St. Thomas never had the slightest doubt: he was always above our pre-Einsteinian illusions that still are maintained by our cosmology manuals; strenuously and consistently he maintained that all events are present to God" (Lonergan 2000, 105).

In a footnote, Lonergan added: "[B]efore time" is "an illusory figment of the imagination" (2000, 105). The basic issue is as always the issue of picture thinking. We accept the assertions of faith—the full humanity of Christ, the action of Christ in the Church and the sacraments, the life of grace, and our understanding of prayer—within a notion of God in which there is fullness of act. All the assertions of the Creed are to be accepted in faith even though it is only "through a glass darkly" that we can get a glimmer of how it is that these are so. In this process modern science, whose seemingly contrasting assertions also baffle the mind, can give us something of an analogy. Speaking of Athanasius' quasi-theoretical rule for Trinitarian theology that "whatever is said of the Father can also be said of the Son, except that the Father is the Father and the Son is the Son," Lonergan notes: "[T]he terminal notion of consubstantiality not only transcends all imagery but in some way also transcends all intelligibility grasped in an image. For just as Maxwell's equations for the electro-magnetic field were derived from images yet without any image corresponding to them, so also the rule laid down by Athanasius refers only to concepts and judgments. For the same things are said of the Son as are said of the Father, except that the Son is not the Father. Not only does this rule prescind from images but it also cannot be grasped or understood in anything imaginable" (2009, 197-99).

Still, this whole increasingly critical process of purifying, not only our conceptions of God's actions in time but also our understanding of the natural world in which we live, fits into a perspective of divine love and faith as "the eye" of that love. Faith not only relates us to God but it also transforms and magnifies and glorifies our understanding of the natural world as well.

> Without faith the originating value is man and the terminal value is the human good man brings about. But in the light of faith, originating value is divine light and love, while terminal value is the whole universe. So the human good becomes absorbed in an all-encompassing good. Where before an account of the human good

related men to one another and to nature, now human concern reaches beyond man's world to God and to God's world. Men meet not only to be together and to settle human affairs but also to worship. Human development is not only in skills and virtues but also in holiness. The power of God's love brings forth a new energy and efficacy in all goodness, and the limit of human expectation ceases to be the grave (Lonergan 1996a, 115).

Such power coming from faith and love is needed to address the present challenge of humanly rooted climate change. Such faith sees God's glory reflected in a humanity that is fully alive and striving to make our world a better place for ourselves and for our children.

To say that God created the world for his glory is to say that he created it not for his own sake but for ours. He made us in his image, for our authenticity consists in being like him, in self-transcending, in being origins of value, in true love (ibid.).

Faith as the eye of religious love sees the hand of God beneath the struggles of human existence and without such faith we can become overwhelmed at the magnitude of the challenge.

Without faith, without the eye of love, the world is too evil for God to be good, for a good God to exist. But faith recognizes that God grants men their freedom, that he wills them to be persons and not just his automata, that he calls them to the higher authenticity that overcomes evil with good. So faith is linked with human progress and it has to meet the challenge of human decline. For faith and progress have a common root in man's cognitional and moral self-transcendence. To promote either is to promote the other indirectly. (ibid.)

In this light of faith, can we can say, in spite of the apparently terrifying perspectives opened up by scientific understanding, that we live in "a friendly universe" (Byrne 2010; Dawkins, 2)? The Christian tradition would answer "yes."

Faith places human efforts in a friendly universe. Faith reveals an ultimate significance in human achievement; it strengthens new undertakings with confidence (ibid., 117).

Faith even sets technological and scientific progress in a positive light for such progress realizes human and natural potentialities. Faith reveals that the human person exists to bring about an ever fuller achievement in this world; and that achievement because it is for the good of the human person, is also for God's glory.

> It is not propaganda and it is not argument but religious faith that will liberate human reasonableness from its ideological prisons. It is not the promises of men but religious hope that can enable men to resist the vast pressures of social decay. If passions are to quiet down, if wrongs are to be not exacerbated, not ignored, not merely palliated, but acknowledged and removed, then human possessiveness and human pride have to be replaced by religious charity, by the charity of the suffering servant, by self-sacrificing love" (ibid.).

This basic "change of mind" involves repentance and life-long conversion; for humans are sinners, from Lonergan's perspective: "If human progress is not to be ever distorted and destroyed by the inattention, oversights, irrationality, irresponsibility of decline, men have to be reminded of their sinfulness. They have to acknowledge their real guilt and amend their ways. They have to learn with humility that religious development is dialectical, that the task of repentance and conversion is life-long" (ibid., 117-18).

Lonergan had no illusion that philosophy could save the world. A correct philosophy, precisely because it was correct, would seem "out of it" to the great run of people. Still, the community of Christ is called to preach to the world the Good News and to do so today in a way that fits into the scientific and historical consciousness of the contemporary world. What is needed, Lonergan felt, is a Christian humanism that affirms the inter-relationships of all things and the dignity of the human person. "So a contemporary humanism is dynamic. It holds forth not an ideal of fixity but a program of change. It was or is the automatic progress of the liberal, the dialectical materialism of the Marxist, the identification of cosmogenesis and christogenesis by Pierre Teilhard de Chardin" (ibid., 93).

As I read Lonergan's works, there is often in the back of my mind the dramatic vision of the universe described by Teilhard de Chardin as a vast upwardly directed dynamism stretching from the depths of matter to the human person, human community, the *noospshere*, and ultimately the "Omega Point" of divine love (Haughey 2009, 108-113). Lonergan's approach was less poetic, but it perfectly complements Teilhard's. Taking the sciences on their own terms as instances of human explanatory understanding, Lonergan wove them together into a world vision of emergent probability in which the human person plays a central role in creating new emergent technological, economic, and political schemes in continuity with, not destructive of, the natural schemes of the universe. Both Lonergan and Chardin spoke to a human need for a vision of history that could both enlighten the minds of people and motivate their hearts to action.

Concretely, such initiatives will involve great processes of collaboration and feedback. Lonergan once spoke to Jesuits about facing the gains of modernity in the natural sciences, and he urged them to work out strategies for dealing with secularist views on religion as well as with concomitant distortions in the very notions of the human. Such strategies will involve creative projects emerging from "a thorough understanding of a situation and a grasp of just what can be done about it" (ibid., 184). Such projects will not be a static proposal, but rather an ongoing project constantly revised in the light of the feedback from its implementation. Finally, for Lonergan: "[I]t will not be some single, ongoing project but a set of them, constantly reported to some central clearinghouse with the twofold function (1) of drawing attention to conflicts between separate parts and (2) of keeping all parts informed both of what has been achieved elsewhere and what has been tried and found wanting. Finally, all such projects must be in Christ Jesus, the work of those who take up their cross daily, who live by the Spirit in the Word, who consecrate themselves to loving, who banish all tendencies to hatred, reviling, destroying" (ibid., 187).

CONCLUSION

The Christian narrative is one of life, death and resurrection: progress, decline and redemption. Bernard Lonergan's analyses of human consciousness, primarily scientific consciousness, opens the way for understanding the structures of human progress as well as the intellectual and ethical biases at the core of the decline uncovered in the climate change debates. May Lonergan's analyses also shed light on the eternal promise of redemption as we work together to contribute to creating a better world. May the essays in this volume effect a change of mind regarding climate change and contribute a framework, not only for personal transformation, but also for authentic public policy.

SOURCES

Brin, David. 2010. "Climate Skeptics v. Climate Deniers." *Skeptic* 15.4: 13-17.

"British Panel Says Climate Center Did Not Distort Data." 2010. *New York Times*, March 31, A11.

Byrne, Patrick. 2003. "Ecology, Economy and Redemption as Dynamic: The Contributions of Jane Jacobs and Bernard Lonergan." *Worldviews: Environment, Culture, Religion* 7.1-2: 5-26.

———. 2010. "Intelligibility and Natural Science: Alienation or Friendship with the Universe?" Lonergan Workshop, Boston College, June 21-25. Privately distributed.

"Climate Fears Turn to Doubts Among Britons." 2010. *New York Times*, May 25, A1 + 8.

Edwards, Denis. 2006. "Resurrection and the Costs of Evolution: A Dialogue With Rahner on Noninterventionist Theology." *Theological Studies* 67: 816-33.

Dawkins, Richard. 2006. *The Selfish Gene*. New York: Oxford University Press. [30th anniversary edition]

Haughey, John C. 2009. *Where is Knowing Going? The Horizons of the Knowing Subject*. Washington, DC: Georgetown University Press.

The Jerusalem Bible. 1966. Garden City, NY: Doubleday.

Liddy, Richard. 2006. *Startling Strangeness: Reading Lonergan's Insight*. Lanham, MD: University Press of America.

Lonergan, Bernard J. F. 1985. *A Third Collection: Papers by Bernard Lonergan, S.J.* Edited by Frederick E. Crowe S.J. New York: Paulist Press.

———. 1992. *Insight: A Study of Human Understanding*. Edited by Frederick E. Crowe and Robert M. Doran. *Collected Works of Bernard Lonergan*, vol. 3. Toronto: University of Toronto Press.

———. 1996a. Reprint. Toronto: University of Toronto Press. *Method in Theology*. Original edition, New York: Herder and Herder, 1974.

———. 1996b. Reprint. Toronto: University of Toronto Press. *A Second Collection*. Original edition, London: Darton, Longman & Todd.

———. 2000. *Grace and Freedom: Operative Grace in the Thought of Saint Thomas Aquinas*. Edited by Frederick E. Crowe and Robert M. Doran. *Collected Works of Bernard Lonergan*, vol. 1. Toronto: University of Toronto Press.

———. 2009. *The Triune God: Doctrines*. Edited by Robert M. Doran and H. Daniel Mansour. Translated by Michael G. Shield. *Collected Works of Bernard Lonergan*, vol. 11. Toronto: University of Toronto Press.

Newman, John Henry. 1967. *Apologia pro vita sua: Being a History of His Religious Opinions*. Edited by Martin J. Svaglic. Oxford: Clarendon Press.

"Open Letter: Climate Change and the Integrity of Science." 2010. *Manchester Guardian*, May 6. Accessed from http://www.guardian.co.uk/environment/2010/may/06/climate-science-open-letter/print on November 5.

Rockström, Johan et al. 2009. "A Safe Operating Space for Humanity." *Nature* 461, September 24, 472-75.

"Scientists Becoming Active in Defending Climate Data." 2010. *New York Times*, March 3, A11.

HANS URS VON BALTHASAR &
DEEP ECOLOGY

TOWARDS A DOXOLOGICAL ECOLOGY

Anthony C. Sciglitano

To join the Swiss Catholic theologian Hans Urs von Balthasar with the twenty-first century issue of human-caused climate change is not a simple task. Balthasar would have little to offer, at least *qua* theologian, on such policy issues as the latest IPCC report, the Kyoto accords, or the relative merits of public and private approaches to reducing our carbon footprint. Thus ecological or environmental reformism addressing issues of legislation would not be a place for a meaningful dialogue here even if one judges statutory action to be our most pressing need in an attempt to alter the current direction of climate change. This does not mean that Balthasar would oppose international legislation on these issues. He was not a Christian anarchist, and he thought that there is a role for Christians to play in the structures of worldly governance. By working in these structures, Christians can participate in God's ongoing work to align the world with God's will and wisdom.

Yet within the realm of environmental philosophy, there is for Balthasar an entirely suitable dialogue partner: Deep (radical) Ecology (DE). For those familiar with Balthasar, the convergences between DE and his work will be sufficiently substantial to justify dialogue here. Balthasar also engages a genealogical critique of Western instrumental reason and anthropocentrism, recommends contemplative, meditative thinking along with poetic discourse, and thinks that the encounter with reality ought to inspire responses as wonder, reverence, and joy. Additional similarities will emerge as this essay continues. Yet Balthasar does not merely mirror what we find in DE. He challenges DE's antipathy to divine transcendence, low esteem for human moral responsibility, and aspects of DE's genealogy of Western intellectual culture. In addition to pointing out these challenges, I argue that Balthasar offers a theological contribution to a serious difficulty within ecological thought generally and DE in particular. More specifically, I argue that what I am calling Balthasar's *doxological ecology* helps address the critical issue of identity and difference that Val Plumwood and other DE proponents have observed in the DE movement while also helping to strengthen DE's opposition to anthropocentrism without thereby eliminating humanity's special moral responsibilities to the non-human world. This argument neither resolves the current human-caused climate change issue nor does it offer a particular proposal for reducing our carbon footprint. What my application of Balthasar's thinking does accomplish in relation to the climate crisis is support for a communal practice rooted in contemplation that respects the natural world in all its diversity as having both an intrinsic value and a need for human care. Like DE proponents, Balthasar offers a different way of perceiving and thus acting towards the natural world.

My first task in this argument is to develop some of the rudiments of the DE view: an overview of DE proponents' basic critique of instrumental reason, atomistic anthropology, and anthropocentrism; several observations about the sources for Deep Ecology; a brief description of DE's positive proposal; and some criticisms that have been leveled at DE especially by ecofeminists Val Plumwood and Karen Warren. In the second section, I identify some of the resources that Balthasar thinks are available in the Western intellectual tradition's struggles to grasp the intrinsic value of the natural world. Section three discusses Balthasar's theologically informed ontology found in the third part of his trilogy. The next section provides a brief Christological and Trinitarian specification of Balthasar's general ontology in view of the relation between divine transcendence and immanence to the world and his retrieval of Eastern divinization motifs. In my concluding remarks, I draw implications from Balthasar's views for action aimed at mitigating the effects of human-caused climate change.

DEEP ECOLOGY (DE): IDENTITY & DIFFERENCE

The Deep Ecology movement joins a critique of Western anthropo-centric and atomistic anthropology to a positive philosophical pro-gram that advances a radically egalitarian and relational ontology (Sessions and DeVall 1985, 67).[1] DE founder Arne Naess and pro-ponent Warwick Fox label their ecological view "deep" as opposed to "shallow" ecology, in part because DE proponents ask fundamental philosophical questions precisely whereas practitioners of ecological science often decline to do so. As Naess, the founder of DE, explains, "we ask why and how, where others do not. For instance, ecology as a science does not ask what kind of a society would be the best for maintaining a particular ecosystem—that is considered a question for value theory, for politics, for ethics" (ibid. 65, 74). DE "goes beyond the so-called factual scientific level to the level of self and Earth wisdom" and "beyond a limited piecemeal shallow approach to environmental problems and attempts to articulate a comprehensive religious and philosophical worldview" (ibid.). From the DE perspective, "shallow" ecology is the ecological movement that looks to the very causes of our current predicament, namely, anthropocentrically-oriented modern science and its concomitant political processes, tends toward utilitar-ian calculation, and uses instrumental short-term reasoning as the so-lution to environmental problems. For DE proponents, shallow ecol-ogy is akin to hoping Dr. Frankenstein will do a better job next time.

We have already noticed that the DE platform contends for a radi-cally egalitarian and relational view of the nature-human relationship. For Naess these fundamental goods and normative criteria are ar-ticulated in his notions of "self-realization" and "identification" (1989, 8-11; Sessions and DeVall 1985, 70-73). The self is essentially rela-tional and can never be grasped outside of its relation to all other enti-ties. Self-realization comes in and through identifying in some way with other beings, usually starting from intimate human relations and expanding outward to others. In other words, the self deepens as it re-alizes its own inherent possibilities for relation, not by dominating or standing apart from humans and other beings. Such domination and alienation would destroy potentials for relationship and thus hinder

---

1   Rachel Carson and Aldo Leopold are renowned catalysts for the Deep Ecology movement, while Naess is considered the founder of the move-ment itself.

self-realization. Moreover, the self in question here is not the atomistic self of Lockean political theory who putatively chooses to be in relation with others, to be in society as it were, but rather the inherently relational self who transcends towards a greater and truer Self only by way of relations with others (Naess 1989, 8-10). Important for our purposes is that Naess works with the notion of form (*gestalt*) to describe the inherent relationality of all beings. In a *gestalt*, no being stands apart from the total form; instead, all parts contribute to the form and all parts of the form depend upon one another and upon the whole for their significance. Naess uses a number of examples, including the parts of a musical score and a triangle of dots, to explain this notion (1989, 57-63; Diehm 2002, 27). If we take a triangle of dots and add a dot to the three and the meaning of each other dot changes for they are no longer parts of a triangle. Remove one and the same point holds. Notice that not only the dot that is added or subtracted undergoes change; the other dots do as well. So, change any one of the dots and the whole changes; the whole changes and the individual dots take on different meaning. His use of *gestalt* helps make clear that self-realization can never be an isolated affair, and that all elements or parts of the total form have their own value in relation to all the other parts. Humans do not create value by themselves, and non-humans are not valuable solely because they serve human use. What is important to notice on Naess' account is that the human is not separate from the total *gestalt* of the natural world; humans are a part of the world. The potential of nature deeply conditions the potentials inherent to human existence and vice-versa. For Naess, all beings gain intrinsic value by participating in the total *gestalt*. Equally important, and certainly central to Balthasar's discussion of aesthetic form, is that *gestalt* thinking challenges the fact-value divide as a form of abstraction that, while it has its uses, also fails to describe the world as experienced and living (Naess 1989, 60). It is in this very divide that allows the ecological scientist to forego the types of question DE advocates prioritize.

Karen Warren, Val Plumwood, and other critics have noted the tendency of Naess and Fox's notion of identification to stress sameness and to elide difference (Plumwood 2002, 197-98). Christian Diehm has taken up this and related criticisms and has given a nuanced account of their different applications to deep ecology thinkers. Warwick Fox in particular seems vulnerable to this charge when he prioritizes cosmological and ontological identification over against the particularities of concrete personal relationships. While Fox cannot be accused of excluding what he calls the "felt nearness" of human beings from his account of human development, it is nevertheless true, as Plumwood, Warner, and Diehm observe, that the realm of concrete, local, human relation takes a secondary place to the universal and impersonal (or at least "trans-personal") (see Diehm 2002). Diehm puts up a reasonable defense of Naess on this same count, but he thinks that, in the end, Naess' position needs supplementation in the direction of human difference, that is to say, responsibility or responsiveness, to the natural world (Diehm 2002, 33). Unlike Fox, Naess conceives of identification as moving from intimate human relations of such quality that they allow for deep sympathy with wider and wider circles of being. For Naess, personal attachments play a more fundamental role and, thus, he is less vulnerable to Plumwood's critique than Fox, at least on this point. Yet Plumwood still pushes her critique further and observes that Naess' notion of personal or trans-personal identification seems to assume a kind of ever-expanding self, whether or not this self grows to subsume all reality or whether it is simply realizing its inherent possibilities for identifying types of relation. In this sense, it does not seem to her that Naess provides a clear road away from anthropocentrism, and, more importantly, it is not clear to her how this marginalization of difference supports human responsibility.[2]

From an opposite angle, the DE writer Peter Reed claims that difference is more fundamental than sameness or identification whether specified as ontological or as interest related. Against his friend, Arne Naess, he makes the point that even an interest related difference will involve someone identifying with another's interest in one direction, whether it is the human identifying with the interest of an endangered animal/species or whether it is the human demanding that the animal identify with their interests, which will more often be the case

2 Working from Warren's and Plumwood's critiques of deep ecology, Diehm notices that Naess does not want to claim that the self is identical to nature as to an aspect of one's ontological self, but rather that there can be an identity of "interests" among the human and non-human. Naess appears to believe that this removes the danger inherent to "identification" of eliding the real differences between the realm of the human and that of others. Plumwood disagrees and, in her sardonic manner, observes that while one can "identify" with the need for food shared by both humans and other creatures, it would be hard to imagine identification on issues of grass selection for, say, the diet of the wombat (Plumwood 2002, 200).

(Plumwood 2002, 199-200). Using Martin Buber's "I-Thou" formulation, Reed begins from a call to recognize and stand in awe of nature as what calls to us and expresses its radical otherness. Reed, however, runs too far toward difference and thus undercuts what Plumwood would view as the importance of solidarity and human agency's ability to contend for the eco-system beyond an appreciation of difference. On her account, Reed's position enacts a kind of "hyper-separation" between human beings and nature while Naess's position fails to acknowledge real difference (Plumwood 2002, 198-99). Transposed into the theological realm, the issue here may be stated in terms of the classical Christological heresies of monophysitism and Nestorianism. Ecological monophysites are incapable of giving distinctness its full import. Ecological Nestorians, on the other hand, fail to adequately account for likeness, unity and thus solidarity in Plumwood's sense. They overplay distance and difference to such an extent that it is not clear any kind of love or care involving human moral agency has a genuine ground. I will argue below that Balthasar's doxological ecology offers a key contribution to the issues of identity and difference. But first we need to develop two more points relevant to this discussion.

In seeking a kind of religious and philosophical wisdom upon which to build their deep ecology, DE writers try to locate some of the more problematic positions in the Western theological and philosophical tradition. We cannot speak of a full-scale genealogy of Western environmental misbehavior here, but several elements of their critique do rise to the top through criticisms they make and through the sources they choose for their own thought. At least for Michael Zimmerman, Martin Heidegger's critique of Western metaphysics has been important. This critique implicates the Platonic and medieval Christian tradition in a reifying conception of Being as *logo-centric*, that is, as known in terms of things that are wholly transparent to human knowledge, use, manipulation, and consumption (Sessions and DeVall 1985, 99-100; Zimmerman 1993, 195-196). On Heidegger's account, Western civilization goes from a primordial Greek experience of Being as emergence (*physis*) that inspires wonder and gratitude to this reification that supports the later Enlightenment and modern exploitative industrial and technological craze (Zimmerman 1990, 173-175). The modern period exacerbates the problem by separating out the knowing subject from the material world viewed as an object exploitable for human benefit. We will see that Balthasar agrees with this assessment of modern philosophical predilections and views the exploitative utilitarian approach to the natural world as nothing less than blasphemy. Christianity, however, will not infrequently be viewed by DE and other environmental writers as pulling the divine and human apart from and raising both above nature and thus supporting the ecologically destructive developments of modernity.

Deep Ecology antipathy to divine transcendence becomes clearer with attention to the writers and traditions they tend to cite as sources for their thought.[3] Spinoza, native American spirituality, the American Transcendentalists, Friedrich Hölderlin, Taoism, and Buddhism are sources listed by DE writers as the minority tradition they find helpful. This does not mean that all Christian sources are out of bounds. Saint Francis certainly makes an appearance and Naess, working from some of E. F. Schumacher's ideas, thinks that Scripture permits varied approaches to ecology some of which he deems positive (Naess 1989, 187-89). A non-DE writer such as Manussos Marangudakis, while trenchantly critical of the Western theological tradition in terms of ecological issues, argues that Eastern Christian theology has much to offer ecological reflection (2001). In general, however, what these sources reveal is that the Deep Ecology movement places a high premium on immanence and also that they find support in a range of discourses that includes the religious and the poetic. Certainly they would find common ground with Balthasar who at times seems to favor poetic to philosophical discourse for its capacity to reflect divine glory.[4] Negatively stated, DE is deeply suspicious of doctrines of divine transcendence and human otherness. With the notable exception of Saint Francis, and elements of Hölderlin's poetry, DE authors opt for patterns of thought that rule out a personal, transcendent divine, including ones that seek to balance it with divine immanence so that God is both Other and Non-Other; DE authors simply overlook and/ or rule out Jewish and Christian theologies of a transcendent God who is at the same time also Emmanuel or God with us. To transpose their language into Balthasarian theological language, we might say that DE authors return to a purely cosmocentric worldview where the sacred gets confined to the world itself. What I want to suggest below

---

3   Peter Reed is, in a sense, an exception to this tendency.

4   Kevin Mongrain argues that Balthasar's use of poetic language and his valorization of poetic discourse is not merely ornamental but essential to the relationship he sees between faith and reason (2007).

is that the exclusion of divine transcendence, rightly understood, is one of the reasons for DE's difficulties with issues of identity and difference. In addition, a strong view of divine transcendence provides a crucial block to anthropocentrism especially if divine transcendence is accompanied by a mediating metaphysics.

RESISTANCE & ADVENT:
RESOURCES FOR A DOXOLOGICAL ECOLOGY

In his thoughtful essay on the importance of poetic discourse to Balthasar's doxological theology, Kevin Mongrain observes that "[r]eestablishing a constructive relationship between theological, cosmological, and anthropological discourses is central to Hans Urs von Balthasar's entire intellectual project" (Mongrain 2007, 381). For Balthasar, one of the major difficulties this task presents concerns how to retrieve the cosmological element without falling into a cosmocentric and fatalistic worldview he considers pre-biblical. Modernity offers resistance in this regard as it splits into two basic approaches to the ancient world. On the one hand, modernity yields a "metaphysics of spirit," especially prevalent in the German Idealist tradition, that forfeits the encounter with glory in the universe, strikes against divine transcendence, and offers a deeply anthropocentric vision. On the other hand, modernity produces poets such as Hölderlin and Goethe who seek to retrieve classical cosmological wonder but are unable to free themselves, at least on this account, decisively from fatalism and anthropocentrism respectively. The key, on Balthasar's reading, is unwillingness on the part of Hölderlin and Goethe to view antiquity as a kind of advent that gets taken up into Christianity (Balthasar 1991, 248, 451). Of course this can be stated from the other side to the extent that Christian theologians have too often sought to throw off rather than integrate the insights of the classical tradition (ibid. 451). In the following we will address precisely what Balthasar thinks the classical tradition has to offer a revitalized contemplative approach to the world that avoids the anthropocentrism of certain modern forms of thought. We will begin, however, with a discussion of what Balthasar sees as the major resistance to a truly contemplative encounter with Being, that is, with an abbreviated genealogy meant to call into question some of the assumptions regarding Christianity and anthropocentrism circulating in ecology literature currently. Much of this must be more nearly gestured to than argued here.

For Balthasar, anthropocentrism, and thus also the devaluation of creation as site of divine glory, takes its origins not so much in the Hebrew Bible, Plato and neo-Platonism, or even in the standard bearers of medieval Christian thought, as in late medieval Nominalism and in Luther's turn to Scripture alone and decisively away from metaphysics (ibid. 18-21, 452). Of course this is only the beginning. Anthropocentrism accelerates exponentially in the eighteenth and nineteenth centuries. Balthasar does agree with ecologically minded authors that the Hebrew Bible or Old Covenant shows a God who is no longer reducible to forces of nature or circumscribed by the cosmos. Instead, God becomes the Creator who freely enters into covenant with free human beings, free, that is, beyond mere fate or the fatalism of cosmocentric myth. Moreover, this liberation of the divine from the cosmos, what Balthasar calls "de-divinization," can seem to warrant human carelessness or exploitation of brute and protean matter (see Balthasar 1990, 177, 395). Indeed, Balthasar thinks that in certain cases the extreme anthropocentrism of modernity draws from the joining of Jewish de-divinization of nature on the one hand and from the Christian proclamation of divine *kenosis* on the other. Whereas the first leaves the cosmos bereft of glory or intrinsic value, the second can be—and has been—read as a definitive end to divine transcendence and a symbol of Promethean human power and liberation. So Balthasar agrees that a certain attenuated rendition of Judeo-Christian revelation can support, and in fact has been illicitly enlisted to support, the anthropocentric Prometheanism of modernity. This rather thin story, however, needs a two-fold correction.

First, Balthasar would observe that Israel's revelation breaks open the closed regimes of mythical and cosmocentric systems in which human beings had only two options: submit fatalistically to the domination of the gods and/or nature, or assert domination over the gods and/or nature. Israel's revelation of a God beyond myth and nature is quite positive in that it bestows on humans the dignity of a dialogical relationship with the divine which affirms and activates their freedom and moral agency. The freedom and responsibility awakened by God's offer of a covenantal relationship carries over into human persons' relationships with nature and other persons as well. The God of Israel thereby becomes the transcendent norm that challenges all forms of slavery to rigid, closed mechanistic functioning of religious, national, social and economic systems. Surely no one would argue that a deterministic cosmic system or mythological system (or market system) necessarily fosters human dignity or justice; such systems make slaves of both those who manage them and who are managed by them. This was the entire point of Israel's prophets in their critiques of "pagan" religious systems and unjust human political, economic, and social systems.

Second, Balthasar would point out that Israel liberates the natural world and its gods from the burden of being the "lords" over a humanity prone to Promethean backlashes against the implacable workings of its heartless masters. Israel is therefore beyond the zero-sum dialectic of resignation to fate versus willful self-assertive projects to secure freedom. Israel perceives and praises the Covenantal God of freedom and dignity throughout Creation, which is now neither master nor slave, but instead simply free to be itself in its own dignity; the gift God bestows on humanity frees it to pass that same gift along to nature. Indeed, the Psalms are filled with praise of God's glorious reflection throughout the created world that fills out God's fundamental affirmation of creation in Genesis and returns again in the Wisdom literature. It takes a massive effort at forgetting to suggest that divine transcendence meant for Israel or for Judaism that divine glory is remote from the world. This forgetting not only relates to Israel's scriptures, but also to Jewish liturgy, as Solomon Schechter points out in his *Aspects of Rabbinic Theology* (Schechter 1961, 21-45). For Balthasar, modern authors, not excluding Christian authors, are particularly adept at "forgetting" in this regard. The consequence of this forgetfulness is that the doxological space of non-competitive, mutually affirming and solicitous unity-in-difference between God and humans, and in turn between humans and nature, is closed. This closing of the doxological space is precisely what makes possible modernity's renewed war on nature. We should not be deceived, Balthasar would caution us, when modern thinkers co-opt and pervert the language of divine transcendence and divine kenosis for Promethean ends fundamentally at odds with the original meaning of this language. Forgetfulness and anti-doxology, however, are not on Balthasar's account necessary. They are contingent features of modernity and can therefore be remembered again through a thicker description of and engagement with Israel's revelation.[5]

Finally, we can notice that Israel's doxological orientation toward God, that is, their praise and gratitude for creation as gift, places a tremendous obstacle to the reduction of the non-human world to the human whether ontologically or in terms of interest. This is no accident. Not incidental to Balthasar's efforts to remember for the ecclesial community a Christian sacramental vision of the cosmos is his massive resistance to what he calls "identity metaphysics." He accomplishes this remembering through a sustained reflection on the relation of the biblical God to the world. For Israel, he argues, it would be strange to conclude that simply because creation is not the Creator it cannot be good or have its own value. What it does mean unequivocally is that human beings cannot be the origin of value for the created world. Rather, all that is, as Augustine observes, has value as gift and reflection of the Creator.

If Balthasar does not allow Judaism or Christianity to go unscathed in his account of the loss of cosmic glory and the descent into anthropocentrism, this does not lead him into simplistic denigrations of divine transcendence or human freedom. It does lead him to argue that late medieval Nominalism is where the human and nature come apart precisely because this is where any notion of shared Being (*esse commune*), and thus metaphysical mediation for the form of revelation, begins to fade from view. The fourteenth century functions for Balthasar, as for John Milbank (2006, 14-16) and Louis Dupré (1993, 3)[6], as a watershed for the doctrine of creation. The continuity-in-difference between human being and nature gets shattered and the way opens to oppose one to the other, human voluntaristic subject to nature as protean object. Central to this discussion, for Balthasar and Dupré, is

---

5   This issue of forgetting and remembering is taken from the work of Cyril O'Regan on Balthasar, much of which is forthcoming in the next year. O'Regan would add the term "misremembering" to designate, in part, a willful form of forgetting that also transforms its initial object.

6   Agreements here are not total. In the second edition of *Theology and Social Theory*, Milbank expresses a somewhat more positive account of medieval and early modern development and a disagreement with what he thinks of as Balthasar's "conservative" account of modernity. Dupré's account adds an assessment of early modern humanism as a key element of the passage to modernity.

a notion of form (*gestalt*) that inspires contemplation and receptivity to the created world perceived as good, true, beautiful, and purposeful. Instead, Nominalism sets up a formless world that yields to force or efficient causality and helps support what Charles Taylor calls the "mutual benefit" society developed during the Enlightenment. Nature comes to be viewed as raw material for human benefit, exploitation, and dominance. Under an eighteenth century hermeneutic regime, nature can no longer be the reflection of a personal, transcendent, *and* deeply intimate divine glory. We are now in a position to ask what classical antiquity might contribute to the two-fold endeavor to block anthropocentrism on the one hand and to encounter the non-human world as having its own value on the other.

On Balthasar's reading, classical antiquity perceived an epiphanic cosmos that interlaced the transcendentals of the Beautiful, True, and Good understood as underlying the appearance of all forms. Interior goodness and truth shine out in radiant form. The cosmos, whether in terms of stars or gods or in more rarified philosophical formulae, called forth human self-transcendence, wonder, awe, humility, and service. In its poetic, philosophical, and religious modes the classical world provided a mediation to the biblical form of revelation up to the fourteenth century. We will see that for Balthasar, and, in his view, for much of the Christian tradition, divine transcendence in no way restricts divine immanence or the value of non-human entities. To the contrary, they are directly rather than inversely proportional. This is also the case for the classical transcendental. All forms in the world both radiate truth, goodness, and beauty while never exhausting these same transcendentals. The transcendentals transcend particulars while inhabiting them. It is for Balthasar both the remarkable plenitude and variety of the natural world that elicits the self-transcendence of human knowing and being.

Our discussion of Balthasar's survey of ancient metaphysics here is necessarily cursory and can only help show that his epiphanic construal of the cosmos and his ec-static theological anthropology play against anthropocentrism and for a deeply positive estimation of worldly being. Homer and Plato will perform the roles of exemplars for a much more nuanced and extensive discussion of Greek and Roman poets, philosophers, and religious thinkers found in *Glory of the Lord* 4 (Balthasar 1989). With respect to Homer, Balthasar observes that his theandric context, which differentiates between human and

divine or immortal and mortal, makes possible genuine dialogical relationship, prayer, wonder, and humility before all that is. It is to the
credit of Hölderlin and Goethe, on Balthasar's view, that they each
assimilate much of this classical wonder and contemplation. Homer
also includes divine immanence in his epics. It is not unusual to have
heroes receive radiant grace bestowed upon them by their god or even
to have a goddess such as Athena take human form to show favor to
Odysseus. It is the poet's vocation to contemplate and proclaim the
events of glory wherever they occur or are discovered. The emphasis in
poetic and narrative discourse—and its particular gift—will be on the
singularity and momentousness of particular glorious events, and it is
just this form of glory that philosophy will find it difficult to represent.
Homer sees glory as the gift of the gods. Certainly human beings have
their roles to play in gaining glory, but glory does not belong to them.
They are to hold fast their obedience to the gods if they wish to attain
it. Indeed, this loyalty or faith on the part of mortals, including not
only the heroes but also the poets, requires looking beyond the self
and beyond the human to the gods and the world as it shows itself to
the subject. Reality must be perceived and represented truthfully.

Philosophy seeks to universalize mythical insight and for Balthasar
the convertible transcendentals of the good, true, and beautiful accomplish just this task. Mongrain has shown that Balthasar recognizes in Plato's Diotima a form of mythical grace and mediation for the
Beautiful intended to generate fruitfulness in the Good. Socrates contemplates the beautiful in all forms on his ascent to the Good which
comes to be grasped as the transcendent source of all radiant forms.
So for Socrates in the *Symposium, eros* or desire for the beautiful leads
him to the Good where he also discovers wisdom or truth. Two points
should be noted here. First, the beauty of particular forms in the world
draws Socrates beyond himself toward the world of objects. This is
the ec-stasis toward the luminous world that is given to us and not determined by us in Plato. For Socrates, wisdom demands holding fast
to and even sacrifice for the truth. Anthropocentrism receives a sharp
blow here every bit as much as it does in the poet's receptivity to the
commands, judgment, grace, and favor of the gods.

Secondly, it should be clear that resources exist in the mainline
Western tradition for attributing value to non-human entities in-
somuch as all entities participate in the good, true and beautiful
apart from human interest or utility. Of course this is by no means a

complete picture of Plato's approach to the natural world. Christians would need to reject, for instance, his negative estimation of corporeality in *The Phaedo*. What it does show is that there is wisdom there that can be received into a doxological ontology. We should notice significant areas of agreement here between Balthasar and DE thinkers. Certainly Arne Naess and Balthasar each call for the retrieval of form, for contemplation, and for recognition of the intrinsic value of the natural world. We can also observe that in contrast to Heidegger's nearly univocal story of fall from the contemplative early Greek thinkers to the industrial techno-crazed age of modernity, Balthasar does not think that all Western intellectual culture is devoid of resources that support a renewed sense of gratitude, wonder, surprise, awe, and joy in relation to Being. Certainly he thinks Goethe and Hölderlin in the modern period offer resources to be enlisted in the cause, but he also reads against Heidegger by noticing elements in Homer, Plato, the Greek tragedians, Vergil, Plotinus, and Pindar that beg for retrieval and contest anthropocentric and the purely immanentist tendencies that support it.[7]

BALTHASAR'S DOXOLOGICAL ONTOLOGY

> The forms and individual characters of living and growing things and of inanimate things and of animals and flowers and all nature, constitute their holiness in the sight of God. Their inscape is their sanctity. (Merton 1986, 25)

The Hopkins-inspired passage above leads nicely into what must be a somewhat abstract discussion of ontology meant to affirm the ontic claims above with a doxological ontology. In speaking of Balthasar's doxological ontology we should take care to observe that for Balthasar form never passes away into pure concept or pure logic where all is transparent to reason and contemplation can yield to manipulation. This emphasis on form is where Balthasar's theology offers ecological thought a way beyond not only anthropocentrism, but also materialism and idealism. The perception of form, for Balthasar, is the perception of a material and external expression of an interior energy, splendor and radiance. However, the interior and exterior of form are not separable as they are in allegory (idealism) where a grasp of inner

---

7   For his summary discussion of ancient metaphysics, see Balthasar 1991, 637-39.

meaning allows the disposal of external form or in materialism where no interiority matters and all is surface. Rather, Balthasar's vision is sacramental. Thus, for instance, Balthasar is clear that Jesus' humanity is no mere sign serving as an instrumental pointer to something else, but is rather the visible expression of the invisible Father for us (Balthasar 1982, 437). But this is also true of Creation. If Creation is God's Creation, it stands to reason that it will in some way reflect divine glory. Because for Balthasar the form of God's self-revelation in Christ provides the measure for our contemplative vision of Creation, it is helpful to begin with some of Balthasar's more formal considerations of Christian revelation prior to our ontological discussion.

For Balthasar, God's self-expression in Christ reveals God's profound love for the world through his incarnate life even through the paradoxical formlessness of Good Friday and Holy Saturday. The Resurrection is what allows this revelation to have shape, and thus become legible, on the other side of death and thus beyond the ken of reason alone. Of course divine fidelity and commitment to the world does not cease at the ascension, but continues in the Spirit and in Eucharist. This profound and perduring divine commitment to creation has as its anthropological corollary that human beings can give full trust, "to the point of self-sacrifice, to the depths of the form of the world" (Balthasar 1982, 444). Christian piety grasps the invisible splendor of Christ's form as self-abandoning love for the world to the glory of the Creator and pursues this same path. Of course this means that faith sees beyond surface images to grasp the interiority of this revelation through its expression, for the surface image would speak to ontological rupture or death as final. Alternatively we might say that faith sees the whole form including the resurrection, and only then can grasp the meaning of the Cross. In analogous fashion, Balthasar believes, when we confront the material world, we grasp the mysterious interiority that both radiates from and remains hidden in its form (in large part because we know ourselves to have both bodily form and interior life) (ibid. 442-44). Here, he thinks it is appropriate to speak of a kind of "natural faith" and "natural piety" (ibid. 447, 451). When Balthasar turns to his discussion of the truth of the world, this operative notion of form remains. Worldly being, of which human being forms a part, exhibits a differentiated relationship between form and splendor at all the different levels of Being from basic matter to free rational spirit (ibid. 442).

In our utterly non-exhaustive discussion below, we will take two approaches to Balthasar's most intensive and sustained ontological reflection, which is found in *Theo-Logic* 1. The first approach concerns what he calls the "freedom of the object" and considers human being within a hierarchy rooted in shared or common Being (*esse commune*) but expressive of difference. The second approach offers a broader vision on his notion of truth at the philosophical level and concentrates on three terms: mystery, *aletheia* (unconcealedness), and *emeth* (trustworthiness). Through this reflection I want to show the way in which Balthasar recommends a relationship between human being and non-human being that has genuine continuities with Deep Ecology along with differences that, I will suggest, recommend themselves as a way to deepen the Deep Ecology movement. Of course this returns to issues of identity and difference. Most important will be to show the profoundly ec-static and doxological character of all creation[8].

If the mutual relation of form and splendor shape all Being, for Balthasar this continuity in Being yields another, namely, the idea that a form's interiority is the ground of its freedom, of a self-expression that cannot be grasped until it is expressed and even then is not fully reducible to its expression. Even inanimate nature, he contends, has an interior energy that does not exhibit itself as such, but in an unpredictable form and in an infinite variety of forms. On Balthasar's account, this fundamental continuity in Being also suggests important differences in the kind of interiority or intimacy present at the different levels of Being. In terms of the DE-ecofeminist debate, Balthasar certainly recognizes the continuity, interrelatedness and inherent value of all Being, but he is not interested in collapsing differences. Indeed, this difference in what he calls the intensity of a being's interiority or intimacy allows for human moral responsibility for Being itself. However, on Balthasar's account, this moral responsibility need not forfeit the wonder, reverence, and joy one takes in creation or lapse into a new kind of anthropocentrism. To see this clearly we need to invoke our second and more extensive approach.

On Balthasar's view, the ontological truth of the world has three fundamental qualities that form the ground for analogy to God's

---

8   Much of the discussion that follows is from the first volume of Balthasar's *Theo-Logic*. Because this discussion also appears in my forthcoming book with Crossroad on Balthasar, *The Cross and the Covenant*, I must thank John Jones for allowing it to appear in slightly altered form here.

self-revelation in Christ: *aletheia* (unconcealedness), *emeth* (trustwor-thiness), and mystery. Balthasar thinks that reality shows itself to a knowing subject in a trustworthy yet excessive manner. Indeed, reality always exceeds our conceptual control as a particular form of existence points to a plenitudinal, always greater ground or essence that gives itself appearance or sense-apprehensible form for us (2000, 50, 217). The world that offers itself to "vision" is no superficial surface, but also essence, ground, interior light, and depth. Science assumes this depth dimension throughout its investigations as it seeks the essential laws of nature underlying the infinite variety of appearances (ibid. 84). Thus Being reflects God's self-revelation in Christ through its self-disclosive (*aletheia*), trustworthy (*emeth*) revelation that permits true and not merely phenomenal knowledge, and also in its mysterious and hidden excess (ibid. 50). When we know a tree, for instance, we never imagine that this is the only possible realization of its essence or that, knowing its essence, we can dispense with the particular and varied realizations. This suggests that intellectually true judgments never end inquiry, but rather open up further contexts and intellectual vistas for human wonder (ibid. 49-50).

We might ask how precisely the knowing subject relates to this epiphanic world. For Balthasar, the relation of human to non-human being will bear analogy to God's relation to the world precisely be-cause humanity bears God's impress in its very being. This is no an-thropocentric claim for Balthasar, but rather a warrant for the love of God's creation as God loves it. In imaging the Creator, human beings make space for and give a voice to creation as both same, in the sense of shared Being, and as other in the sense of free rational spirit. The non-human world gives humanity material for knowing and for enact-ing its freedom; the human voice serves non-human being by helping it fulfill its epiphanic end. In our time this service no doubt means giv-ing voice not only toward the end of praising God, but also by giving a voice to vulnerable non-human being. Such a theological discourse is both poetic and prophetic, but not thereby irrational insofar as ratio-nality begins from a position of Eucharistic gratitude and hospitality. This requires a brief explanation.

Balthasar describes the relation of world and free rational spirit as one of mutual gift. Each moves beyond itself outwards and thus to-wards the other in a kind of erotic ec-stasis or transcendence. From the side of the free rational spirit, this means always already being the recipient of Being and its manifestness for knowing and acting. Balthasar speaks of the "unchosen necessity of being broken open to receive, through the subspiritual gateway of the senses, the life and truth of the other that comes uninvited, pouring into the domain of its intellect" (2000, 39). The erotic going out to creation, being broken open, and mutual self-giving imaged both by natural being and human being images God's "letting-be" a space within the divine life for that which is other than God without this space or freedom suggesting anything like divine indifference. Rather, this "letting-be" is joined to the non-coercive and disinterested love that provides the incarnate measure for human ethical behavior towards one another and towards creation. The act of knowing appears as a kind of paschal or Eucharistic hospitality and service to an epiphanic, self-giving and self-surrendering world, all of which reflects the Creator (2000, 149-153).

CHRISTOLOGICAL & TRINITARIAN SPECIFICATION: UNITY-IN-RELATION AS DOXOLOGICAL GROUND

Up to this point I have stressed the role divine transcendence might play in support of an ecological movement concerned to root out anthropocentrism and to emphasize relationality without collapsing relation into sameness or identity. This discussion bore the name doxological because, in Balthasar's view, all created form bears analogy to the abundant self-abandoning love of the Creator and thus gives rise to wonder, gratitude, awe, and reverence. Humans finds themselves at once members of the created and finite order and called to serve it with the gifts peculiar to the person's kind of interiority. At one point, Balthasar names this relation to the natural world a kind of "natural piety" and he insists that this natural piety means that when Christ comes, he does not come merely under the sign of contradiction. Instead, Christ magnetically draws into himself the grammar of Being (and history) and takes it into his own self-revelation. For Balthasar, then, creation is one form of God's self-revelation, one form of God's loving condescension for the world. All "creaturely Being" for Balthasar is an "indicative utterance about God" (1982, 448). Indeed, there is no clear marking off of the divine splendor as it shines through creation from "what Christian theology calls supernatural revelation" (ibid. 449). Here we have Balthasar adhering to the Chalcedonian Christology that recognizes deep unity without confusion. It might be best to say that after the resurrection, we live in an impure world. Grace penetrates nature too deeply to inoculate one from the other. Transcendence and immanence are in no way opposed here. Indeed, given Balthasar's divinization eschatology, we would have to say that transcendence serves immanence, while divine-human intimacy reveals transcendence.

Of course greater specification of both comes for Balthasar through the revelation of Christ, which takes up all creation, human culture, and, more particularly, Israel's revelation and gives all of this validity in eternity, that is, in the life of the Triune God. Certainly Balthasar's theology has neo-Platonic resonances as an exitus-reditus schema appears to inform his depiction of the Christ-form. Nevertheless, Christ's descent goes well beyond the neo-Platonic circuit through an extreme and personal form of suffering that requires free self-renunciation, service, death, and even the descent to hell so as to retrieve all things for God. Only with this free personal revelation can revelation call upon human beings in their freedom. Balthasar wants to make clear that just these salvific and revelatory actions are for the world and not merely for the Church. Yet the Church reflects precisely this being-for-the-world in its Eucharistic liturgy, its ethical solidarity with creation, with humanity and especially the poor, in its prayer and its saints. With respect to environmental issues, Balthasar makes abundantly clear his view on the merely utilitarian approach to nature: "In the rhythm of the form of plants—from seed to full growth, from bud to fruit—there is manifested an essence and to reduce the laws of this essence to mere utilitarian principles would be blasphemous. And in the totality of beings, as they ascend and maintain their equilibrium, there is revealed a mystery of Being which it would be even more blasphemous and blind to interpret by reducing it to a neutral 'existence'" (1982, 444).

The utilitarian and instrumental view of nature is nothing short of blasphemous, for it ignores the splendor of God's work and instead seeks self-centered exploitation, whether economic, ecological or both. What Balthasar thinks is called for is a contemplative beholding of creation's sacramentality, its disinterested self-giving and fruitfulness, its radiant goodness. It is this creation that human beings are meant to serve with all their intelligence, imagination, and collective action. Christians, then, are not for Balthasar called to ascend in a neo-Platonic contemplation away from the world to the One, but rather descend with Christ into and for the world. Thus contemplation leads to service and even mission, a key category of Balthasar's Christological work in Theo-Drama.

Certainly Balthasar's emphasis on form is rooted in his reflection on the Incarnation. Yet form does not evaporate in the Godhead; instead, Balthasar speaks of the Triune God not as formless, but as "superform" (Balthasar 1982, 432). It is this superform of the Trinity that serves as the positive ground for the multiplicity-in-unity encountered in the world of nature. Stated more concretely, difference-in-unity is divine glory as self-abandoning love, which Balthasar often speaks of in terms of "letting-be." This letting-be of difference and love for what is at once other and same in the Trinity is for Balthasar what provides an eternally positive ground for the temporal world in its difference from the divine and in all its multiplicity and diversity within shared Being. Multiplicity finds a positive ground in the Trinity beyond any accidental fall from the One or natural emanationism that may well end up in a kind of total absorption or identity.

If Christ's relation to the Father and Spirit discloses the immanent Trinity as ontological ground for the diversity of creation, it is also true that, for Balthasar, the doctrine of the Trinity must be rooted in Christ's incarnation, that is, in his deep historical and natural penetration of God's creation. In this sense, while we must still speak of divine transcendence, we must also specify "transcendence" as the capacity for divine immanence beyond even the non-Aliud of Nicholas of Cusa to a radically intimate relation that Balthasar names the "Other in the non-Other." In other words, Creation is both in God and permeated by God's presence without losing its creaturely reality. Of course, poets may carry more adequate linguistic tools than theologians and philosophers when it comes to expressing this mysterious interpenetration of God and world.

CONCLUSION

In his poem "Advice to a Prophet" Richard Wilbur pleads with the crier "mad-eyed from stating the obvious" to cease haranguing reluctant hearers with dire numbers bespeaking our universal demise which, he thinks, lies beyond our imagining. Instead, he tells the prophet to show us what we will be missing. For Wilbur, the loss of the surrounding richness leads to a kind of existential aphasia: "Ask us, prophet, how we shall call/Our natures forth when that live tongue is all/

Dispelled, that glass obscured or broken/In which we have said the rose of our love and the clean/Horse of our courage, in which beheld/ The singing locus of the soul unshelled. And all we mean or wish to mean" (Wilbur 1963, 6-7). For Balthasar, the loss would be mutual. If we lose sight of beauty and the metaphysical richness of the world, we cannot serve the world's epiphanic *telos*. I will heed Wilbur's plea by not stating the obvious about the losses that are and will be occurring due to climate changes, since all the statistics and descriptions can be found in the IPCC reports and other scientific documents. Nevertheless, a few points are significant.

Although some aspects of climate change and climate change reports have been exaggerated in the popular media, the facts of climate change and the effects of emissions on the global climate are grounded in solid evidence. Even with an optimistic view of what will take place in climate summits, such as the one at Cancun that is taking place as I write, or with a hopeful estimation of wealthier nations' emissions reductions, we can anticipate an approximately 2° C (3.6 °F) warming above pre-industrial temperature norms. We have all heard of the melting polar ice caps, consequent rising of sea levels, and, in other areas, desertification. What is perhaps every bit as disturbing as the loss of biological diversity, and perhaps the loss of entire species, is the asymmetrical ecological relation between rich and poor countries. Rich countries produce far greater greenhouse gas emissions while the requirements for adapting to new ecological circumstances such as hotter summers, colder winters, drought and flooding decidedly favor the rich over the poor (*The Economist* 2010, 15, 85-88, 105).

There is little doubt that technical expertise is absolutely necessary to support agricultural developments, to predict the actual changes that will take place and supply corrective and adaptive strategies, and to grasp the economics surrounding occurring changes. In such a crisis, I think it is small-minded to denigrate human know-how. Yet none of this will be sufficient without an underlying and disinterested sensibility for the sacramentality (be it supernatural or "natural") of the created world and our responsibility to the poor so a reason other than mere survival (or even comfort) in the moment forms the inspiration for change. It seems to me that Balthasar contributes to ecological issues by helping us gain a new vision in which the natural world no longer appears as mere fodder for our next project, but as living testimony to divine glory as self-giving love. On this vision, humanity receives a call to the realm of natural being of which we are a uniquely responsible part. This is, indeed, a difference that matters and that stirs a kind of ontological subsidiarity. It is true, of course, that different species take care of their own needs in their own environments to a certain extent. However, it is also the case that human freedom, for better and for worse, must take responsibility for a much wider range of existence precisely as shepherds, or, as a friend puts it, as "good gardeners" who seek not so much their own interest as to serve others, and to make of the natural world not our servant, but our friend.

SOURCES

Balthasar, Hans Urs von. 1982. *The Glory of the Lord: A Theological Aesthetics, Vol. 1, Seeing the Form.*, Translated by Erasmo Leiva-Merikakis. Edited by Joseph Fessio, S. J. and John Riches. San Francisco and New York: Ignatius Press and Crossroad.

——. 1989. *The Glory of the Lord: A Theological Aesthetics, Vol. 4, The Realm of Metaphysics in Antiquity.* Translated by Andrew Louth, Brian McNeil, C. R. V., John Saward, Rowan Williams and Oliver Davies. Edited by J. Riches. San Francisco, CA: Ignatius Press.

——. 1991. *The Glory of the Lord: A Theological Aesthetics, Vol. 5, The Realm of Metaphysics in the Modern Age.* Translated by Oliver Davies, Andrew Louth, John Saward and Martin Simon. Edited by J. Riches and B. McNeil. San Francisco, CA: Ignatius Press.

——. 2000. *Theo-Logic, Vol. 1, Theological Logical Theory: The Truth of the World.* Translated by Adrian J. Walker. San Francisco: Ignatius Press.

——. 1990. *Theo-Drama: Theological Dramatic Theory, Vol. 2: The Dramatis Personae: Man in God.* Translated by Graham Harrison. San Francisco: Ignatius Press.

DeVall, Bill and George Sessions. 1985. *Deep Ecology: Living as if Nature Mattered.* Salt Lake City, UT: Peregrine Smith Books.

Diehm, Christian, 2002. "Arne Naess, Val Plumwood, and Deep Ecological Subjectivity: A Contribution to the "Deep Ecology-Ecofeminist Debate." *Ethics & The Environment* 7.1: 24-38.

——. 2007. Identification with Nature: What it Is and Why it Matters. *Ethics and the Environment* 12.2: 1-22.

DiZerega, Gus. 2001. "Deep Ecology and Liberalism: The Greener Implications of Evolutionary Liberal Theory." *The Review of Politics*: 699-734.

————. 1995. "Empathy, Society, Nature, and the Relational Self: Deep Ecology and Liberal Modernity." *Social Theory and Practice* 21.2: 239-70.

Dupré, Louis. 1993. *Passage to Modernity: An Essay in the Hermeneutics of Nature and Culture.* New Haven: Yale University Press.

Marangudakis, Manussos. 2008. "On nature, Christianity and deep ecology—a response to W. S. Helton and N. D. Helton." *Journal of Moral Education* 37.2: 245–48.

Merton, Thomas. 1986. *Seeds of Contemplation.* New York: New Directions Publishing.

Milbank, John. 2006. *Theology and Social Theory: Beyond Secular Reason.* Oxford: Blackwell Publishing.

Mongrain, Kevin. 2007. "Poetics and Doxology: Hans Urs von Balthasar on Poetic Resistance to Modernity's Turn to the Subject." *Pro Ecclesia* 16.4: 381-415.

Naess, Arne. 1990. *Ecology, Community, Lifestyle: Outline of an Ecosophy,* trans. David Rothenberg. Cambridge: Cambridge University Press.

Plumwood, Val. 2002. *Environmental Culture: The Ecological Crisis of Reason.* London and New York: Routledge.

Pojman, Louis P. and Paul Pojman. 2010. Environmental Ethics Readings in Theory and Application. Boston, MA: Wadsworth.

Schechter, Solomon. 1961. *Aspects of Rabbinic Theology: Major Concepts of the Talmud.* New York: Schocken Books.

Sessions, Robert. 1991. "Deep Ecology Versus Ecofeminism: Healthy Differences or Incompatible Philosophies?" *Hypatia* 6.1: 90-107.

Wilbur, Richard. 1963. *The Poems of Richard Wilbur.* New York and London: Harcourt Brace Jovanovich.

Zimmerman. 1990. *Heidegger's Confrontation with Modernity: Technology, Politics, Art.* Indiana: Indiana University Press.

Zimmerman, Michael E. 1993. "Rethinking the Heidegger-Deep Ecology Relationship." *Environmental Ethics* 15: 195-224.

13

THOMAS BERRY AS THE GROUNDWORK
FOR A DARK GREEN CATHOLIC THEOLOGY

Peter Ellard

Over the past twenty years, Catholic theologians have responded to the environmental crisis. Leading figures such as Denis Edwards, Elizabeth Johnson, and Rosemary Radford Ruether have been joined by many calling for a reevaluation of Christian doctrine in light of the current destruction of earth's eco-systems. Furthermore, conferences of Catholic Bishops in many countries have written powerful statements challenging the established malaise, and both John Paul II and Benedict XVI have called for a renewed awakening to our responsibility toward creation. There is even a Catholic Coalition on Climate Change stressing stewardship and care for creation on the parish level. Many of these efforts have been progressive and truly reflective of a greening of religious institutions and theology (see Forum on Religion and Ecology 2010; Hessell and Ruether 2000).

However, much of what has been published has attempted to add various amounts of "green" to established theology. Sacramental, covenantal, and stewardship approaches dominate. It is my belief that, in a time of crisis, we are called to go deeper and darker in establishing the contours of an ecological theology. Though Thomas Berry is not a theologian, he offers as a self-identified "geologian" the groundwork for building a dark green Catholic theology for a planet in crisis.

This essay outlines key aspects of Berry's thought and demonstrates how they serve as the basis for a dark green Catholic theology. My use of the modifier "dark" seeks to stress the radical nature of Berry's views. This term was inspired by Bron Taylor's work on dark green religion, though my usage is not meant to signify full agreement with Taylor's analysis of Berry.

THE CONTEXT FOR A
DARK GREEN CATHOLIC THEOLOGY

Throughout his prolific career, Berry consistently presented a twofold context to his ideas regarding the crisis before us. These are, first, the need for recognition of the magnitude of the crisis and, second, the need for all contemporary endeavors to be grounded in contemporary science. This dual context also serves as the foundation of any theology based in Berry's work.

The Magnitude of the Current Crisis

Recent reports issued by the Inter-Governmental Panel on Climate Change, World Watch, and other scientists have been universally stark. Berry consistently mined this information to accentuate in clear terms the gargantuan nature of the climate crisis, and he stressed that the changes taking place were so massive that they must be viewed on a geological scale. He compares the level of change with the last glacial period and with the Cretaceous extinction. He notes that there is, in truth, "no historical parallel" (1999, 3). The overall picture he sometimes paints is indeed bleak.

Although the data and images of the devastation can paralyze us, it is essential that we express the order of magnitude as a context so we realize that any solution must come with the same level of response. Berry cautions that we need "something more than ecology, more than Deep Ecology" (1988, 2). We also need to do more than add the environment to a long list of concerns of the Church. Berry holds that an honest picture of the level of crisis should be embraced because it is precisely out of these conditions that transformative moments occur and that the "great spiritual traditions of humankind emerged out of confrontation with terror" (2009a, 1). Nothing short of great spiritual traditions—or current traditions greatly transformed—are in order in response to the current terror. Finally, Berry notes that this "is not a situation that can be remedied by trivial or painless means" (2006, 19). The change required on the part of our industrial economy will bring about great hardship. There will also be great change within religious institutions, and reluctance, fear and entrenchment will occur.

With respect to the causes of the current climate crisis, Berry presents an indictment of the petrochemical industrial extraction and consumption culture of the modern world (1988, xii). He offers an important reminder that many of the "positive" advancements in medicine, democracy, technology, life expectancy, globalization, genome research and that which we call progress have come at the price of ecosystem devastation (1999, 2). He describes our entrancement with progress and the industrial world as a "deep cultural pathology" and insists that none "ever invented could be so perverse and so devastating to the delicate balance of life and existence on the planet" (2009b, 47; see also 1988, 1).

Berry is sympathetic to Lynn White's narration of the causes behind the devastation (Berry 1988, 80), and he singles out anthropocentrism, androcentrism (1999, 181), and population pressures on the planet (1988, 72) as each playing their part in creating the crisis. Still, the real cause he points to is a "particular mode of consciousness" which has alienated humans from a true awareness of their identity and connection to and with the rest of the natural world (1999, 4). The solution requires, therefore, "a vast turn in human consciousness" (1988, 38). We need to be awakened from our technological entrancement (1988, 37) and we need to create a new vision of "mutually enhancing human- earth relationships" (1988, xii). Like addicts, we need to begin our recovery within the context of the devastation we have caused. Our recovery must also begin within the context of our primary guide forward. This guide and ultimate resource is the revelation of modern science.

Science & the New Story of an Evolutionary Universe

Berry opens his seminal essay in 1978 with the words: "It's ALL a question of story" (1978, 1). He contends that the story of the universe, from cosmology to geology and from evolutionary biology to quantum physics, is the single most important discovery of the 20th century (1999, 163). Scientifically, the new story reveals that approximately 14 billion years after the "big bang," the universe is still in the process of creation. We exist in a cosmogenesis. Everything that exists in the universe today has existed from the beginning in a different form. From the primordial point of power and singularity, the vast diversity of the universe has sprung and continues to creatively transform itself by its own self- governing mechanisms. Evolutionarily speaking, everything in the universe is related—genetically, chemically, and sub-atomically—to everything else. The universe really is a "uni," a single time developmental event or "being." Everything *in* the universe *is* the universe.

One problem, as Berry sees it, is that we do not understand the meaning of the new science. Even some scientists, as recalled by Brian Swimme in his essay "Cosmogenesis," come away from the sciences with a decreased sense of meaning (2000, 239). Berry notes that Christians are stuck in a mindset developed from the older biblical story of creation, and this story no longer functions cogently. However, since we have failed to absorb the meaning of the wonders that modern science reveals about our existence, we are stuck between stories. Much work has been done in recent years to mine the Bible for support for an ecological theology (Horrell 2010), but what is still missing, Berry insists convincingly, is the realization that the new "scientific account of the universe is the greatest religious, moral and spiritual event that has taken place" in recent years (Berry 1998, 98).

This brings us again to the question of the need for a theological response that is a darker shade of green. For, if we fully embraced the new science, it would radically alter our entire worldview and change our consciousness. The new story of the universe has the potential to function as the single story for all the world's peoples, religions, and cultures. This story need not negate the value of other "creation" stories, but it does replace them in terms of revealing what "happened" and, as importantly, what continues to be happening. Moreover, the most practical value of this story is providing us with the answers we seek in our moment of crisis. It functions as the guide for how to act in an age of global climate change (Berry 1988, xi) and, I believe, it will help us produce a theology that can function in the face of bio-cide.

This new story is both a scientific and a religious "text." There is, of course, a long history within the Christian tradition of reading the "book of nature." The difference here is that the book is written, in part, by scientists. One of the pitfalls of the last few hundred years, as outlined by Elizabeth Johnson in her essay "Losing and Finding Creation in the Christian Tradition" (2000), is that theologians and religious institutions have given over the interpretation of the scientific data to the scientist alone. As Ian Barbour, John Haught and other scientifically-informed theologians have recognized, theologians can and must reenter the conversation with science. We also need to mythologize scientific findings as a way to assist us in parsing these truths. As Berry contends, we need to be story tellers, artists, choreographers, and liturgists to help us enter into greater communion with the universe through its story. This is precisely what Berry and Swimme sought to present in *The Universe Story* (1992).

The first thing Berry wants us to know about the universe is that "it is primary" while "the human is derivative" (Berry with Clarke 1992, 97). This primacy extends to every category we can conceive. Whenever we seek to understand something, anything, about ourselves, about the divine, about Earth, or about theology, our understanding needs to take place within the context of the universe.

THE UNIVERSE & SCIENTIFIC REVELATION

Berry believes that science is an aspect of divine revelation. More emphatically, he insists that the universe is "the primary revelation of the divine" (2009b, 31). This primacy gives the universe a status unparalleled by any single part of it or expression of it. The most important thing in the phenomenal world is the universe (2009b, 47). This holism is a key part of the Berrian worldview because it leads us to recognize that our identity is much more than this separate corporeal/spiritual unit that I call "myself." This holism also helps us to begin to realize the interrelatedness of all that is, and that everything we experience —including divine revelation—takes place in, and is a product of, a self-governing and self—emergent process (2009b, 83). Revelation is natural and reflective of the fact that the universe itself, as a whole and in all of its parts, is a mode of divine presence (Berry with Clarke 1992, 81).

A Mode of Divine Presence

One of the most consciousness-altering statements Berry presents is the following: "If we lived on the moon our sense of the divine would reflect the lunar landscape" (2009b, 42; see also Berry with Clarke 1992, 9). This is a profound realization. All of human understanding and creativity, both interior and exterior, is a product of our experience of the natural world. Our sense of the divine comes from the natural word (1988, 81). In fact, one could argue that every moment and every reality is a sacred presence, a way that the divine reveals itself in and through creation. The divine is not confined to creation as in pantheistic thinking; for Berry, creation is the only way we can experience the divine. In this sense, every moment is theophanic. In truth, although Berry does not call it so, there is a deep strain of mysticism that runs within his thinking that has affinities to both Nicholas of Cusa and John Scotus Erigena. However, and more specifically in debt to Teilhard de Chardin, Berry says that the universe is a mode of divine presence not only because it reveals divine truth, but also because the very nature of the universe itself is a spiritual, as well as, a physical reality.

The Psychic/Spiritual & Material/Physical Nature of the Universe

If everything in the universe comes from the universe itself and if there is consciousness or spirit or love or incarnation in the universe, then these are a product of evolution and have been present in the universe from the start. Berry writes: "[T]he evolutionary process of the universe has from the beginning a psychic-spiritual as well as a material physical aspect" (2009b, 29). Again, it is important to note that this is not just latent potential spirit. All material interactions before humans arrived had a psychic component, a mind component, a soul component in them, which much later flared forth in a particular way in humans, including in Jesus (2006, 55). This psyche/mind/ spirit/ soul aspect of all material things remains in all things. Ultimately, since the universe is one, "the spiritual and physical are two dimensions of the single reality that is the universe itself" (1999, 49-50).

The Spiritual Revelation of Gravity

It might be helpful if we take a moment and look at how Berry interprets one scientific discovery, gravity, as he seeks to identify "the spiritual dynamics of the new cosmological story given to us by science" (2009b, 5). Newton's understanding of gravity tells us that every object in the universe pulls on every other object and the magnitude of that "pulling" is dependent on the respective mass of the object. Its range is infinite and its effects universal. An Einsteinian expansion suggests that gravity actually bends the space. It warps space. Thomas Berry looks at this and sees the mystery of gravity and divine revelation. We know what it does, but in reality, we have no idea what it is. All any scientist can say is that it is a fundamental force originating with the big bang. But again, they will only tell you what it does. Gravity is, therefore, a marvelous mysterious force that holds the whole universe together in such a way so as to allow for its expansion and to set the conditions for stars, planets, and eventually humans. One of the things that Berry wants us to ask is: When was the last time we thanked gravity for this act? When was the last time we celebrated the feast day of gravity in our churches? Without it, we would not exist. We thank God for our existence, but we miss the self-emergent vehicle that made it possible, and this omission leads to our alienation.

For Berry, gravity is part of divine revelation and reveals that "every being in the universe is ultimately present to and influencing every other being in the universe" (2006, 23; see also 1999, 53). We exist in relationships, and this creative balance is grace active on the universal scale. Berry refers to this bonding force as "the Great Compassionate Curve." This curve is a mode of being in the universe. It too has a psychic/spiritual aspect (2009b, 55). We might also contemplate it as being in the image and likeness of the divine. Berry is asking for us to expand our consciousness and our approach to the material and forces of the universe. This is exactly the kind of transformed understanding that must take place in order to make the human species viable. It is also in recognizing the particular subjectivity of the forces and "things" in the universe that we come to understand more fully who we are, and how we should act.

A COMMUNITY OF SUBJECTS

Perhaps the most quoted words of Thomas Berry are found in the phrase: "The universe is a community of subjects, not a collection of objects" (with Clarke 1992, 20). This, he says, we learn from science and primal peoples (1988, 91), and this is what we must relearn if we are to heal our relationship with Earth. We have already noted that everything in the universe is materially kin to everything else and that everything has an effect on everything else. Here, however, Berry seeks to help us form a more intimate relationship with other modes of being in the universe by recognizing their subjectivity. The universe is a community, a single and sacred community where everywhere is the center and where everyone is part of and an expression of the universe itself. Each aspect has its own vitality and center of being (1988, 134).

This realization is not something that Berry feels he is adding to science, but one that science itself reveals. This interior center or "self" of all subjects is also a spiritual self and a mode of divine expression. He believes that this primordial tendency—an inner subjectivity that, along with the tendency toward differentiation and then bonding, reveals some fundamental truth about the universe. Within this tripartite schematic Berry sees another way of viewing the Trinity (2009b,

81). This is also part of the revelation that solidifies the earth-human relationship.

The Primacy of Earth & the Human Response to It

Earth is an amazing place, an outcome of 14 billion years of creative evolutionary energy. Crossing the globe, we experience marvel after marvel. The problem, again, as Berry notes, is that we fail to see these marvels. We fail to recognize the grandeur and wonder of the planet, so we destroy it. We see it as a collection of objects for human use. What is needed, Berry claims, is a land mystique similar to that proposed by Aldo Leopold. Berry points to Leopold and similarly to John Muir, Rachel Carson, and Fredrick Church as modern prophets who have had this realization in the past and to whom we can look for inspiration about how we should act. These prophets call us to an intimacy with the land (2006, 36). This intimacy is highly important for Berry because we do not need a human answer to the problem of global climate change. Nor do we need a divine answer. We "need an earth answer to an earth problem" (1988, 35).

Thinking and acting as beings separate from Earth, humans created the crisis, and this mode of being cannot solve the crisis. There is also a danger of alienation in seeking divine intervention through prayer, hope, or faith. What is needed is Earth intervention. Earth, Berry says, is a "numinous material principle out of which all life emerges" (2009a, 75). It is the mode of divine expression that creatively evolved to a point where it could express itself in human form. One of the reasons that this is important, as Mary Evelyn Tucker points out in her preface to Berry's *Evening Thoughts*, is that this helps us find our way back home to Earth (2006, 10). It also helps us more faithfully define who we are.

Reinventing the Human: Anthropology in a Universe Context

Berry has determined that our cultural coding has overridden our genetic coding to disastrous effect (1988, 194). He feels that we need to re-image, re-invent, what it means to be human. We exist not primarily as individual separate physical, spiritual, and mental beings. That is only superficial appearance. If we could "see" on the subatomic level, we would not see the separateness. Indeed, we have many layers of "self." On some fundamental level, science reveals that humans are Earth conscious of itself. We do not live *on* Earth. We *are* Earth as it has expressed itself in a unique way, an amazing way—self reflective and aware. Whatever sin and salvation mean for humans, they need to be understood within this context.

We can extend this context a step further and see the universe as our larger "macro" selves (Berry and Swimme 1992, 27). In humans, as far as we know, the universe itself has come into this level of consciousness for the first and perhaps only time. Berry uses the term "Great Self" (1999, 168) to describe this layer of human being. Moreover, this consciousness has been present in the universe in some form since its inception, since the primordial flaring forth. For Berry, any conversation of a human "soul" or "spirit" must move away from one with exclusive focus on the micro self. It is all or nothing. We live and die, are graced, and are saved communally or not at all. We have a unifying principle "known traditionally as the soul," but this is not unique to the "micro self" human (1999, 79). Thus, if there is punishment for sin, it is collective. If there is salvation, it is universal.

Humans have unique intelligence, unique centers or spiritual selves, and the additional burden of responsibility for our actions, Berry insists, and all emerged out of the evolutionary process (2006, 114). For Berry humans are "that being in whom the universe reflects on and celebrates itself in conscious self-awareness" (2009b, 30). This understanding of humans is foundational for altering our consciousness and foundational of any anthropology within a dark green Catholic theology.

Often, after a few classes on cosmology and its significance, I ask my students, "how old are you?" It takes a few minutes, but if I have done my job well, they offer the correct answer—around 14 billion years old. We were there at the beginning. It is not that we were *at* the primordial flaring forth. We were the fireball, the initial forces, the first hydrogen atom, the first gaseous stars, the first galaxies. Everything about us was there and we have been evolving ever since. This is a remarkable thing and awareness of it should have deep therapeutic resonance. Digging deeper, perhaps darker, we come to an awareness that not only are we the universe, but at "the heart of the individual is everyone else" (Berry and Swimme 1992, 134). More than the fact that we are cousins to everything else, we are everything else. Everything else is part of our "Great Self" identity (Berry with Clark 1992, 22). This is the extent of our bonds of community. We could look at this as a form of mysticism. Berry looks at it as a scientific fact.

This also takes us to an awareness that, while we have a unique role to play in the universe—allowing it to think of itself, we are not "the superior mode of being in the universe" (Berry 2006, 41). We are special in our own way, but derivative and reverent to the whole at all times. The scientific story of the universe tells us about who we are, where we are, and the relationships within which we exist. Knowledge of this story reveals, heals, guides and disciplines us (1988, 124). Earth calls to us and gives us the very roadmap we need to address the climate crisis.

The Great Work: Human-Earth Relations

If the universe, and on the local level Earth, is a community of subjects, then our central interaction should be one of relationship between subjects. For Berry, the primary task of the human endeavor at this moment is our "great work"; to be "present to the planet in a mutually enhancing manner" (2009b, 47).

Our being present to the planet in this way will help steer us away from the current destruction. At its core, this entails a remaking of the human as a species. If every "thing," every subject, is intimately present to everything else, we need to come to awareness of and foster this intimacy. In doing so, we must recognize that we have a limited role with limited rights (2009b, 50). We must, as Berry says, "live on the planet on the planet's terms" (Berry with Clark 1992, 47). This is our task. It is the root of our spirituality. It is the root of our anthropology. It is the root of our theology. To think differently is the root of what Berry calls the "counterproductive, addictive, paralyzing, manifestation of a deep cultural pathology" (1988, 32; see also 2009b, 123).

Berry is hopeful that we can crawl out of this abyss and he points to many individuals and communities who have begun to live in communion with the planet. Part of the answer lay in humans spending more time away from the human made world of cement, clay, wires, and glass, and spending more time in and with the non-human made world. This may seem romantic, but it is not. It may seem like the luxury of the rich at the moment, and perhaps it is. Living in communion with Earth means, however, that there is one more responsibility to add to the calls for eco-justice from Rosemary Radford Ruether, Larry Rasmussen, and other scholars. Economic systems must allow for the poor and marginalized to experience intimacy with the natural world. This is difficult to do in a sweat shop.

Ultimately, the solution, for Berry, is not found by going back and restoring the truth of the past. The solution is found in "re-ordering the human in its relationship with the planet" and in its understanding of itself (2009b, 28). In coming to awareness of our place as the universe conscious of itself, we will come to community with the numinous quality within all and the source from which all things come (2009a, 118).

THE DIVINE WITHIN THE CONTEXT
OF THE UNIVERSE STORY

In most of his works, Berry does not use the word God much, and he believes the word is overused (Berry with Clark 1992, 10) to the point that it currently has little meaning. He insists that whatever this divine source is, it is "a mystery too vast for human comprehension" (1988, 198) and an "awesome, stupendous presence that cannot be explained adequately in human words" (Berry with Clark 1992, 11). What then can be said? Berry prefers terms like "originating power" (2009b, 116), "originating source" (1999, 56), "final term of reference" (2009a, 29), "mysterious forces" (2009a, 2), "all pervasive mysterious power" (Berry with Clark 1992,11), and "that numinous reality whence the universe came into being" (2009b, 49). Berry also refers to it as "the ineffable pervasive presence in the world about us" (Berry with Clark 1992, 1) and the "numinous reality" underlining it all (2009b, 57). Finally, Berry presents a view in line with Alfred North Whitehead and Gordon Kaufman when he explains that to "spontaneity as the guiding force of the universe can be thought of as the mysterious impulse where by the primordial fireball flared forth" (1988, 197).

Notable in this array is a lack of focus on categories within a theistic framework. The divine, in a Berrian system, will be largely devoid of these theistic underpinnings. In fact, Berry articulates a certain dependency with regard to the divine. With the universe as divine manifestation, in a certain way the divine is dependent on the universe to reveal itself for the "divine cannot be its own manifestation" (2009a, 44). This is how the divine reveals itself.

Even when we see ourselves as an aspect of divine manifestation, we need our bodies, souls, and minds—all aspects of universe—for this revealing. For Berry, we have no capacity to commune with the divine except that which is mediated through senses or consciousness

(2009a, 44). All of this is part of the evolutionary unfolding of the universe, so anything we can say of God and God's activities takes place within that sequence (2009a, 125).

Furthermore, according to Berry and supported by his interpretation of Aquinas (ST 1.47.1), the divine reveals itself most fully in the diversity of the entire universe (2009a, 19). Berry and Swimme note that the "story of the universe is a story of majesty and beauty as well as of violence and disruption" (1992, 7). Violence is one of the ways that the universe creates and it is part of the context. This means, of course, that, just like the universe, the divine is both wonderful and violent. The divine is life-giving and life-taking. The divine is made manifest through destruction, through cancer, and through plague. We need to take comfort in this. Within our current mode of consciousness, taking comfort in Berry's understanding of the divine is very difficult. Still, to answer the question, "What is God like?" we need to ask "What is the universe like?" That is how we come to know what God is like. Not surprisingly, there is little talk of intrinsic evil within a Berrian system. In a real sense, there is no room for it.

This sense of the divine is, therefore, fundamentally evolutionary. The world is in constant change and transformation and so is our experience of the divine. This is what the revelation of science presents to us as part of the ground of a dark green Catholic theology.

A Dark Green Catholic Theology

I think it is helpful, and in keeping with Berry's sources in the Buddhist traditions, to present here a short retelling of what is known as the poison arrow story. There was a man shot by a poison arrow. His friends came to his aid to remove the arrow before the poison could do its damage. The injured man stopped them and asked a series of questions: What kind of arrow was it? Was kind of poison was it? Who was the man who shot the arrow? All the while, his friends tried in vain to get him to allow them to remove the arrow, but the injured man refused until at last, not surprisingly, he died. The point of the story within the Buddhist tradition is to remind the listeners that they should be focused always and foremost on the immediate task at hand; the end of suffering and its causes. The other issues may have some importance, but until you take the poison arrow out, they have little immediate value.

Berry feels this way about many doctrines and ideas within Christianity. In one provocative statement in *Befriending the Earth*, he writes: "I suggest that we give up the Bible for a while, put it on the shelf for perhaps 20 years. Then we might have a more adequate approach to it" (Berry with Clark 1992, 75). Was he serious? Could we really do this with a wide swath of Christian doctrine?

Put It on the Shelf...for a While

Berry found many aspects of his beloved Catholic tradition to be presently problematic. He rarely talks about Jesus and says that our "[e]xcessive concern with the historical Christ is presently just not that helpful" (Berry with Clark 1992, 75). He says that a central focus on "an interior intimacy with the divine," following "Christian life disciplines," performing "spiritual and corporeal works of mercy toward others," or focusing "our lives on the gospel" (ibid.) leaves us missing our central task. As long as this is our primary concern, he says, "any concern about the universe or planet Earth has no great urgency, no overwhelming relevance to the spiritual process" (ibid.). Our preferential option or love cannot be exclusively for the human poor, it must be for Earth—which, of course, includes the poor.

With regard to our image of God, Berry notes that the first point that distracts us from acting to heal Earth "is a transcendent, personal, monotheistic creative deity" which "tends to desacralize the natural world" (2006, 25). Continuing, he contends that the "[e]mphasis we place on the spiritual soul of the human, a soul created immediately by God, a soul that establishes the human in some manner above or outside of rest of creation" sets us apart in a manner that is not true and is hurtful to the rest of the natural world (2009b, 39). Berry's indictment is sweeping. He even believes that "[t]he Ten Commandments are no longer adequate" and "[n]or does the Apostles' creed give us any direction" because, in memorizing these, "we learn nothing significant about the natural world" (Berry with Clark 1992, 47).

Some of these ideas may need to be "put on the shelf" for a while, as Berry urges. Only after we have awoken from our alienation, only after we have absorbed the truths revealed by science, and only after we have healed our deep cultural pathology will we then be able to return to some of these traditions and interpret them anew in the context of the universe and its evolving story. Berry offers a metaphor that is analogous to the poison arrow story when he describes our world as a sinking boat. When a boat is sinking, no matter what the difficulties within the boat, the health of the boat must be addressed first. There is indeed value in the Bible and tradition, but I fear he is correct when he warns that "[w]e will drown reading the book" (ibid.). The life systems of the planet require a much more radical approach.

Two specific items of Christian doctrine come under particular attack by Berry in several of his writings. These are what he perceives to be the over occupation with transcendence and redemption. He sees both as promoting a detachment from the world, ourselves, and the divine. The very idea that we can transcend the universe and the idea that we can be redeemed apart from the world contribute to the problem. Ideas lead to actions. Berry insists that our "excessive emphasis on transcendence is leading us to destroy the planet" (Berry with Clark 1992, 19). Transcendence comes in many forms. Berry highlights our image of the transcendence of God, of the spiritual aspect of humans, of the mind, and of our historical and millennial destiny (2006, 25-27). While there is value in the idea of divine transcendence, our deliberations on the idea are just not helpful right now.

Berry's critique of focusing on redemption is the emphasis given to its transcendence of Earth. That we seek to be redeemed "from a flawed world" and "long for our true home in some heavenly region" (2009b, 39) are, according to Berry, mistaken beliefs. We may need to be redeemed, but our redemption is not individualistic. It is communal. It is universal. We know this to be true because the "universe story and the human story are a single story" (2009b, 83). If there is redemption, it takes place for the whole universe, not just this micro part of it that we call humanity.

At some point, these ideas will need to be rethought within the new context of the divine revelation of the universe story. There is sin. There is grace. There is salvation. However, all of these theological notions must be evaluated from the perspective that the rest of the natural world needs to be included in each. When we sin, the world sins—not just because our actions affect the world, but because on a fundamental level, we are the world. Grace and salvation therefore are required for our larger selves as well as our smaller selves. There is no more a sense that we can be saved or graced while the rest of the universe passes away or is left behind, then we can walk to the store and leave behind our mind, our hearts or our legs.

Resources for a Dark Green Catholic Theology

The two primary resources for a dark green Catholic theology are the natural world and our scientific study of it. Science reveals the manner by which the divine has created and sustains the world. Theology needs to utilize the tools and discoveries of science not primarily to validate already pre-conceived ideas, but to rethink them and to discover new ones. Furthermore, it is the role of theology to interpret, with an appropriate sense of awe and wonder, the content that science reveals. We should, as Berry says, "situate theology in terms of the understanding the meaning of science" (Berry with Clark 1992, 26). This is also one of the points that John Haught makes in *Is Nature Enough?*

Science reveals our most sacred story, but what is equally needed as a resource is our direct experience of the earth. This connects us with the numinous reality. Berry cites the experience of Aldo Leopold with the "fierce green fire dying in the eyed of the wolf" and Jane Goodall's communion with chimpanzees (2009a, 149) as examples of the kind of intimacy we could and must access if we are to build a functional future. A more suburban, but equally valuable, example is offered by Brian Swimme where he narrates his encounter with a bug under a street lamp (1994, 241).

Although Berry feels that the Bible cannot help us in this moment of crisis (2006, 57) and that the Christian tradition is "inadequate" (2006, 101) and "no longer sufficient" (2009b, 71) to supply the cultural therapy required to respond to the imperiled Earth, he nonetheless does offer support from these sources. Berry occasionally references the biblical tradition but he largely looks to the "Cosmic Christ" of Johanine and Pauline traditions (2009a, 22). He believes that the universe has a Christic dimension from its inception and this aspect remains in every part. He also looks to the Celtic monastic tradition and that of St. Francis, as well as, Hildegard of Bingen, Scotus Eriugena, Aquinas, Roger Bacon, and Duns Scotus for additional support within the Catholic tradition.(2009b, 86-7). Finally, and most importantly, though not exhaustively, he looks to the writings of Pierre Teilhard de Chardin in whom he finds the best modern guide.

Berry believes that a functional Catholic theology must not only be in dialogue with other great religious traditions of the world, but it must have its content influenced by them. Each religious tradition exists in its own proper mode and each is complete in its own mode. He makes as many references to Confucian traditions and those of native people as he does to the Catholic tradition. From his perspective, they teach us a kinship with nature, with other creatures, and with the "powers" of the earth and universe that is sorely lacking in the Catholic tradition. He even notes that "the salvation of Christians lies in the unassimilated elements of paganism" (Berry with Clarke 1992, 21). In truth, any source that leads to developing a communion with and compassion for the suffering Earth is valid for theology in an ecological age when humans are forcing changes to the global climate that can be so devastating.

CONCLUSION

For over two thousand years, theologians have made proclamations on the nature of God and God's actions in the world. Our confidences in the authoritative and complete nature of our understanding have led to a loss of wonder and loss of connection to the natural world. We are very skilled, we think, at explicating the nature of the divine and at how best to talk about it. This too, says Berry, is part of our current pathology and delusion (2009a, 150). The central reason why the truths arrived at are flawed, is that our theology has not come from "thinking or acting within this accepted story of the universe" (2009b, 26).

As noted above, within a Berrian system, the explicit goal is to create a "mutually enhancing mode of human-earth relations" (2006, 20). To do this, a dark green Catholic theology must re-envision sacraments, liturgy, ritual, Christology, and Trinitarian faith seeking understanding. This is in line with what Larry Rasmussen calls "doing our first works over" (2009, 1). Everything must be rethought and redone within this new context.

The new understanding will clearly not be the Vatican II Catholicism of my mother, but my mother's Catholicism cannot meet the challenge of global climate change that we are facing today. Nor are any of the proposed stewardship models strong enough. The Church and its theology will change, or it will become irrelevant. So too will the human species change, or it will become prematurely extinct.

There is much work to be done in explicating in more depth what a Berrian dark green Catholic theology might look like. Issues associated with theodicy and personal responsibility for sin are just a few of the issues that need to be addressed. Still, it seems clear that re-engineering what it means to be human, healing our relationship with Earth, and recognizing the revelation of modern science are warranted. In keeping with Berry's position, our task as theologians is to talk about the divine sparingly. This may seem counter intuitive to the nature of "the-ology," but it does seem like the proper place to start. The entire theistic framework needs to be "put on the shelf." Although Berry's writings are in a very limited sense in conversation with contemporary theologians, a strong affinity can be seen between his efforts and the work of Elizabeth Johnson (1992; 2000) and Sallie McFague (1993; 2008). Speculation about what or who God is in God's own being without reference to the universe is not fruitful at the present time. We need to transform our consciousness. We need to remove the arrow.

If we do not respond appropriately, Earth will. Earth's response will coerce a new balance in which we realize our place. There will be an Earth answer to the Earth problem. Whether or not this answer is one that will transform humans or extinct them depends on our actions now. A Berrian inspired dark green Catholic theology can help us choose transformation.

SOURCES

Benedict XVI, Pope and Woodeene Koenig-Bricker. 2009. *Ten Commandments for the Environment: Pope Benedict XVI Speaks Out for Creation and Justice*. Notre Dame: Ave Maria Press.

Berry, Thomas. 1978. "The New Story." *Teilhard Studies*, Number 1.

———. 1988. *The Dream of the Earth*. San Francisco: Sierra Club Books.

———. 1994. "Ecological Geography." In *Worldviews and Ecology: Religion, Philosophy and the Environment*, ed. Mary Evelyn Tucker and John Grim, 228-37. Maryknoll: Orbis Books.

———. 1999. *The Great Work: Our Way into the Future*. New York: Bell Tower.

———. 2000. "Christianity's Role in the Earth Project." In *Christianity and Ecology: Seeking the Well Being of Earth and Humans*, ed. Dieter Hessel and Rosemary Radford Ruether, 127-34. Cambridge: Harvard University Press.

———. 2006. *Evening Thoughts: Reflection on the Earth as a Sacred Community*. Edited by Mary Evelyn Tucker. San Francisco: Sierra Club Books.

———. 2009a. *The Sacred Universe: Earth Spirituality and Religion in the Twenty-First Century*. Edited with forward by Mary Evelyn Tucker. New York: Columbia University Press.

————. 2009b. *The Christian Future and the Fate of the Earth.* Edited by Mary Evelyn Tucker and John Grim. Maryknoll: Orbis Books.

Berry, Thomas, with Thomas Clark, S.J. 1992. *Befriending the Earth: A Theology of Reconciliation Between Humans and the Earth.* Mystic, CT: Twenty-Third Publications.

Berry, Thomas and Brian Swimme. 1992. *The Universe Story: From the Primordial Flaring Forth to the Ecozoic Era—A Celebration of the Unfolding of the Cosmos.* San Francisco: Harper.

Boff, Leonardo. 1997. *Cry of the Earth: Cry of the Poor.* Maryknoll: Orbis Books.

Bouma-Prediger. 1995. *The Greening of Theology.* Atlanta: Scholar Press.

Catholic Bishops Conference of the Philippines. 1988. *What Are We Doing to Our Beautiful Land?* SEDOS 4, 112-15. Accessed from http://www.aenet.org/haribon/bishops.htm on April 17, 2010.

Catholic Bishops of Alberta Canada. 2008. "Celebrate Life: Care for Creation." Accessed from http://www.inee.mu.edu/documents/24CelebrateLifeCareforCreation-Alberta_000.pdf on April 17. 2010.

Catholic Bishops of the Watershed Region. 2001. *The Columbia River Watershed: Caring for Creation and the Common Good.* Seattle: Columbia River Watershed Project. Accessed from http://www.thewscc.org/files/pastoral-english.pdf on April 17, 2010.

Christiansen, Drew S.J. and Walter Grasher, eds. 1996. *And God Saw That It Was Good–a Catholic Theology and the Environment.* Washington DC: United States Catholic Conference of Bishops.

Conradie, Ernst. 2006. *Christianity and Ecological Theology: Resources for Further Research.* Stellenbosch: Sun Press.

Delio, Ilia. 2008. *Christ in Evolution.* Maryknoll: Orbis Books.

Edwards, Denis. 1995. *Jesus the Wisdom of God: An Ecological Theology.* Maryknoll: Orbis Books.

————. 1992. *Made from Stardust.* North Blackburn, Victoria, Australia: Collins Dove.

Forum on Religion and Ecology at Yale University. 2010. "Publications" and "Resources for Educators." Accessed from www.yale.edu/religionandecology on May 26.

Fragomeni, Richard. 1994. *The Ecological Challenge: Ethical Liturgical and Spiritual Responses.* Collegeville: Liturgical Press.

Gardner, Gary. 2002. *Worldwatch Paper #164: Invoking the Spirit: Religion and Spirituality in the Quest for a Sustainable World.* Washington, DC: Worldwatch Institute.

Haught, John. 2006. *Is Nature Enough? Meaning and Truth in the Age of Science.* Cambridge: Cambridge University Press.

Hessel, Dieter and Radford Ruether, Rosemary, eds. 2000. *Christianity and Ecology: Seeking the Well Being of Earth and Humans.* Cambridge: Harvard University Press.

Horrell, David and Cheryl Hunt, eds. 2010. *Ecological Hermeneutics: Biblical, Historical and Theological Perspectives.* New York: T & T Clark Publishers.

Intergovernmental Panel on Climate Change. 2007. Accessed from http://www.ipcc.ch/ on May 10, 2010.

John Paul II, Pope. 1990. *The Ecological Crisis: A Common Responsibility.* Washington, DC: United States Catholic Conference of Bishops.

Johnson, Elizabeth. 1993. *Women, Earth, and Creator Spirit.* New York: Paulist Press.

———. 2000. "Losing and Finding Creation in the Christian Tradition." In *Christianity and Ecology: Seeking the Well Being of Earth and Humans,* ed. Dieter Hessel and Rosemary Radford Ruether, 4-21. Cambridge: Harvard University Press.

Kaufman, Gordon. 1992. "Nature, History and God: Toward an Integrated Conceptualization." *Zygon* 27: 379-401.

Kerns, Laurel and Catherine Keller. 2007. *Ecospirit: Religion and Philosophers for the Earth.* New York: Fordham University Press.

Kwok Pu-lin. 1999. *Christology for an Ecological Age.* New York: Continuum.

Lonergan, Ann and Caroline Richards, eds. 1987. *Thomas Berry and the New Cosmology.* Mystic, CT: Twenty-Third Publications.

Matthews, Clifford, Mary Evelyn Tucker and Phillip Hefner, eds. 2002. *When World's Converge: What Science and Religion Tell Us about the Story of the Universe and Our Place in It.* Chicago: Open Court.

McDougall, Dorothy. 2003. *The Cosmos as Primary Sacrament: The Horizon for an Ecological Theology.* New York: Lang.

McFague, Sallie. 1993. *The Body of God: Toward and Ecological Theology.* Minneapolis: Fortress Press.

—. 2008. *A New Climate for Theology: God , the World and Global Warming.* Minneapolis: Fortress Press.

Radford Ruether, Rosemary. 1992. *Gaia and God: An Ecofeminist Theology of Earth Healing*. San Francisco: Harper Collins.

————. 2000. "Ecofeminism: The Challenge to Theology." In *Christianity and Ecology: Seeking the Well Being of Earth and Humans*, eds. Dieter Hessel and Rosemary Radford Ruether, 97-112. Cambridge: Harvard University Press.

Rasmussen, Larry. 1997. *Earth Community, Earth Ethics*. Maryknoll: Orbis Books.

————. 2009. "Doing our First Works Over." *Journal of Lutheran Ethics* 9.4 (April). Accessed from http://www.elca.org/What-We-Believe/Social-Issues/Journal-of-Lutheran-Ethics/Issues/April-2009/Doing-Our-First-Works-Over.aspx on April 20, 2010.

Schaefer, Jame. 2009. *Theological Foundations for Environmental Ethics: Reconstructing Patristic and Medieval Concepts*. Washington, DC: Georgetown University Press.

Swimme, Brian. 1985. *The Universe is a Green Dragon*. Santa Fe: Bear.

————. 1994. "Cosmogenesis." In *Worldviews and Ecology: Religion, Philosophy and the Environment*, ed. Mary Evelyn Tucker and John Grim, 238-42. Maryknoll: Orbis Books.

Taylor, Bron. 2009. *Dark Green Religion*. Berkeley: University of California Press.

Tucker, Mary Evelyn. 2006. "Preface." In *Evening Thoughts: Reflection on the Earth as a Sacred Community*, ed. Mary Evelyn Tucker, 9-14. San Francisco: Sierra Club Books.

Tucker, Mary Evelyn and John Grim, eds. 1994. *Worldviews and Ecology: Religion, Philosophy and the Environment*. Maryknoll: Orbis Books.

United States Conference of Catholic Bishops. 1991. *Renewing the Earth*. Washington, DC: USCCB.

United States Conference of Catholic Bishops. 2001. *Global Climate Change: A Plea for Dialogue, Prudence and the Common Good*. Washington, DC: USCCB. Accessed from http://www.usccb.org/sdwp/international/globalclimate.shtml on April 17, 2010.

Waldau, Paul and Patton, Kimberley. 2006. *A Communion of Subjects: Animals in Religion, Science and Ethics*. New York: Columbia University Press.

World Council of Churches. 1990. *Now is the Time*. Geneva: World Council of Churches.

FEMINIST PERSPECTIVES

RUSHING WINDS & RISING WATERS

SEEKING THE PRESENCE OF GOD IN A RADICALLY CHANGING WORLD

Colleen Mary Carpenter

We have always lived in a dangerous world. The fishermen of Breton have traditionally (and humbly) prayed, "Oh God, the sea is so vast and my boat so small." The psalmist cries, "Save me, God, for the waters have reached my neck … the flood overwhelms me" (Ps 69:2, 3). In the Lorica, St. Patrick asked for "God's shield" to protect him from a litany of evils, both natural and personal, including burning, drowning, poisoning, and wounding. Yet in that same prayer, St. Patrick invoked the light of sun, brilliance of moon, splendor of fire and stability of Earth as God-sent protectors from harm (Redmont 1999, 115-16)). Here we see that the world is both dangerous and an obvious source of God's goodness, protection, and care for us. Like St. Patrick, Christians have traditionally known, clearly, that the world was good; we also knew equally clearly that the world was vast, powerful, and not always safe. Even as we were reassured by the repeated statement in Genesis that God saw that the world was good, we understood that "good" was not the same as controllable, or easy to manage. True goodness—God's goodness— is neither small nor humanly "manageable." C.S. Lewis captured this dimension of the goodness of God in *The Lion, the Witch, and the Wardrobe*, in his depiction of Aslan, the great lion and the son of the Emperor Beyond the Sea. When the children ask nervously if Aslan is "quite safe," Mr. Beaver scoffs at such foolishness. "Safe? 'Course he isn't safe. But he's good" (Lewis 1950, 80).

Like Aslan, our world is good, but not safe. Or at least: it used to be good. Current scientific understandings of what is happening to our planet—what we have done to our planet—suggest that our future will be one of rising seas, shortages of fresh water, unpredictable extreme weather (more droughts, floods, and hurricanes), and baffling new challenges for agriculture as new pests invade and new weather patterns disrupt the growing season. It is an ugly, ugly picture, and it is no surprise that many people respond to it with terror, despair—and denial. It is an especially difficult picture for us as Catholics to grasp, because we are a sacramental people, and we have traditionally taken great comfort in finding the spirit of God in the world, in the land and water and wind and fire that surrounds us. Rivers and lakes remind us of the waters of our baptism; a cool breeze on a hot day is a symbol of the breath of God moving all around us, always. We do not identify God with the world, but we see creation as revelatory of God, God's goodness, and God's presence. We live in an sacramental world, a world where God is not always far away and beyond our reach, but where God is both beyond us and all around us, as near to us as our own breath while still being greater than the fires of all the stars we see on a clear night.

This sacramental sense of God's presence in the world around us is threatened by changes we are forcing on the global climate. If we have damaged God's creation so deeply that it no longer supports us but instead seems to oppose our lives and health at every turn, and if we learn to fear the winds and waters and to mistrust the land that we have poisoned, how will that change our sense of God's presence in the world? Will the world around us come to be seen as revelatory not of God's goodness, but of God's anger, even God's punishment? Will we come to see God as arbitrary, dangerous, and unpredictable, instead of as the Merciful and Steadfast One who causes the rains to fall on the good and bad alike? These questions point us to the fact that climate change is not just an ecological problem; it is a theological one—a problem of the sacramental imagination. How do we place our new experiences into a pattern that makes sense? What aspects of our long and complex tradition will we choose to emphasize? What story will we tell ourselves about what is happening? We face difficult choices about how we will choose to understand this new world we are bringing into being, and we need to be careful about how we choose to interpret our experience. The choices we make at this point in our history will dramatically shape the kind of religious experience of creation that our children and our children's children will have available to them. This is true not simply in the practical sense of our decisions today having an impact on how dramatically climate change plays out, but also in the directly imaginative sense that we will be the ones offering a framework for interpreting a changed creation; we will be the ones to choose the images and stories from our tradition that will attempt to make sense of what is happening all around us.

This creative theological practice of mining our tradition for help in understanding a dramatically new situation is not a novelty; it is a deeply-rooted Christian practice. It goes all the way back to the first days of Christian history, when the followers of Jesus turned to the psalms and prophets in an effort to find a way to make sense of Jesus' death and resurrection. Today, as theologians struggle with the new issue of climate change, we are finding again that our ancient tradition offers us a wealth of resources. Interestingly, current theological work on interpreting our relationship to a changed creation has revived a long-neglected aspect of our tradition—pneumatology. After centuries of neglect, the study of the Holy Spirit has recently become an area marked by creative proposals and lively disputes, by a retrieval of tradition as well as an exploration of new themes. Finding the right way to speak of the Holy Spirit in the world seems to be the key to how we will imagine, and thus experience, God's presence in the midst of climate change. There is not yet a consensus on how to speak properly of the Spirit of God and climate change; in fact, two of the most prominent alternatives are diametrically opposed to one another. Should we understand the Spirit in our world today as wounded, traumatized, and in agony? Or do we see the Spirit as a vivifying presence that gives us hope in the midst of the death we have unleashed on the world?

Choosing between those two approaches is not easy. Describing the Holy Spirit of God as wounded and traumatized is disturbing, and yet it is also disturbing to leap glibly over the agony of creation today with a smiling focus on resurrection hope. Do we turn our gaze to the radically new extent of sin and death in the world, or does that deny the fundamental Christian conviction that death leads to resurrection? On the other hand, do we see an insistence on life and hope as an encouraging path to the future, or a misguided attempt to avoid reality? In this essay, I will examine both of these alternatives—conceiving of the Spirit today as either the Wounded Spirit or the Spirit of Life—and weave together the insights offered by each framework into a vision of how we might most thoughtfully, honestly, and humbly continue to find the presence of the Spirit of God in the world.

CLIMATE CHANGE AS A
GENUINELY NEW PROBLEM

The world is not a safe place; it never has been. Floods, earthquakes, and deadly storms are not new: we read about such devastating natural occurrences in the Bible, in ancient historical records across the globe, and of course in the newspaper today. So why is today any different? Are we really at a crisis point, both environmentally and theologically?

The answer to that question is a decisive "yes." The Fourth Assessment Report of the Intergovernmental Panel on Climate Change (IPCC), issued in 2007, presents a terrifying, shocking, and "very conservative" picture of the current scientific consensus about changes that are already being seen in the world due to climate change (McFague 2008, 11). While many people think of climate change as something that might happen in the future (and too many think of it as something that is not likely to ever happen), it has become clear that our world is already being remade. The oceans are already becoming more acid; storms are increasing in intensity; the Rocky Mountains are blighted by millions of acres of dead pine trees; only 25 of the 150 glaciers in Glacier National Park remain; the Himalayan Glaciers, which supply more than half of the drinking water for 40% of the world's population, are disappearing (McFague 2008, 10-17). Even more disconcertingly, we are now seeing changes that are not simply a direct result of human activity, but a result of positive feedback loops in which a (relatively) small initial change produces conditions that accelerate and reinforce the change. For example, melting permafrost does not just alter the landscape of Siberia: it also releases tons of methane into the atmosphere. Methane is "a greenhouse gas twenty times more potent than carbon dioxide," and the amount that scientists now estimate is being released by melting permafrost could "double atmospheric levels of the gas, leading to a 10% to 25% increase in global warming" (Sample 2005). The process that we unwittingly started is no longer under our control: it is accelerating on its own (McFague 2008, 12-13).

Environmental activist Bill McKibben argues that the changes have been so sudden and severe that we no longer live on the same planet that nurtured us when we were children. The world of the past ten thousand years existed in "the sweetest of sweet spots," and its regular temperature and predictable weather enabled us as human beings to thrive. However, we no longer live in that world; we live, McKibben argues, on a different planet (2010, 1-2). This new planet is marked by drought, glacier melt, an acidifying ocean, and strange new phenomena such as "enormous sheets of a mucus-like material" forming in the seas, killing fish and other animals that come in contact with it (26-7). Climate change has altered the world so thoroughly that we are starting to experience second and third-order effects, most completely unexpected. For example, melting of sea ice changes reflectivity in the Arctic, but we had not counted on the death of phytoplankton, nor do we have any real sense of how that will play out across the world's oceans.

In the "old Earth," before the industrial revolution and the dramatic rise of greenhouse gas emissions, the atmosphere held about 275 ppm of carbon (McKibben 2010, 13). Today that number is 390, and scientists believe that the highest level at which our civilization is sustainable is much lower than that—350. Above that level, we "threaten the ecological life-support systems that have developed in the late Quaternary environment, and severely challenge the viability of contemporary human societies" (Rockstrom 2009, 473). Moreover, it is not just "human societies" at risk today, but a great part of plant and animal life, both land- and sea-based, throughout the globe: the rate of species extinction "is estimated to be 100 to 1,000 times more than what could be considered natural" (Rockstrom 2009, 474), and "up to 30% of all mammal, bird, and amphibian species will be threatened with extinction this century" (474). We are, Bill McKibben argues, transforming our planet, "sabotaging its biology, draining its diversity, affecting every other kind of life that we were born onto this planet with. We're running Genesis backward, de-creating" (2010, 25).

De-creating. Human activity is "de-creating" God's creation. It was once unimaginable that we could do so much damage; it is now becoming unimaginable that we will escape it.

IMAGINATION, INTERPRETATION, &
THE SPIRIT OF GOD

In arguing that how we *imagine* the Spirit in relationship to the world today is a significant theological issue, I am making use of particular understandings of sacrament and sacramentality, the role of the imagination in faith, and the relationship between imagination and reality. Accordingly, before advancing to a discussion of the Spirit, it is necessary to spend some time with these linked notions of sacrament, imagination, and truth.

In *Sacraments and Sacramentality*, Bernard Cooke lays out an understanding of the basic sacramentality of the Christian life that begins with a consideration of what, exactly, it means to be human. According to Cooke, to be human involves at least three characteristics: to be *conscious*—to be aware of what is going on around us; to be able to reach out to those around us and to *love*; and, to know *freedom* and to have the power to shape ourselves and our future. "To live this way," Cooke tells us, "alert, aware, concerned, and loving and open to others, free and self-determining—does not come easily. It is a challenge…[and] our Christian faith tells us that this goal would be beyond us if it were not for the personal help of God" (Cooke 1994, 13).

Cooke begins with this understanding of the human—conscious, loving, and free—in order to explore what it might mean that our humanness is transformed in sacrament. Sacraments are for people, he reminds us, and they are meant to make God's presence, God's grace, known to us. A sacrament, according to Cooke, is "a specially significant reality that is meant to transform the reality of 'the human' by somehow bringing persons into closer contact with the saving action of Jesus Christ" (10). This definition of sacrament is not limited to "certain formally religious actions;" rather, it "touches everything in our life that is distinctively human" (2). After all, Cooke argues, "Christians' faith is meant to embrace and transform all aspects of their being and activity. What it means to be human, what it means to grow as a human, what it means for people to share life as humans—these have been changed forever by what happened in Jesus' life and death and continues to happen in the mystery of his resurrection" (11).

Thus, a sacramental understanding of our lives is one through which we are *awake* to the presence of God in our lives, a presence that embraces us, changes us, and brings us to union with Christ. In order to gain that kind of understanding of our lives, we need to live reflectively, to work to interpret our experience correctly, to allow ourselves to be shaped by the Christian story. Cooke argues that we need to attend carefully to the ways in which we interpret our experience, since we usually interpret it not simply according to the Christian story but according to the myths and models of our family, community, and nation, and some of those myths and models are more appropriate than others. How accurate, he asks, is the story in which we have been raised? "How adequate, or misleading, are the models according to which we have been educated to view reality?" (33). In order to judge the lens through which we are interpreting our lives, Cooke says, we need a hermeneutic of experience, "a set of principles, insights, and critical judgments that equips us to interpret our experience in a more accurate and more profound way" (33).

Interestingly, Cooke's definition of a hermeneutic of experience dovetails quite neatly with biblical scholar Sandra Schneiders' definition of the imagination. Schneiders defines the imagination as "our constructive capacity to integrate our experience into dynamic and effective wholes which then function as the interpretive grids of further experience" (1986, 16). In other words, it is our imagination that enables us to take the raw data of our experience, turn that into a story or model, and then use that model as a reference point as new things happen to us. If our imagination is "dynamic and effective," as Schneiders argues (or, in Cooke's terms, if we are using an appropriate hermeneutic of experience), our interpretation of the world *does* respond to new data, and with more and more data, our interpretation of the world comes closer and closer to helping us understand what is really going on.

And that is what we are trying to figure out in our lives: what is really going on? Do we really live in a world where power is the measure of all things? Which is stronger, a tank or a single person standing in its way? Is human evolution a winner-take-all story of cut-throat competition, or is it the story of communities that cared for their sick and elderly and tended one another so that the many survived where the individual could not? *What is really going on?* Christians answer that question by putting their own experience in conversation with the life and death and resurrection of Jesus. Who is my neighbor? What is the kingdom of heaven like? When did I see you hungry, Lord, and feed you? My God, my God, why have you forsaken me?

Our imagination—our interpretation of our own experience and of
the world around—is shaped by how Jesus answered these questions,
and how we slowly slowly learn to live those answers in our own lives.

In order to live sacramentally, we need to attend carefully to how we
see the world, how we are interpreting the world. We need to educate
our imaginations; we need to develop the insight and judgment to understand what is happening around us. In particular, as we face the
crisis of climate change, we need to rethink our understanding of the
relationship between the Holy Spirit of God and the created, damaged world around us.

THE WOUNDED SPIRIT & THE SPIRIT OF LIFE

How shall we imagine the Spirit today? What elements of our tradition will help us formulate a hermeneutic of experience adequate to
our new situation? Here I will consider the pneumatological proposals
of Mark Wallace and Elizabeth Johnson, both of whom turn to the
Spirit in their discussions of the ecological disaster unfolding around
us. Within the context of "de-creation" described by McKibben and
others, it is perhaps not surprising that Mark Wallace argues that we
need to see the Creator Spirit as being harmed by our devastation of
the planet. This portrayal of the Spirit of God as injured and helpless
in the face of human sin is shocking, even horrifying. Johnson, on the
other hand, describes a joyous, vivifying Spirit whose healing presence
gives us hope. In the end, I believe that Johnson's imaginative portrayal
of the Spirit of Life surrounding and supporting us corresponds more
faithfully to the totality of our experience of God and the world, but
I remain haunted by Wallace's vision of the Wounded Spirit, and cannot forget that grief and lament have an important place in any Christian account of the history of our world.

Theologian Mark Wallace writes with passion about the theological problems raised by our current environmental crisis, which he
terms "ecocide" (Wallace 2000, 313). Ecocide is in the end, he argues,
a spiritual disease, an addictive pattern of destructive behavior that we
recognize as deadly even as we find ourselves powerless to escape it
(313-14). Ecocide is, simply put, sinful, and the fight against sin is not
something humans can take on alone. Defeating sin is ultimately the
work of God; thus, Wallace turns to tradition, to pneumatology. He
claims that "hope for a renewed earth is best founded on belief in the
Spirit as the divine force within the cosmos who continually indwells

and works to sustain all forms of life" (314). A thoroughly traditional understanding of the Spirit as "the Lord, the Giver of Life" will be the bulwark against human de-creation. Wallace is setting the Spirit of Life over against the ecocidal human pursuit of Death. However, Wallace recognizes that many contemporary Christians experience the Spirit only "as the forgotten member of the Trinity, the shy member of the Godhead, the left hand of God" (314). In order to combat this oversight—especially since a robust understanding of the Spirit's life-giving power is central to a Christian response to ecocide—Wallace "reenvisions" the Spirit. He offers us not a "ghostly" or "disembodied" Spirit, but "an earth-centered model of the Spirit as the 'green face' of God who sustains the natural order" (314). The key phrase here is "earth-centered": the earth-centered Spirit is an embodied Spirit, even an embodied "being," and "suffers loss and pain whenever the biotic order is despoiled through human arrogance." Here we reach the point where Wallace's new vision of the spirit is truly new, truly a departure from tradition: "Because God as Spirit is enfleshed within creation, God experiences within the core of her deepest self the agony and suffering of an earth under siege. *The Spirit then, as the green face of God, has also become in our time the wounded God*" (315).

Elizabeth Johnson also has contemplated the relationship between pneumatology and the ecological crisis we now face. In her 1993 Madeleva lecture, *Women, Earth, and Creator Spirit*, she opens with reflections on ecocide, and goes on to reflect on both the life-giving and feminine symbolism associated with the Spirit. Where Wallace claims that the association of Earth and the Spirit should lead us to see the Spirit as wounded, vulnerable, and at risk of death, Johnson instead points out that this association between matter and spirit has a radiantly hopeful character: "A theology of the Creator Spirit overcomes the dualism of spirit and matter…[and] leads to the realization of the sacredness of the earth. The Spirit of God dwelling in the world with quickening power deconstructs dualism and draws in its place a circle of mutuality and inclusiveness. Instead of matter being divorced from spirit and consigned to a realm separate from the holy, it is an intrinsic part of the cosmic community, vivified, indwelt, and renewed by the Creator Spirit. The Spirit creates matter. Matter bears the mark of the sacred and has itself a spiritual radiance. Hence the world is holy, nature is holy, bodies are holy.… About the Creator Spirit this can be said: loves bodies, loves to dance. The whole complex, material universe is pervaded and signed by her graceful vigor" (59-60).

Graceful vigor...or unending agony. These two visions of the relationship between Earth and the Spirit are both rooted in a clear-eyed assessment of the grave danger we face. In the end, then, we have a choice about how we see the Spirit and the world today: we can see the Spirit enfleshed in the world as cruciform, suffering, and deeply wounded by our destructive actions—or we can see the same wounded world being tended by the vivifying, restoring Spirit. The choice is not and should not in fact be an easy one, and in the end it rests not in theological argumentation but in our imaginations: in how and what we choose to see. How are we going to integrate our experience of ecological devastation with our experience of God—and what interpretive grid are we going to shape in order to help us understand further experiences of the world and of God?

Wallace's understanding of a God wounded by ecocide is not, as it seems at first glance, a proposal born of despair. Rather, it is a proposal that seeks to emphasize the power of the Holy Spirit, and a proposal that finds hope in the Spirit's life-giving nature. Although I believe it fails in the end, this understanding of the Spirit begins in very traditional, and very hopeful, territory. Wallace opens his consideration of the Spirit by recalling Biblical expressions of the Holy Spirit as vivifying breath, healing wind, divine dove and living water. He argues that the Spirit "reveals herself in the biblical literatures as an earthly lifeform who labors to create, sustain, and renew humankind and otherkind" (316). The choice of the word "lifeform" is unusual, but his point that we find the Spirit active in the world is both important for his argument and absolutely standard in pneumatological reflection. There is no disputing that "of all the activities that theology attributes to the Spirit, the most significant is this: the Spirit is the creative origin of all life. In the words of the Nicene Creed, the Spirit is *vivificantem*, vivifier or lifegiver" (Johnson 1993, 42). Wallace would certainly agree with Johnson that "as the continuous creative origin of life, the Creator Spirit is immanent in the historical world. 'Where can I go from your presence,' sings the psalmist, "and from your Spirit where can I flee?' ...The Spirit fills the world and is in all things." (Johnson 1993, 42).

From here, however, Wallace and Johnson decisively part company. Whereas Johnson moves immediately from an acknowledgement of the Spirit's presence to a reminder of the Spirit's transcendence,

Wallace rejects this and instead moves deeper into considering the immanence of the Spirit. "As once God became human in the body of Jesus," Wallace argues, "so continually God enfleshes Godself in the embodied reality of life on earth. Quintessentially, then, both Spirit and
earth are life-givers: the Spirit ensouls the earth with the quickening
breath of divine life and the earth enfleshes the Spirit as it offers spiritual and physical sustenance to all living things" (Wallace 2000, 320).
The upshot of this dramatic and disturbing claim is that: "[I]t appears
that God as Spirit is vulnerable to serious loss and trauma just insofar
as the earth is abused and despoiled. In an earth-centered model of
the Spirit, God is a thoroughgoing incarnational reality who decides
in freedom, and not by any internal necessity, to indwell all things.
In making this decision, however, the Spirit places herself at risk by
virtue of her coinherence with a continually degraded biosphere. God
then, is so internally related to the universe that the specter of ecocide
raises the risk of deicide: to wreak environmental havoc on the earth
is to run the risk that we will do irreparable harm to the Love and
Mystery we call God" (320).

Here Wallace's argument breaks down. The idea that we are doing
"irreparable harm" to God cannot stand. It gives human beings (and
human sin) far too much power; it neglects absolutely the character of
the Spirit as lifegiver, as the "divine creative power [that]…renews the
face of the earth." Johnson reminds us that we sing of the power of that
renewal every Pentecost: "The damaged earth, violent and unjust social
structures, the lonely and broken heart—all cry out for a fresh start. In
the midst of this suffering the Creator Spirit, through the mediation
of created powers, comes, as the Pentecost sequence sings, to wash
what is unclean, to pour water upon what is drought-stricken, to heal
what is hurt, to loosen up what is rigid, to warm what is freezing, to
straighten out what is crooked and bent" (1993, 43).

The Spirit does not struggle in agony over what is hurt; the Spirit
heals what is hurt. The Spirit does not find herself contorted and bent
over, but straightens what is crooked and bent. The Spirit is not a victim of deicide; the Spirit is Life! Wallace, however, sees the Spirit as
cruciform, as suffering the agony of crucifixion—and not just once
long ago in the death of Christ, but every day. "The Spirit's suffering
from persistent environmental trauma engenders chronic agony in
the Godhead," he claims (2000, 321). Yet trauma is a specifically human experience, one that includes not just suffering but an interior

response of being overwhelmed, being unable to resist and unable to cope with what is happening (Jones 2009, 13). Certainly there has been a strong movement in several areas of theology over the past few decades to speak of a suffering God, especially through discussion of the suffering of Christ, but to discuss trauma in the Godhead makes no sense. A victim of trauma suffers from "a loss of a sense of self, a breakdown in normal knowing and feeling, and a paralyzing lack of agency" (ibid. 15). Climate change, however much pain it inflicts on Earth and on Earth's people, cannot inflict such harm on the Source of Life. What is needed here is a more nuanced understanding of the suffering of God, one that clearly separates the activity of "suffering" from the related but different activity of "suffering with." We as humans suffer; we have come to recognize that Earth suffers; but God is the One who suffers with us.

Elizabeth Johnson offers a way forward: "The Love who is the Creator Spirit participates in the world's destiny...the Spirit is in the groaning and in the midwifing that breathes rhythmically along and cooperates in the birth. In other words, in the midst of the agony and delight of the world the Creator Spirit has the character of compassion. In multifaceted relationships she resists, reconciles, accompanies, sympathizes, liberates, comforts, plays, delights, befriends, strengthens, suffers with, vivifies, renews, endures, challenges, participates, all the while moving the world toward its destiny" (1993, 59).

"Suffering with" offers a much broader scope of action, as can be seen in Johnson's plethora of verbs describing the Spirit's activity. This broad scope of action, rooted in the core notion of compassion, is faithful to the myriad ways we experience the Spirit in the world today. A faithful, sacramental imagination perceives the world, even in the midst of climate change, as a place of both agony and delight—even as we face the depths of agony, we cannot forget the delight.

Since Wallace links the suffering of the Spirit to the suffering of Christ on the cross, we cannot forget that the Christian story does not end on Good Friday. Yes, we acknowledge the reality of Christ's passion and the reality of the grief of his friends and disciples, but we also celebrate Easter Sunday, the day when grief and pain was answered decisively by God's love and life. In Wallace's discussion of an Earth-Spirit, no resurrection is anticipated or imagined—just further trauma, further pain. The story does not resolve. It is entirely possible, of course, that Wallace intended that this shift from focusing on the

suffering Spirit rather than the suffering Christ is meant to emphasize that there is no "happy ending." However, if this is the case, the
Wounded Spirit is not a source of hope for us as we face changes in
the global climate, but is instead a source of the deepest despair. If in
the end we are in the midst of murdering God, how can God save us?

Despite these serious objections, I do not wish to completely dismiss
Wallace's pneumatological imagination. Its powerful insistence that
we are faced with a situation of unimaginable grief points us to two
significant strands within our tradition: lament and prophetic grief.
One way to look at Wallace's proposal of the Wounded Spirit suffering in the damaged Earth is that he is identifying Earth itself as crying
out in lament. It is not just the people of God who cry out in pain, and
who cry out in need of God's justice, but Earth itself. As a tradition,
lament enables us to mourn, to grieve, to protest against wildly, cruelly,
desperately awful situations. "How long, O Lord? Will you forget me
forever?" cries the psalmist (Ps 13:1). In the words of South African
theologian Denise Ackermann, lament is "a coil of suffering and hope,
awareness and memory, anger and relief, a desire for vengeance, forgiveness, and healing that beats against the heart of God. It is our way
of bearing the unbearable" (Ackermann 2003, 111). Despite our faith
in God's triumph over sin and in the victory of goodness over evil,
we as human beings too often experience real pain, real setbacks, real
agony. The words and actions of lament face up to this pain, and give
voice to it. Ackermann points out that lament is "the language both of
suffering victims as well as of penitent people" (110) and speaks to the
complexity of our responses to climate change. Earth itself is in agony,
and we are responsible; we are both suffering and sorry, and lament
is the best way to express our sorrow and suffering. Any attempt to
ignore our pain, to insist only on God's triumph, is not only disre-
spectful to those in pain but ends by becoming an attempt to lie about
God. And when we refuse to face the truth, when our hermeneutic of
experience breaks down into self-deception, our sacramental imagination has failed and we are no longer speaking about God.

Speaking truthfully about God is often painful. The practice of lament is closely related to another form of speech we need to remember today: the prophetic practice of calling numbed, frightened people
into mourning, into a recognition that they are living a lie. In *The Prophetic Imagination*, Biblical scholar Walter Brueggemann speaks of
the necessity of grief in our broken world, and of the prophetic voice

that forces us to mourn what needs to be mourned. "The task of the prophetic imagination," he argues, "is to cut through the numbness, to penetrate the self-deception, so that the God of endings is confessed as Lord (Brueggemann 2001, 45). Surely we are at a time in our history when our self-deception needs piercing, when our collective, numbed refusal to engage climate change needs to be dramatically shaken. In this respect, Wallace's emphasis on the wounded God is helpful; he is reminding us, with the force and vision of a prophet, that there is need to mourn. After all, "only those who mourn will be comforted…those who do not mourn will not be comforted and those who do not face the endings will not receive the beginnings" (Brueggemann 2001, 57).

However, despite the fact that Wallace's vision of the Wounded God calls us to prophetic mourning and lament, we must recognize that both these traditions embrace more than the experience of pain and loss. Instead, the tradition of lament incorporates both mourning and praise, and the tradition of prophetic grief incorporates both grief and amazement. "Lament is never an end in itself," Ackermann argues (2003, 111). "It is undergirded by the hope that God not only can but that God *will* hear the cries of the suffering *and* the penitent and *will* act with mercy and compassion" (ibid.). Wallace's vision of an impending deicide offers neither hope nor new life. He presents us with a grief as stark as Jeremiah's, but there is no sense that the grief is necessary to make room for God to do a new thing, for God to amaze us once we let go of what we must release. Instead, Wallace's wounded God has no strength left for a new thing, let alone an amazing thing; this God is dying. Thus Wallace's talk of deicide moves his work decisively out of the tradition of either lament or prophecy, neither of which envision the possibility that God cannot and will not respond to our heartfelt cries.

THE RENEWING SPIRIT IN A WOUNDED WORLD

With Wallace, we must mourn; yet with Johnson, we must not let go of hope and trust. Indeed, it is the Spirit of Life who mourns with us, and who leads us to new life. Climate change brings us into a situation in which we must lament: we must cry out as a suffering people, a penitent people, and we must find ways to hold together both loss and hope. We must attend carefully to the shifts and changes in our sacramental imaginations as our experience of creation is shaken by climate change, and we must find ways to *imagine* a future for the planet and for ourselves. This is an enormous task, as great as the practical task of finding ways to heal Earth and mitigate climate change. This imaginative task is one that must be embraced by the church in its liturgy, by all of us in our daily lives, and by theologians in their work to refine and focus the images of God's Spirit that are arising out of the depths of our experience of a planet in crisis. Great and difficult as it is, the work before us is undergirded by an ancient tradition of trusting in God through terrible endings, because God is the author of new beginnings. And we shall see those new beginnings only if we first find ways to imagine them: "How can we have enough freedom to imagine and articulate a real historical newness in our situation? That is not to ask, as Israel's prophets ever asked, if this freedom is realistic or politically practical or economically viable. To begin with such questions is to concede everything to the royal consciousness even before we begin. We need to ask not whether it is realistic or practical or viable but whether it is *imaginable*" (2001 Brueggemann, 39).

What can we imagine? What will our future be? Our disruption of the global climate today confronts us with these questions in a dramatic new way. Yet the Spirit who groans over our wounded planet is also the Spirit who inspires our understanding, our intellect, our imagination. As we work to develop our sacramental imaginations in these new conditions, and as we attend to the Spirit in that work, perhaps we will discover that one route forward is to focus our attention neither on the Wounded Spirit nor on the Creator Spirit, but instead learn to see and appreciate and imitate the Renewing, Liberating Spirit in our midst. God is not wounded nor dying in the suffering of the planet, nor is it enough to focus on the beauty and power of untouched creation. Rather, the Spirit of God is present wherever life is blooming out of death, and wherever human hands act to heal and restore what we have damaged. It is a lovely thing to see the presence of God in the flow of a mighty river, but I suggest that what we need to do is to see the presence of God in the volunteers who dedicate an afternoon, several times a year, to River Clean-Up Days in or near their hometowns. The Renewing Spirit scrambles over muddy riverbanks seeking trash much as the woman in the parable scrambled through her house searching for the lost coin (Luke 15: 8-9). I hope that we will learn to recognize the Renewing Spirit in the world around us and will act in ways that enable us to recognize this Spirit in more and more places in our lives. We can—indeed, we must—move beyond the climate-disruptive and planet-damaging behaviors that have become all too "normal" in our lives, and begin to live in a new way. May God the Renewing Spirit guide our hands and our hearts as we seek the healed, renewed world that we are still only beginning to imagine sacramentally.

SOURCES

Ackermann, Denise. 2003. *After the Locusts: Letters from a Landscape of Faith*. Grand Rapids: Eerdmans.

Brueggemann, Walter. 2001. *The Prophetic Imagination*. 2nd edition. Minneapolis: Fortress Press.

Johnson, Elizabeth. 1993. *Women, Earth, Creator Spirit*. New York: Paulist Press.

Jones, Serene. 2009. *Trauma and Grace: Theology in a Ruptured World*. Louisville, KY: Westminster John Knox Press.

Lewis, C. S. 1950. *The Lion, the Witch, and the Wardrobe*. New York: Harper&Row.

McFague, Sallie. 2008. *A New Climate for Theology: God, the World, and Global Warming*. Minneapolis: Fortress Press.

McKibben, Bill. 2010. *Eaarth: Making a Life on a Tough New Planet*. New York: Henry Holt.

Redmont, Jane. 1999. *When in Doubt, Sing: Prayer in Daily Life*. New York: Harper Collins.

Rockstrom, Johan et al. 2009. "A Safe Operating Space for Humanity." *Nature*, 24 September, 472-75.

Sample, Ian. 2005. "Warming hits 'tipping point.'" *The Guardian*. Accessed from http://www.guardian.co.uk/environment/2005/aug/11/science.climatechange1on December 2, 2010.

Schneiders, Sandra. 1986. *Women and the Word*. New York: Paulist Press.

Wallace, Mark. 2000. "The Green Face of God: Christianity in an Age of Ecocide," *CrossCurrents* 50: 310-31.

TREES, "LIVING SYMBOLS OF PEACE & HOPE"

WANGARI MAATHAI & ECOFEMINIST THEOLOGY

Anne M. Clifford, C.S.J.

I n 2004 Wangar Maathai, founder of the Green Belt Movement in Kenya, became the first African woman and the first environmentalist to receive a Nobel Peace Prize. The Green Belt Movement is a Non-Governmental Organization that has mobilized poor women to plant thirty million trees in fifteen countries in Africa over a span of three decades (Maathai 2004a, x). The Norwegian Nobel Committee's official announcement of their decision to give a Peace Prize to Maathai describes her as a woman standing "at the front of the fight to promote ecologically viable social, economic and cultural development in Kenya and in Africa." The committee also praised her for her "holistic approach to sustainable development that embraces democracy, human rights and women's rights in particular" (Maathai 2004a, ix).

In her response to the Norwegian Nobel Committee, Maathai thanked its members for placing "the critical issue of environment and its linkage to democracy and peace before the world…. Our work over the past thirty years has always appreciated and engaged these linkages" (Maathai 2004b). In the *The Challenge for Africa*, Maathai describes how the three have been linked in her work using the symbol of the traditional African stool, which is comprised of a seat and three legs (Maathai 2009, 56-58).[1] The first leg represents democratic space, where rights both human and environmental are respected. Of particular importance is equity, including the balance of power between women and men. The second leg symbolizes sustainable management of natural resources in a manner that is just and fair, especially for

---

1 For another treatment of the "three legged stool," see, Maathai's *Unbowed* 2004a, 294.

people on the margins of society. The third leg stands for forming "cultures of peace," characterized by the virtues of compassion, forgiveness and justice. The three legs support the seat, which in her conception represents "the milieu in which development can take place" (Maathai 2009, 58).

Early in her years with the Green Belt Movement, Maathai recognized that a strong and sturdy program (her "three legged stool") would not just appear by magic. "Things will not just happen.... Women must make them happen" (quoted in Gaard 1993, 3). In her decision to focus on women planting trees, an impetus was the rampant deforestation in Kenya, which had reduced the country's indigenous forest by over 90% in a fifty-year period. This was very serious because Kenya was economically dependent on forestry, which underpinned agriculture, wildlife, tourism, and energy. One of the consequences of deforestation that had an emotional impact on Maathai was that women and girls were spending hours every day, walking long distances in search of wood for cooking fuel.

In her post-Nobel Peace Prize reflections, Maathai expressed the personal importance of the Green Belt Movement, "Trees have been an essential part of my life and have provided me with many lessons. Trees are living symbols of peace and hope. A tree has roots in the soil yet reaches to the sky. It tells us that in order to aspire we need to be grounded, and that no matter how high we go it is from our roots that we draw sustenance" (Maathai 2007, 293).

When one considers Wangari Maathai life history, especially her tribal family roots, her reception of the Nobel Peace Prize is extraordinary. Born in Nyeri, near Mount Kenya in 1940, while Kenya was under British rule, Maathai experienced first hand the effects of colonization on her family. Her father, a member of the Kikuyu tribe and a peasant, worked at a British owned plantation that exemplified the colonizers' desire to domesticate wild Kenya and dominate its land, resources, and people. Her mother was the second of her father's four wives, and, like many African women of her generation did subsistence farming. From her mother and her maternal grandmother Maathai learned many important life lessons that contributed to her attunement to the effects of the patterns of domination, especially of women, started by the British and continued by the Kenyan government after 1963, when her homeland gained independence.

Maathai's recognition of the plight of women was heightened by the United Nations' Decade for Women (1976-1986), which prompted her to invite women to work with her in launching a movement to plant trees in 1977 (Maathai 2007, 125). From the start she not only emphasized the need for reforestation and sustainable management of fragile forest ecosystems of her eastern African equatorial homeland but also set about remedying the disempowerment of women, captured poignantly in the title of her memoir, *Unbowed*.

GLOBAL CLIMATE CHANGE & AFRICA'S FORESTS

A question that begs for attention in a book with "climate crisis" in its title is: "What do trees have to do with global warming and why and what difference does it make that women planted them in Kenya and in other parts of Africa? A further question of importance is how is climate change with a focus on tree planting a matter of theological importance? Four insights help answer these questions.

First, climate change is not necessarily a problem within itself, but it does contribute to other environmental problems, such as weather changes that have negative ecological effects. Global warming over the past fifty years has resulted in large measure from human choices. A major factor in planetary climate change is the escalation of the level of greenhouse gasses, such as carbon dioxide, which continues to grow in a fossil fuel world economy. Long present in Earth's atmosphere, CO_2 has both positive and negative effects. As the principle greenhouse gas, CO_2 traps the heat needed for sustaining the planet's biodiversity, especially plant life. Carbon dioxide, along with other greenhouse gasses, provides Earth with a heat-trapping blanket resulting in a planetary average temperature of 57.2 °F, a temperature that allows life to flourish (Godrej 2001, 37-43). The climate change problem we must face today is not due to the presence of greenhouse gases but to their escalated increase in Earth's atmosphere that caused the 1990s to be the warmest decade of the twentieth century. The rise in the concentration of carbon dioxide, which not only allows the sun's rays into Earth's atmosphere as heat energy but also blocks the heat from escaping, is that too much of a normally good thing is dangerous. Clearly, the climate change problem is primarily anthropogenic. Fossil fuel burning industrialization is on a collision course with life on our planet that will be especially devastating for poorer nations in the southern hemisphere. Since human choices contribute to this escalation, land-use decisions must be made that will create and maintain life-sustaining global air temperatures (Pojman and Pojman 2008, 568).

Second, where forests are concerned, a person could argue that since forests trap heat from the sunlight they absorb, deforestation could have a net cooling effect on the planet. This is the case because cleared land surfaces, such as deserts, reflect more solar radiation into the atmosphere than forested land. This position, however, ignores the complex interdependent relationships that are basic to a life-sustaining planet. The world's forests are its lungs (Maathai 2009, 258). Trees absorb carbon dioxide for the photosynthesis that enables them to grow, releasing oxygen while holding vast reserves of carbon. Indigenous trees, particularly in the forests of the tropics, help cool the planet and have a positive effect on rainfall patterns (Maathai 2009, 240-242). Indigenous forests protect watersheds and enrich soil with nutrients, thereby enabling other plant species to flourish and a variety of animal species to have life sustaining habitats. Indigenous forests also help to prevent soil erosion, resulting in a positive impact on waterways. If rivers in mountainous regions, such as the five found in Kenya, are not silt-choked, they can provide suitable sites for dams that can generate pollution free electricity. Put simply, the presence of trees, especially of indigenous forests, reduces global warming. Conversely, "if we lose forests, we lose the fight against climate change" (Maathai 2009, 259).

Third, sustainable forests enabled by constraining land exploitation are essential to human survival in Africa and other parts of the world. The establishment of privately owned monocultural plantations of exotic species of trees and other non-indigenous crops by the British conflicted with traditional African wisdom and cultural patterns regarding agriculture (Maathai 2004, 46). The women of African society are the ones who are most affected by local environmental degradation, especially the loss of forests (Maathai 2009, 242). Lack of kindling for cooking and the loss of previously plentiful potable water needed for drinking, food preparation, bathing, and farming are things to which women of Africa are attuned. Loss of indigenous fruit bearing trees is especially serious because they provide vitamin rich food-reducing malnutrition—a major problem in Africa. Furthermore, all trees in the tropics offer welcome shade for humans and provide fencing material and fodder for domestic animals, especially cattle and goats.

Fourth, maintaining sustainable ecosystems by limiting anthropogenic climate change requires motivation. Most Americans and western Europeans know about climate change and its causes. There are no lack of reminders about the "inconvenient truth" of where the responsibility for escalated global warming lies. However, awareness of ethical "oughts" does not easily translate into the will to take the necessary steps to end global warming. Religious faith traditions with theologies that emphasize the gift-like character of creation and dependence on a Creator-God, such as Christianity, can provide the fundamental "why" for ecological action—*love*. Jesus Christ linked love of God with love of neighbor. The fundamental reason why Christians choose to love our neighbors, especially those most vulnerable, is that God loves us. The divine love that embraces each of us and all creation is the basis for our abandoning self interest and performing acts of charity for other persons and for engaging in care of nonhuman creatures.

TREES OF GOD, SOURCES OF LIFE

For generations African peoples, including the Kikuyu, lived lives of ecological balance. In contrast, the conquering Europeans treated the African territories as possessions to exploit for economic gain in the name of attaining "progress." The colonizing countries had a worldview that set them apart from and in control of the land and its indigenous peoples who had long had a culture of intimacy with its forests and savannahs. Maathai learned about the importance of indigenous trees at her mother's knee long before concern about levels of greenhouse gasses and climate change became a global issue. One identity forming religious narrative especially impacted her life—the beloved story of the fig trees of Kenya, called *mĩgumo*.

From her mother Maathai learned that the *mĩgumo* were the sacred trees of God. In her memoir, Maathai recalls with appreciation an area teeming with life that surrounded a large indigenous fig tree near her home. Beneath the shade of the tree's branches a stream of crystal-clear water bubbled to the surface, creating a stream along which banana plants, sugar cane and other food crops grew. Her mother stressed that the Kikuyu loved and reverenced the fig tree (Maathai 2007, 45). Because the native fig trees were God's, no one should ever violate the *mĩgumo*. Neither the fig tree nor its branches could be cut. The Kikuyu religious narrative regarded *mĩgumo* as not only a beautiful tree, reflecting God's goodness, but also as a source for the miraculous wonders of life that God provided the Kikuyu people.

From her early experience of contemplating the local fig tree, Maathai acknowledged the sacredness of the indigenous trees of her homeland. She drew attention to the tragic decision of the British colonial government to burn indigenous forests in order to start tree farms of soft wood pines imported from the northern hemisphere and eucalyptus trees (known in one Kenyan dialect as *munyua mai* meaning "water guzzler") from the southern hemisphere (Maathai 2007, 39). Maathai grieved over the loss of the native trees and recognized, even as a child, that this decision was wrong.

To encourage the Kikuyu to create exotic monocultural tree farms, the British provided local farmers with pine and eucalyptus seedlings, stressing their commercial value. These actions reflect an attitude that regards the elements of the natural world, such as indigenous trees, as accidental, impersonal material. The colonizers patriarchal land ethic promoted the idea that land exists to fulfill human needs and feed human greed. Later she would recognize the importance of native trees, such as the *m_gumo*, for the survival of the Kikuyu people. The British-managed exotic tree farms destroyed healthy ecosystems. In the wake of the destruction of indigenous forests, the rain ran rapidly downstream, taking with it rich top soil resulting in a marked decrease in surface and underground water levels (Maathai 2007, 39).

From childhood, Maathai was ever eager to learn. She received scholarships to attend Catholic boarding schools run by orders of women religious, the first run by Consolata Missionary Sisters from Italy (*Unbowed*, 53) and the second, the only Catholic high school for African girls in Kenya, run by Loretto Sisters from Ireland, (Maathai 2007, 69). At age twelve, before beginning high school she decided on her own to become a Catholic. At these schools, she recalls that she learned that to serve God, one served one's fellow human beings and that volunteerism must be guided by concern for the common good and a deep sense of justice (Maathai 2007, 60, 70).

Upon her completion of high school, Maathai became a recipient of a Kennedy Airlift scholarship, a program that President Kennedy developed with the goal of providing promising young Africans an education that would enable them to become leaders in their chosen fields when African nations gained independence. Maathai's scholarship took her to another Catholic school, Mount St. Scholastica College in Kansas run by Benedictine sisters. There she majored in biology and minored in German (Maathai 2007, 79-85). A recipient of yet another scholarship, Maathai was able to earn a master's degree in Biology at the University of Pittsburgh (1966). After teaching for a while at the University of Nairobi, she pursued a Ph.D. in Germany (1971), making her the first Kenyan woman to earn a doctorate in science.

Maathai's education led her to discover the connection between the fig tree's root system and underground water reservoirs. Because fig tree roots burrow deep into the ground, breaking through the rocks beneath the surface soil and into the water table, water traveled upward along the roots until it found a weak spot in the ground and then gushed forth as a spring of pure clean water. Wherever mature "trees of God" grew, there would likely be a stream, lush foliage, and an abundance of frogs. Reverence for the fig tree, handed on from mother to daughter, a spiritual practice with deep roots in her culture, contributed to Maathai's commitment to the conservation of biodiversity and to her decision to gather women to plant trees (Maathai 2007,44-46).

In Maathai's scientific assessment, deforestation was the root cause of the shrinkage of Kenya's watersheds, the depletion of its soil, and the growing poverty of its people. She concluded that unless there was reforestation on a massive scale, especially of indigenous trees, the people of Kenya would grow poorer and be more malnourished.

On the surface, the Green Belt Movement seems to be about rural women planting bands of trees near human settlements. This initiative, however, is about so much more. In Kenya, like most of sub-Saharan Africa, it is women's role to collect the wood needed for cooking nutritious foods for their families. Wood sacristy requires women to walk long distances to find kindling. In times of wood scarcity, girls often are taken out of school to collect wood or care for younger children as their mother's workload increases (Maathai 2007, 127). As Maathai points out, the Green Belt Movement undertaken by women not only provides villages with nearby trees thereby promoting environmental rehabilitation and conservation; it also contributes to gender equality and women's empowerment (Maathai 2009, 242).

ECOFEMINISM & AFRICAN CHRISTIANITY

Ecofeminism, a term coined by Françoise d'Eaubonne (1974), brings together insights from feminism and ecology, especially deep ecology. From feminism, it draws attention to the effects of systemic patriarchy, principally sexism, but also racism and classism. From ecology, it takes the understanding of Earth and its systems as a network of interdependent relationships. From deep ecology, it builds on the insight that humans assuming that they are the hierarchical apex of Earth's entire ecosystem is inimical to life on the planet. Ecofeminism brings the three together with the goal of ending discrimination against women and subjugated men and treatment of nonhuman nature as if it is a "thing" that exists solely for human benefit.

D'Eaubonne called for a revolution in thought and action. Developments in Maathai's life would evoke both forms of revolution. At the University of Nairobi Maathai experienced gender-based discrimination and resistance to the contributions she wanted to make. She challenged the idea that "a woman could not be as good or better than a man" (Maathai 2007, 117). She recalls that she was vilified as a female who was not following the tradition of a "good African woman" (Maathai 2005). Of necessity, she became a feminist and set about resisting sexually based oppression and seeking liberation for women and also men. The latter is important because she recognized that patriarchy not only oppresses women; it also discriminates against nonwhite and non-Western men.

Although Maathai does not use "ecofeminism" in reference to the Green Belt Movement in her writings, it is clear that she is mindful of the interconnectedness of the undervaluing of women and the domination of Earth. She recognizes that impoverishment of the land and the human poverty that accompanies it affects women more severely than most men. This is the case because women, particularly African women, are the poorest of the poor and because, along with nonhuman nature, women are the primary sustainers of society. Maathai's ecofeminism is not an academically oriented theory as it often is for women living in the Northern hemisphere. Her ecofeminism is grassroots critical engagement of human-Earth and inter-human relations. Another Kikuyu woman, Teresia Hinga provides further content to the significance of Maathai's position. She notes that in the pre-colonial era, there was gender mutuality in Kikuyu society. As agriculturalists, the Kikuyu, like most African tribal societies, had developed divisions of labor without valuing the role of one gender over the other

2   Hinga was a senior lecturer in the Department of Religious Studies of Kenyatta University in Nairobi and a lecturer in religious studies at De Paul University in Chicago when her article was published in *Women Healing Earth*.

with both women and men having "usufructuary rights" to tribal land. Patriarchal colonialism undermined these cultural patterns, disempowering women (Hinga, 1996, 179-180).

Not a theologian by education, Maathai does believe that when people use biblical religion to legitimate the domination of creation, this is morally wrong. She believes that when we degrade the planet, we also degrade ourselves. Conversely, when people work to heal Earth they experience their own spiritual healing (Maathai 2010b). Although she is open to the good found in the religions of the world, Maathai's religiosity has been most strongly influenced by the traditional beliefs of the Kikuyu and the teachings of Catholicism. Regarding the latter, she has cast a critical eye on the manner in which Catholicism and other forms of Christianity were presented to Africans. She faults European Christian evangelization for many of Africa's problems, including widespread ecological damage that contributes to global warming. Without in-depth knowledge of African traditional religion and its role in the lives of the people, European missionaries judged African traditional beliefs to be demonic. Hinga provides content to the demonizing of African religions, noting that the African sense of respect for Earth as a source of life led Africa's European colonizers to regard Africans disparagingly as nature worshipers, branding them as idolaters (Hinga 1996, 172).

With recognition that Christianity has been impacted by many cultural belief systems during its long history, Maathai accepts that Christianity has been a voice for justice and freedom. But she also finds it necessary to apply a hermeneutics of suspicion to the ways in which Western Europeans presented Christianity to Africans. She draws attention to how Christianity was used to implant a skewed sense of European superiority over Africans, which Africans uncritically internalized (Maathai 2004a, 48). Put simply, Christianity mediated the presence of the sacred and at times obscured it.

Maathai has assessed the historical roots of the Christianity that Europeans brought to Africa. No matter how well-meaning the Christian missionaries of the nineteenth and early twentieth centuries may have been, their efforts to spread the "Good News of Jesus Christ" through simplistic catechesis led Africans to assume that God favored them less. They concluded that "God had decided not to reveal himself to them directly but only to others—the Europeans—who were now offering them God's messages" in a holy book, the Bible (Maathai 2009, 38).

In addition, the Catholicism presented to Africans, influenced by the Council of Trent's teaching that the greatest of graces promised by God was eternal life, emphasized salvation—getting to heaven, the dwelling place of God—while neglecting creation and God's abiding presence there. This prompted African Christians to set their eyes on the afterlife, devaluing life in the present. The pre-Christian reverence for Mount Kenya by the Kikuyu illustrates how the deprecation of African religious culture is linked to the continuing destruction of its ecosystems (Maathai 2009, 167). Mount Kenya had been a sacred place for Kikuyu people who regarded it as the dwelling place of God. The people believed that the rains, clean drinking water, green vegetation, and crops, all of which had a central place in their lives, came from God of the Mountain (Maathai 2009, 173).

European Christian missionaries suppressed this belief with their emphasis on heaven as God's home. Yet, if the mountain were to be given the reverence the Kikuyu culture accorded it, people would not allow illegal logging and clear-cutting in the forests, nor would they plant non-indigenous species. Cultural revival, therefore, could contribute to environmental conservation. Maathai believes that a revival of traditional perspectives need not conflict with Christian teachings. Heaven is not a distant place above us; it is right here, right now. The Kikuyu people were not wrong when they said that God dwelled on the mountain because, if God is omnipresent, as sound Christian theology teaches, then God is indeed present on Mount Kenya giving the people what they need to lead them through life (Maathai 2009, 39).

Reflecting on the religious culture of her early childhood, Maathai stresses that religion gives people identity and moral character. Conversion to Christianity contributed to the people's loss of appreciation for their culture and its rich religious mythology. African people accepted that they were destined to be followers of and collaborators with their colonizers (Maathai 2009, 41). They acquiesced to the racism that trivialized their cultural mores. They tried to emulate the Western pattern of acquiring ownership of material things, including taking land by force. Before colonization, land ownership was alien to Africans (Maathai 2009, 227). A family may have acquired a certain level of "exclusive" use of a piece of land, but no family or individual was its owner. Persons living on a tract of land were merely the trustees (or *Muramati*) of God's creation (Hinga 1996, 176). The new concept of private land ownership disenfranchised women who could not own land, but could instead access land at the pleasure of their fathers or husbands (Maathai 2009, 228).

Maathai's ecofeminism is evident in her criticism of European patriarchy and the ways in which it promoted private ownership of land to grow cash crops like fast growing pine trees and coffee, leading to the neglect of indigenous trees and foods such as millet, yams and green vegetables, and medicinal plants. The outcome was and continues to be malnutrition and disease (Maathai 2009, 234). It is not surprising, therefore that Maathai draws attention to indigenous spiritually and its role in the culture of Africa. She is an advocate for spirituality that affirms Earth as sacred. She stresses the importance of indigenous spiritual values because they link people with their roots, God, one another, and the environment (Maathai 2004a, 48).

Although Maathai is critical of the European colonizers' disregard for the spiritual values of Africans, she approves the post-colonial Africanization of Christianity during the past five decades by churches in Africa. She draws attention to Pope John Paul II's visit to Nairobi in September 1995 and his apology to Africans for the sins committed by missionaries who regarded their traditional religious practices to be demonic. She praises the pope's positive response to the incorporation of African traditions in song and dance that he experienced in the Eucharistic liturgy celebrated in Uhuru Park in Nairobi (Maathai 2009, 178-179).

Maathai interprets Pope John Paul's message to the people of Kenya as telling African Catholics that it is up to them to decide what they wish to take from other cultures, to claim what is good and retain it, and to decide what impedes their development and abandon it (Maathai 2009, 179). In this regard, a hermeneutics of remembrance is key. Maathai recognizes that Africans cannot return to where they were prior to European colonization. However, she argues that it is imperative that they abandon their negative assessment of their cultures filtered through the prism of colonial racism and reclaim their heritage. Inculturation of Christianity by Africans can yield an African way of being Christian that incorporates traditional Kikuyu spirituality and its respect for human and nonhuman life.

TREES, SYMBOLS OF HOPE

Maathai is both a realistic and hopeful woman. Her sense of hope is rooted in her belief that "all of us have a God in us, and that God is a spirit that unites all life" (quoted in Suzuki 1997, 227). Since close to 85% of Kenyans are Christians (Maathai 2004, 127), the God they recognize as dwelling in them is most likely the God of Jesus Christ. In the Christian tradition, hope is one of the three theological virtues. As a virtue, the objective of hope is attaining what is truly good. Hope-filled persons set life directions in anticipation of future possibilities that include deliverance from the evils that rob them of the goal of improving their lives.

The virtue of hope is evident in Maathai's poignant words about what trees symbolize for her, "The tree for me is hope; it is the future. When I look at the tree, I see promise. It starts from a seed. Eventually it becomes a huge tree. It becomes an ecosystem in itself" (Maathai 2004b). These words apply to more than trees. They apply to the women who painstakingly planted them. For African women perceived to be subordinate and socialized to feel disempowered, hope can often be in short supply. Initially, skeptics argued that tree planting should be left to the professional foresters and not to uneducated women. But Maathai saw promise in them. To encourage women to trade their fear for hope, Maathai invoked the traditional Kenyan "Harambee" spirit. *Harambee* means, "Let us pull together!" as a community of purpose. To make the identity of the community evident, she called her initiative for tree planting the "Save the Land Harambee" (Maathai 2004a, 20).

Maathai initiated the Harambee under the auspices of the National Council of Women of Kenya. She recalls that the women who agreed to join the Harambee quickly became innovators. They substituted broken pots for seedbeds to keep the seeds and seedlings from domestic animals. Through trial and error, the women learned how to nurture seedlings in their own compounds and then plant them in suitable places on public lands, church property, and the boarders of farms. They soon became "foresters without diplomas" (Maathai 2004, 29, 128). If the seedlings took root, the women were compensated at the rate of $.04, a modest amount but symbolically significant to women who previously had no way of earning any money.

The first tree-planting ceremony took place in Nairobi on World Environment Day, June 5, 1977. The participants recited a "committal," "Being aware that Kenya is being threatened by the expansion of desert-like conditions; that desertification comes as a result of the misuse of land and by the consequent soil erosion by the elements; and that these actions result in drought, malnutrition, famine and death we resolve to save our land ... by planting of trees wherever possible. We make this commitment to save our country from actions and elements which would deprive present and future generations from reaping the bounty which is the birthright and property of all" (Maathai, 2004a, 21). With the renewal of this commitment at every tree-planting ceremony, communal ripples of hope for a better future are sent forth, encouraging more tree-planting and other forms of "pulling together" for community action.

Because interest in planting trees grew, the "Save the Land Harambee" title was replaced in 1987 by the "Green Belt Movement" (GBM) to draw attention to the planting of long rows of trees, often a thousand or more, in "belts" near human populations and in mountainous regions where indigenous forests previously grew (Maathai 2004a, 28). One of the unique attributes of the GBM movement is that it is a value-driven organization committed to the principles articulated by the United Nations' Millennium Goals (Maathai 2009, 239-243). Achieving each of the eight goals depends heavily on healthy ecosystems, which makes the seventh goal, *to ensure environmental sustainability* the pivotal one. In Africa the achievement of the third goal, *gender equality* is enhanced by the achievement of the seventh. However, without women taking up the mantle of equality and contributing their energy to environmental sustainability, the likelihood of achieving goal seven is greatly diminished.

Maaathai, a well-educated former university professor, recognized that the major key to gender equality was education. Mindful that poverty resulted in the inability for girls to even complete primary school, Maathai engaged the women in communal problem solving by holding educational seminars guided by basic questions, such as:

1. What problems do we see in our community?

2. Where do these problems come from and what are their causes?

3. What are their solutions; what actions do we need to take to effect needed change? (Maathai 2009, 167-171).

The method of engagement used by Maathai is similar to the "See, Judge, Act" dynamic of the Catholic Action model traceable to the Belgium priest, Joseph Cardijn, who in 1924 founded the Young Christian Workers in Belgium (Cardijn Community International n.d.). The GBM seminars however also have a unique dimension. The participants are urged to be attentive to which mirror they are using as they address the questions posed. Is it that of the colonizers, the current oppressive government officials, their husbands, or is it their own mirror? For this reason, the seminars are called *kwimenya* (self-knowledge) seminars (Maathai 2009, 170-171). These seminars enable individual and communal self-empowerment, a core value of the Green Belt Movement (Matthai 2010, 16).

Through the education seminars, women become aware that planting trees or advocating for saving forests is part of a larger mission to create a society that respects democracy, decency, adherence to the rule of law, human rights, and the rights of women. According to Maathai, by "clothing the naked Earth," Kenyan women are laying the groundwork for equitable economic development for their country and modeling what is possible in the continent of Africa and beyond. Planting trees requires women to bend to Earth, but in reality they are rising up and walking into a better future (Maathai 2007, 295).

TREES, SYMBOLS OF PEACE

During the time of her leadership of the National Council of Women of Kenya, Maathai set about empowering women to participate in conflict management and peace building. In her Nobel Peace Prize acceptance speech, Maathai drew attention to the symbolism of peace trees: "In time, the tree also became a symbol for peace and conflict resolution, especially during ethnic conflicts in Kenya when the Green Belt Movement used peace trees to reconcile disputing communities.... Using trees as a symbol of peace is in keeping with a widespread African tradition. For example, the elders of the Kikuyu carried a staff from the thigi tree that, when placed between two disputing sides, caused them to stop fighting and seek reconciliation. Many communities in Africa have these traditions" (Maathai 2004b).

Commitment to "peace making" has accompanied the GBM at significant junctures of its history. In 1989, Maathai exhibited a hermeneutics of suspicion directed toward Kenya's single party government. By gathering women to engage in the planting of indigenous trees on public land, Maathai directly challenged the continuation of the pattern of patriarchal dominance of women and of non-human nature, traceable to the British colonial system and furthered by the second President of Kenya, Daniel Toroitich arap Moi and his political party, the Kenya Africa National Union (KANU).

In Maathai's assessment, President Moi and his conspirators behaved as if they were the colonizers of their own country (Maathai 2009, 166). It became increasingly evident that Moi was determined to take land from the Kikuyu living in the Rift Valley so that he could redistribute it to members of his own tribe, the Kilenjin, and to other loyal supporters. He was not only continuing the colonial pattern of environmental racism, he was escalating it. Moi's government attempted to subjugate the women of the Green Belt Movement by invoking an old colonial law that made it illegal for more than nine persons to meet in one place without a government issued license. At that time, the GBM educational seminars often attracted fifty or more women (Maatnai 2007, 180). Moi's patterns of behavior were destructive of African society and had the potential of sparking tribal conflict and even civil war. When he announced his plan to build a sixty-story tower in Nairobi's Uhuru (freedom) Park, Maathai felt compelled to respond (Maatnai 2007, 184-205). The proposed building would be the tallest in Africa and would house the headquarters of Moi's political party, business centers, and a shopping mall. Financing the building would require a $200 million loan. When completed, its visitors would be greeted by a huge statue of President Moi. This building, however, would be of little benefit to the average Kenyan. In addition, its proposed location meant sacrificing most of Uhuru Park and its indigenous trees.

To stop this project, Maathai's wrote to United Nations and British government officials and others in Europe and North America. She made appeals to editors of Kenya's newspapers and instigated non-violent protests. In the midst of this struggle, more and more people began to advocate for democratic elections and many people, especially young men, were imprisoned and tortured. Maathai, herself was arrested (Maathai 2007, 213-215). In a 1992 demonstration in Uhuru Park dozens of demonstrators were killed, hundreds more were injured, and still others were taken to prison. In response, many women of the Green Belt Movement formed the Release Political Prisoners group and went to Freedom Corner in the park to protest the government's actions, calling for the end to the imprisonment of their sons and grandsons.

The government retaliated, brutally beating the women. In a show of defiance, some of the women bared their breasts, a response usually reserved for an offending son and a reminder—if you had not suckled at my breasts you would not be alive. Many of the government forces backed off and the women escaped, taking sanctuary in the nearby All Saints Cathedral, the major Anglican Church in Kenya (Maathai 2007, 222). To prevent their own imprisonment, the women stayed in the cathedral for a year, during which time they engaged in nonviolent protests by a rotating hunger strike. The military surrounded the church periodically, but after a year passed, due to pressure from around the world, the political prisoners were released and the women ended their protest.

Maathai was repeatedly arrested for publicly advocating an end to a single-party state and the creation of a true democracy. When the Kenyan government finally allowed new political parties to form, Maathai founded the Mazingira Green Party. Finally, at the end of 2002, after twenty-four years of struggle against dictatorship, a democratic election was held and Maathai was elected to parliament with an overwhelming 98% of the vote. She was subsequently appointed by President Mwai Kibaki to be the Assistant Minister for Environment, Natural Resources and Wildlife (Maathai 2007, 289). This appointment enabled her to facilitate the creation of policies to protect Kenya's remaining mountainous indigenous forests, which she speaks of as the nation's "water towers,"[3] which are needed to help to ensure that streams continue to flow and rainfall is reliable. She spearheaded policies for more tree planting and for vetting the introduction of non-indigenous plants, especially genetically modified hybrids to determine whether these plants will have negative effects on the already existing biodiversity and/or on Kenya's economy. Her motivation for these policies is simple. It is not possible for people to live in peace with each other if they do not manage the environment responsibly and accountably.

Since her reception of the Nobel Peace Prize, Maathai has helped found the Nobel Women's Initiative, along with five other women

---

3   Kenya's five mountains, Mt. Kenya, Aberdares, Mau, Cherangari, and Mt. Elgon, are commonly referred to as "water towers," accessed at www. kenyaforestservice.org.

Nobel peace laureates. Her participation in the NWI took her to Sudan in 2008 with a multifaceted mission: (1) to bring awareness of the massive violations of women's rights there; (2) to advocate for participatory governance in Sudan; (3) to give encouragement to women's groups working for peace and reconciliation; (4) to stand in solidarity with all who are working for peace in Sudan; and (5) to call upon citizens throughout the world to take individual and collection action to build a sustainable peace with justice in the region (Maathai 2009, 187).

Maathai stresses the need for engendering cultures of peace in Africa with recognition that this will not be possible without protection of the environment (Maathai 2009, 249). The tragic struggle for control over resources is made evident by what she describes during her flight over Sudan and neighboring nations. She describes a landscape with "remnants of abandoned villages: as many as fifty huts spread out in a circle and, all around them, desert" (Maathai 2009, 249). The tragic state of affairs in western Sudan, especially Darfur, although sometimes attributed to religions and ethnic conflict, is really due to the relentless expansion of the Sahara Desert. Conflict in Darfur is an example of climate-change induced warfare. A major contributing factor is the increase in the temperature of the Indian Ocean and the accompanying 40% decline in rainfall that has brought extreme drought to Darfur. When water was more plentiful, Darfur's farmers shared it with the nomadic herders. In 2003, violence erupted when the farmers sought to protect their land and its water from the Arab pastoralists with whom they had previously coexisted peacefully (Stewart 2010, 49).

MAATHAI &
CONTEMPORARY CATHOLIC TEACHING

As noted above, Maathai has been very critical of the colonial implantation of Christianity in Africa yet also appreciative of African enculturation of Catholicism supported and encouraged in the post Second Vatican Council era. Although an author of several books, Maathai's life energies are invested primarily in activism—planting trees to change lives, advocacy for controlling human-forced changes to the climate, and support for the development and dignity of women.

A question poses itself, how are statements about the environment by recent popes related to her actions? Do papal writings provide resources for a theology that the Green Belt Movement can live with?

Maathai's life commitment to peaceful change in response to the ecological crisis resonates with Pope John Paul II's 1990 ground-breaking World Day of Peace message, *Peace with God the Creator, Peace with All of Creation* (1989). Climate change and its spurring of warfare, especially in Africa, was not on Pope John Paul II's radar in 1990. His emphasis was on awareness of the ecological crisis and its relationship to the ever-present search for peace. He states, "In our day, there is a growing awareness that world peace is threatened not only by the arms race, regional conflicts and continued injustices among peoples and nations, but also by a lack of due respect for nature, by the plundering of natural resources and by a progressive decline in the quality of life" (#1).

Recognizing that Earth is suffering, Pope John Paul stressed that there was an urgent need for a new solidarity among all persons of good will to care for creation for the sake of the common good (#10). He especially associates the ecological crisis with the structural forms of poverty that exist throughout the world. He notes that rural poverty and unjust land distribution in many countries have led to subsistence farming and to the exhaustion of the soil. Once their land yields no more food, many farmers move on to clear new land, thus accelerating uncontrolled deforestation, or they settle in urban centers that lack the infrastructure to receive them (#11). Because of these growing problems, we must care for creation within the broader context of *the search for peace* (#15).

Pope John Paul II's successor, Pope Benedict XVI, has given more attention to ecological problems in his social encyclical, *Caritas in Veritate* ("Charity in Truth") in which he devotes chapter four to the environment. Pope Benedict calls for a conception of authentic development that affirms the ethical significance of global interdependence and the common good. At the core of the encyclical is the connection Pope Benedict makes between the moral duty to live in accord with the demands of charity in a complex globalized world and the need to anchor this love in the truth about the human person and the ethical requirements of contemporary economic life (#6-7).

One of those requirements is attention to the environment. Articulating a position that is very important to Maathai, Pope Benedict speaks of the environment as "God's gift to everyone" (#48). He stresses that the natural environment is more than raw material to be manipulated at our pleasure. It is a wondrous work of the Creator and as such we must treat it responsibly—with mindfulness that we have "a responsibility towards the poor, towards future generations, and towards humanity as a whole" (#48). He calls for "inter-generational justice" and a "responsible stewardship of nature, in order to protect it, to enjoy its fruits and to cultivate it in new ways," including restraining consumption, developing renewable energy technology, and helping the poorest people and countries to deal with environmental challenges (#48 and 50).

In *Caritas in Veritate*, Pope Benedict also gives attention to desertification and the decline in productivity in agricultural areas. He stresses that "[p]eaceful agreement about the use of resources can protect nature and, at the same time, the well-being of the societies concerned" (#50). He does not directly address climate change in this encyclical but does include it in his 2010 World Day of Peace message, *If You Want to Cultivate Peace, Protect Creation*. He begins by stressing the theme of seeing creation as God's gift. Here he argues that this helps us to understand our vocation and worth as human beings (#1). He argues that for the sake of peace we cannot remain indifferent before the problems associated with such realities as climate change, desertification, the deterioration and loss of productivity in vast agricultural areas, and the deforestation of equatorial and tropical regions. Nor can we disregard the growing phenomenon of "environmental refugees," people who are forced by the degradation of their natural habitat to forsake it in order to face the dangers and uncertainties of forced displacement. Pope Benedict cautions, "passivity in the face of actual and potential conflicts involving access to natural resources" is not a viable option today (#4). Certainly, Maathai would agree.

Like Maathai, Pope Benedict calls for cultural renewal to "rediscover those values which can serve as the solid basis for building a brighter future for all" (#5). The present crisis requires us to rethink the path that we are traveling together and to engage in communal discernment and in strategic planning. Green Belt Movement's seminars are obvious exemplars. To protect the environment and to safeguard the climate, both Maathai and Pope Benedict call for commitment of the international community to establish clearly defined rules to combat misuse of the environment and to safeguard the climate. Maathai has stressed the same message again and again since speaking in 1985 at the UN's Global Woman's Conference held in Nairobi. Pope Benedict argues that refraining from environmental damage must be done especially by the richer nations due to the solidarity they owe to those living in the poorer areas of our world (#7). Pope Benedict also speaks directly to the less developed countries and the emerging nations (likely with Africa in mind), reminding them that they also are not exempt from their own responsibilities with regard to creation, for the duty of gradually adopting effective environmental measures and policies is incumbent upon all (#8). He invites all peoples and nations to see the ecological crisis as offering an historic opportunity to develop a common plan of action aimed at orienting the model of global development towards greater respect for creation and for an integral human development. He calls for this development to be inspired by the values proper to charity in truth: the centrality of the dignity of the human person, the need for changes in life-style, and prudence, "the virtue which tells us what needs to be done today in view of what might happen tomorrow" (#9).

Where Maathai's advocacy for the development and dignity persons, especially of women is concerned, Pope Benedict gave attention to this topic when he was Cardinal Ratzinger. The issue of the dignity of persons, especially of women, was addressed in a 2004 letter from the Congregation for the Doctrine of the Faith (CDF) to the Catholic bishops of the world under Cardinal Ratzinger's signature that was sanctioned by Pope John Paul II. The letter's title is "Collaboration of Men and Women in the Church." It begins with the declaration that depicts the Church as "expert in humanity." As such, the Church is said to be "called today to address certain currents of thought which are often at variance with the authentic advantage of women" (#1). Among the currents highlighted is women responding to subordination by antagonism, seeking "power for themselves" (#2). Women in the adversarial role of seeking power is described as a process that "leads to opposition between men and women, in which the identity and role of one are emphasized to the disadvantage of the other, leading to harmful confusion regarding the human person, which has its most immediate and lethal effects in the structure of the family." The letter draws from Pope John Paul II's 1998 Apostolic Letter *Mulieris dignitatem* (On the Dignity and Vocation of Women) in which Mary, the Virgin Mother of God, serves as a model of gender complementarity that emphasizes the mothering and self-giving capacities of women, who though incapable of representing Christ's act in initiating redemptive life as ordained priests, symbolize in their own bodies the self-giving response to Christ incumbent on the Church, his Bride (#26).

Though the 2004 CDF letter has collaboration in its title, how men are to collaborate with women to their mutual benefit as persons for the common good is neglected. Motherhood is presented as women's primary responsibility in the domestic sphere, but this does not rule out a woman's having a position of responsibility outside the home in which she might contribute to the transformation of society.

What is missing in the CDF letter is an acknowledgment that women's reaction to abuse of power is valid. Patriarchy, at times expressed by men in violence toward women and less overtly in attempts to keep women in "their place"—subordinate to men—is not acknowledged. Instead, women "seeking power" is judged to be undertaken to "make themselves the adversaries of men," resulting in lethal effects on the structure of families (#2).

One would expect that Maathai would question a letter written exclusively by men (all the members of the CDF are male) for men (the bishops) with the opening claim that the Church hierarchy is "expert in humanity." She also would question the articulation of women seeking power for themselves. For her, empowering women and facilitating self-knowledge about their gifts and potential to make a difference is foundational for the Green Belt Movement, which she does not believe is an anti-family organization. On the contrary, her grassroots feminism is one that regards motherhood to be central to women's activism seeking to support healthy family life, which at a minimum requires nutritious food and potable water. Her motivation is not antagonism of men but rather service to survival, a service which has given her life meaning. Her words confirm this, "I myself am at my happiest and my best when I am serving" (Maathai 2005).

CONCLUSIONS

Humanity stands at a crossroads of the climate crisis. The problem is serious but the story is not all gloom and doom. Wangari Maathai and the Green Belt Movement provide ample evidence that committed individuals and communities of purpose, comprised largely of women, can have a positive effect on the problem. The GBM has grown from humble beginnings to an influential force in many global forums. At the World Summit on Sustainable Development held in Johannesburg in 2002, the GBM received recognition for its good forestry practices. This honor provided a context for the decision to designate 2011 the "International Year of Forests." The accompanying rationale gives attention to the serious problems related to deforestation and drought and calls for "sustainable management, conservation and sustainable development of all types of forests for the benefit of current and future generations" (United Nations General Assembly 2002).

Maathai will likely continue to use the platform that U.N. recognition and the Noble Peace Prize have provided her to demonstrate that women bring important elements to remedying the negative effects of climate change. In a 26 November 2010 statement given in conjunction with the approaching international Climate Change conference in Cancún, Maathai stressed, "If we truly want to tackle climate change, poverty and conflict we need to think holistically. We need to...think big, connecting the dots between poverty, energy, food, water, environmental pressure and climate change. Focusing on only one dot means that we lose sight of the bigger picture" (Maathai 2010a).

Maathai herself has certainly connected the dots and has done so in ways that recognize the high price of hierarchal dualism where women, non-human nature, the peoples of Africa (the so-called "dark continent"), and theological notions of God are concerned. She recognizes that the colonial patriarchal order that attributed ultimate value to God and his heavenly home, rendering Earth of no intrinsic worth has had a negative impact on African culture and ecosystems. Climate change issues are religious-cultural issues.

The story of the Green Belt Movement is more than a strategy to encourage reforestation. It is ecofeminism making a difference from the bottom up. One of its most significant aspects that sets the GBM apart from other forms of advocacy for ecological sustainability is its communal discernment seminars in which women can become aware of their capacity to challenge abuse of the environment for short-term economic gain and make a positive long-term difference. Maathai and her Green Belt sisters illustrate well the wisdom of ecofeminism and the transformative power of thinking and acting *with* one another and *with* God's creatures. The response to Wangari Maathai and the Green Belt Movement is surely one of gratitude for an action-oriented ecofeminism that is *buoyed by hope* and *committed to peace*.

SOURCES

Benedict XVI, Pope. 2009. *Caritas in Veritate*. Accessed from http://www.vatican.va/holy_father/benedict_xvi/encyclicals/documents/hf_ben-xvi_enc_20090629_caritas-in-veritate_en.html on March 15, 2010.

————. 2010. *If You Want to Cultivate Peace, Protect Creation*. Accessed from http://www.vatican.va/holy_father/benedict_xvi/messages/peace/documents/hf_ben-xvi_mes_20091208_xliii-world-day-peace_en.html on March 15.

————. (Joseph Cardinal Ratzinger). 2004. *Letter to the Bishops of the Catholic Church on the Collaboration of Men and Women in the Church and in the World*. Accessed from http://www.vatican.va/roman_curia/congregations/cfaith/documents/rc_con_cfaith_doc_20040731_collaboration_en.html on December 28, 2010.

Cardijn Community International. n.d. "Joseph Cardijn." Accessed from http://www.cardijn.info/cardinal-joseph-cardijn on October 15, 2010.

d'Eaubonne, Françoise. 1974. *Le Féminisme ou la Mort*. Paris: Pierre Horay.

Gaard, Greta. 1993. "Living interconnections with Animals and Nature." In *Ecofeminism, Women, Animals, Nature*, ed. Greta Gaard. Philadelphia: Temple University Press, 1993.

Godrej, Dinyar. 2001. *The No-nonsense Guide to Climate Change*. Oxford: New Internationalist Publications.

Hinga, Teresia. 1996. "The Gikuyu Theology of Land and Environmental Justice, in *Women Healing Earth, Third World Women on Ecology, Feminism and Religion*, ed. Rosemary Radford Ruether. Maryknoll, NY: Orbis Books.

John Paul II, Pope. 1988. *Mulieris Dignitatem*, Apostolic Letter "On the Dignity and Vocation of Women on the Occasion of the Marian Year." Accessed from http://www.vatican.va/holy_father/john_paul_ii/apost_letters/documents/hf_jp-ii_apl_15081988_mulieris-dignitatem_en.html on December 28, 2010.

————. 1989. *Peace with God the Creator, Peace with All of Creation*, Message for the 1990 World Day of Peace December 8. Accessed from http://www.vatican.va/holy_father/john_paul_ii/messages/peace/documents/hf_jp-ii_mes_19891208_xxiii-world-day-for-peace_en.html on October 17, 2010.

Maathai, Wangari. 2010a. "Cancún Must Be About More Than Climate Change," *The Guardian*, November 26. Accessed from http://www.guardian.co.uk/commentisfree/cif-green/2010/nov/26/cancun-climate-change-conference on December 17.

————. 2009. *The Challenge for Africa*. New York: Pantheon Books.

————. 2004a. *Green Belt Movement, Sharing the Approach and the Experience*. New York: Lantern Books.

————. 2010b. *Replenishing the Earth: Spiritual Values for Healing Ourselves and the World*. New York: Doubleday.

————. 2004b. "Nobel Peace Prize Acceptance Speech," 10 December 2004. The Green Belt Movement's Official Website. Accessed from http.//wangarimaathai.or.ke/ on October 21, 2010.

————. 2007. *Unbowed, A Memoir*. New York: Random House.

Maathai, Wangari with D. Gilson. 2005. "Root Cause: An Interview with Wangari Maathai." *Mother Jones*. Accessed from http://www.motherjones.com/print/18611 on December 28, 2010.

Pojman, Louis P., and Paul Pojmam. 2008. *Environmental Ethics, Readings in Theory and Application*. Belmont, CA: Thompson Wadsworth.

United Nations General Assembly, "Resolution on International Year of Forests, 2011" adopted at the World Summit on Sustainable Development (2002). Accessed from http://www.un.org/en/events/iyof2011/resolution.shtml on October 28, 2010.

Stewart, William. 2010. *Climate of Uncertainty, A Balanced Look at Global Warming and Renewable Energy*. Flagler Beach, FL: Ocean Publishing.

Suzuki, David. 1997. *The Sacred Balance: Rediscovering our Place in Nature*. Vancouver, BC: The Mountaineers.

"Wangari Maathai, Founder of the Green Belt Movement." 2005. Accessed from http://www.nobel-prize-archive.com/wangari_maathai_bio.htm on October 28, 2010.

CATHOLIC SOCIAL TEACHING
PERSPECTIVES

16

"WITH GENEROUS COURAGE"

PROMISE & POIGNANCE IN THE LEGACIES OF POPE JOHN PAUL II & POPE BENEDICT XVI

Connie Lasher & Charles M. Murphy

On 22 September 2009, Pope Benedict XVI addressed the United Nations Summit on Climate Change via videostatement, offering his "support" to those leaders of governments and international agencies who were about to meet to discuss "the urgent issue of climate change." In his address, Benedict encouraged participants in the summit to enter into their deliberations "constructively" and "with generous courage" (2009b).

A few months later, the December 2009 global summit on climate change in Copenhagen failed to achieve consensus. The developing nations were not going to be forced to reduce carbon emissions and hamper their prospects for a higher standard of living. The developed nations were not going to change their lifestyles or tax themselves to the level required for more socially responsible energy use. This impasse at Copenhagen and the ongoing polarizations that dominate media portrayals of the question of human-forced climate change evince neither "generosity" nor "courage" of the sort that Benedict had sought. As the consequences of the global economic crisis linger—a specter of the vulnerabilities and vicissitudes of globalization within which the complexity and scope of climate change must be confronted—the social commentator, Tony Judt, captured the tenor of our times in noting the need for a moral narrative, a vocabulary not based upon materialism but "a moral critique of the inadequacies of the unrestricted market or the feckless state" (2010). It is precisely this kind of "moral narrative" and "critique" that the social magisterium of the Roman Catholic Church offers, not simply *ad intra*, but moreover, "to the people of

our time, [the Church's] travelling companions" on the human journey through history (Pontifical Council for Justice and Peace 2004,
1). While the fragility of that common journey has always been the
message of "The Age of Ecology" (Worster 1985, 339), the growing
evidence and understanding of the dynamics of human-forced climate
change present yet another test of the Church's ability to confront "new
things" in a manner that is at once *comprehensive* and *comprehensible*.

The *Compendium of the Social Doctrine of the Church* presented the
interpretive framework for its overview in terms of "an integral and
solidary humanism" by means of which "the Church intends to offer
a contribution of truth to the question of man's place in nature and
in human society" (Pontifical Council for Justice and Peace 2004, 5).
This integral humanism structures the analysis of all social issues, including the chapter on "Safeguarding the Environment" which states
in its opening paragraphs: "The relationship of man with the world
[non-human nature, creation] is a constitutive part of his human identity" (197). While these statements may be greeted as welcome evidence of the successful incorporation of environmental awareness into
the official teachings of the Catholic Church, they may equally arouse
consternation that the magisterium of the Church remains apparently
impervious to the classic environmentalist indictment of Christianity's intractable "anthropocentrism." However, the judgment of ongoing
anthropocentrism warrants more nuanced analysis. It may be that the
Church's insistent anthropological hermeneutic becomes not simply
the "problem" of Christianity bequeathed to the world, but rather, an
offering of its own distinctive potential for a new millennium of global
community and crisis.

In their 1993 essay "Creation and an Environmental Ethic," Michael and Kenneth Himes assert: "The exploration of [a] relational
anthropology is the basic contribution theologians can make to the
environmental movement" (106). The vigor of theological efforts
in this regard has been as impressive as any in the history of social
Catholicism. However, this contribution has been attenuated to the
extent that theologies of the environment are produced or evaluated
only as instances of post-conciliar "trends." Of particular importance is
the fact that the Second Vatican Council unfolded as a seminal event
in the Church's existence in the modern world at approximately the
same time that contemporary environmentalism came to the fore as

a moment of *renewed* social engagement in an *already existing* modern environmental tradition in the West.

The thinkers whose work preceded and informed the Council and who remained influential after the Council were participants in an era of dynamic intellectual renewal within a twentieth century Catholicism whose varieties, whether characterized as *ressourcement* or *aggiornamento*, sought to engage precisely the meaning of the human from a wide range of disciplinary perspectives. These developments expressed prior and ongoing renewals in the doctrine of grace, which sought to recover a Christian understanding of God's presence to the world in a manner adequate to the questions and crises of the age. These studies frequently included profound and holistic analyses in which the question of technology, of human "mastery" over nature, were formative features. These developing, *de facto* "relational" anthropologies—be it the personalist orientation of a Karol Wojtyla, varieties of Transcendental Thomism, or the *ressourcement* orientation of a Hans Urs von Balthasar—all were contending with the consequences of the modern "turn to the subject," ensuing streams of continental thought, and subsequent developments which were also the focus of nascent environmental philosophies, historical analyses and "genealogies" of the ecological crisis (Lasher 2004). In short, both Pope John Paul II and Pope Benedict XVI matured as thinkers in this era of Catholic philosophical and theological renewal, in the midst of this wider intellectual and cultural "conversation" from which contemporary environmentalism itself emerged.

Each of these popes has approached the issue of environmental concern in a manner that reflects explicit commitment to and interpretation of the Second Vatican Council. In this regard, each has fully engaged the task of reading the signs of the times, doing so in a spirit of openness to ecclesial self-examination and reform. And consistent with the Council, both popes have offered a renewed vision of the meaning of the human in terms of the deposit of faith: Belief in Christ. Therein we encounter the perennial Christian paradox which must be considered in any treatment of the theological legacies of Pope John Paul II and Pope Benedict XVI: The need to speak *comprehensibly* requires a universal, solidary humanism which takes its place as one religious contribution in a world characterized by "pluralism and difference at every level: in ways of thinking, moral choices, culture, religious affiliation, philosophy of human and social development" (PCJP

2004, 6). Yet, in its ecclesial expression, this universal dimension must emerge from the "heart" of Christian particularity as a foundational theological renewal from which the requirement of *comprehensiveness* is fulfilled.

In this essay we offer a theological analysis of the legacies of Pope John Paul II and Pope Benedict XVI in which we consider features we deem both noteworthy and promising in confronting the crisis of human-forced climate change. We examine a cross section of their more formal teachings, supplemented with a few texts of a more "experiential" or informal type which we believe offer insights into the character of each pope's understanding of the human relation to nature and environmental concern. Our aim is to illustrate the manner in which each pope is cognizant of the environmentalist critique of Christian anthropocentrism, how each seeks to correct the "anthropological error" of "absolute dominion" not by de-centering the human, but by re-affirming human distinctiveness within a renewed relational theological anthropology. In each, we find a deeply felt Catholic attunement to the reality of creation as "gift," encountered as the "truth" of God's love made known in Christ.

POPE JOHN PAUL II (1978–2005): THE "TRUTH" OF THE "GIFT"

"The Redeemer of man, Jesus Christ, is the center of the universe and of history." With these words, John Paul II began his first encyclical, *Redemptor hominis* (1979), in which "for the first time, redemption emerges as the primary theological idea of a pontificate" (Miller 1996, 32). We find here in summary form the contours of the pope's legacy: "redemption as a new creation" (John Paul II 1979, #8), the meaning of the human, and of the universe itself, seen in christocentric renewal. In *Crossing the Threshold of Hope*, John Paul described his first encyclical as reflecting "what I had already been living on the threshold of the papacy" (1994, 48). For a pope who was a poet attentive to nature's beauty and remarkable in his well-known love for outdoor pursuits (be it mountain excursions, skiing, or kayaking), his lifelong experience and contemplation of the human relation to nature is not without significance. The renewal he sought finds its locus in what he referred to as that "stupendous text" (John Paul II 1979, #9), *Gaudium et spes* 22, which would appear in virtually all of his major writings to come:

"The truth is that only in the mystery of the Incarnate Word does the mystery of man take on light . ...Christ the new Adam, in the very revelation of the Father and of his love, fully reveals man to himself and brings to light his most high calling" (#8).

David L. Schindler characterizes the significance of *Gaudium et spes* 22: "[H]istory will recognize the heart of [John Paul II's] interpretation to be a new sense of the *imago Dei* resulting from the integration of trinitarian Christology and anthropology" (2000, 94). Schindler's essay points to an important convergence in its analysis of Joseph Ratzinger's 1969 commentary on Chapter One of *Gaudium et spes*, in which the future pope considers the apparent tension between paragraph 12's more "theistic" rendering of the *imago Dei* and its eventual Christological resolution in paragraph 22. Ratzinger upholds the "Augustinian" aspect of the text, in which "dominion over the world is only the consequence, not the content, of likeness to God.... That dominion must ultimately consist in bringing things into man's glorification of God" (Ratzinger 1969, 121-122). Considering the content of paragraph 22, he continues: "In accordance with the whole composition of the text, the chapter on the dignity of man culminates in Christ who is now presented as the true answer to the question of being human" (159). Thus, the anthropology of the chapter "does not lead to an unacceptable form of anthropocentrism but, by taking man seriously, recognizes him as the being who is constituted to be not merely in himself, but above and beyond himself, and who is only in full possession of himself when he has gone forth from himself: Abba, Father" (163).

In *Redemptor hominis*, John Paul II contemplates the dilemma of human sin and "dominion" in its relevance to the question of the environment, and he then turns to a restatement of the Christological resolution in *Gaudium et spes* #22. As his analysis unfolds, we find an extended meditation on disordered relations, including a telling consideration of the human relation to nature in which "exploitation of the earth" and "uncontrolled development of technology" threaten the natural environment:" [T]he uncontrolled development of technology outside the framework of a long-range authentically humanistic plan often bring with them a threat to man's natural environment, *alienate him in his relations with nature and remove him from nature.* Man often seems to see *no other meaning in his natural environment* than what serves for immediate use and consumption. Yet it was the Creator's will that man should communicate with nature as an intelligent and noble 'master' and 'guardian,' and not as a heedless 'exploiter' and 'destroyer' (John Paul II 1979, 15, italics added). Within this christocentric orientation, in which redemption in Christ redeems all creaturely relations, a renewed theological anthropology begins to emerge in which the relation to non-human nature is further integrated with social analysis.

This element emerges with greater force in the 1987 social encyclical *Sollicitudo rei socialis*, which reconsidered the question of development for a changing world in light of the significance, "originality," and "breadth of outlook" found in Paul VI's *Populorum progressio* (1967). In *Sollicitudo rei socialis*, John Paul affirms Paul VI's basic insight about development: "[T]hat the very concept of development, if considered in the context of universal interdependence, changes notably. True development cannot consist in the simple accumulation of wealth and in the greater availability of goods and services, if this is gained at the expense of the development of the masses, and without due consideration for the social, cultural and spiritual dimensions of the human being" (John Paul II 1987, #9).

After an overview of "the actual situation of development," including the negative aspects of the growing, not diminishing, gap between rich and poor unforeseen by Paul VI, John Paul includes among the newer, more positive signs "a greater realization of the limits of available resources, and of the need to respect the integrity and the cycles of nature and to take them into account when planning for development, rather than sacrificing them to certain demagogic ideas about the latter. Today this is called ecological concern" (John Paul II 1987, #26). In the context of discerning "authentic development" in the contemporary context, John Paul laments the rise of what he calls "super-development," with all its wastefulness, a consumerist mentality that becomes a "throw-away" society (#28). The contrast between "being" and "having" ("the cult of having," the "civilization of consumerism") reflects Paul VI's appraisal of the true vocation of the human—a refrain which will become a familiar corollary to GS 22 in John Paul's legacy. For it is the spiritual dimension of the human person, the *imago Dei*, which is the measure for development, even as God the Creator "imposes limits upon [human] use and dominion over things" (# 29).

In a now classic text, John Paul considers in *Sollicitudo rei socialis* the moral nature of authentic human development, which must include "respect for the beings which constitute the natural world, which the ancient Greeks—alluding precisely to the order which distinguishes it—called the 'cosmos'" (John Paul II 1987, #34). He first affirms: "[T]he appropriateness of acquiring a growing awareness of the fact that one cannot use with impunity the different categories of beings, whether living or inanimate—animals, plants, the natural elements— simply as one wishes, according to one's economic needs. On the contrary, one must take into account the nature of each being and of its mutual connection in an ordered system which is precisely the 'cosmos' (ibid.). He then notes the urgency of realizing that natural resources are limited; some are not renewable. "Using them as if they were inexhaustible, with absolute dominion, seriously endangers their availability not only for the present generation but above all for the generations to come." Thirdly, he points to "the consequences of a certain type of development on the quality of life in industrialized zones…the pollution of the environment, with serious consequences for the health of the population." This section culminates in his assertion that "the dominion granted to man by the Creator is not an absolute power, nor can one speak of a freedom to 'use and misuse', or to dispose of things as one pleases…. [W]hen it comes to the natural world, we are subject not only to biological laws but also to moral ones, which cannot be violated with impunity" (ibid.).

In a subsequent section of *Sollicitudo rei socialis* entitled "A Theological Reading of Modern Problems," John Paul II considers both personal sin and "structures of sin" as necessary for understanding "the reality that confronts us" (#36). "For Christians," he continues, "as for all who recognize the precise theological meaning of the word 'sin', a change of behavior or mentality or mode of existence is called 'conversion' to use the language of the Bible (Mk 13:3,5; Is 30:15). This conversion specifically entails a relationship to God, to the sin committed, to its consequences and hence to one's neighbor, either an individual or community" (#38). This allusion to sin and selfishness is followed by an extended exposition of the Christian moral virtue of solidarity, defined as "a firm and persevering determination to commit oneself to the common good; that is to say to the good of all and of each individual, because we are all really responsible for all" (#38). As an application of the virtue of solidarity, John Paul cites another principle from Catholic social doctrine: the universal destination of the things of Earth. With these themes of sin and of conversion to solidarity, of social analysis which includes redemption of the human relation to nature and restoration of nature's integrity, *Solicitudo rei socialis* became the document which placed the theme of the environment legitimately within the "concerns" of Catholic social thought (Murphy 1989).

Pope John Paul II's legacy continued to unfold in volume, scope, and depth his theological analysis of the human relation to the natural world. A watershed moment came in December 1989 when he issued "the first Vatican statement devoted entirely to the ecological question," entitled *Peace with God the Creator, Peace with All of Creation*, his message for the January 1, 1990 World Day of Peace (Christiansen 1996,1). In this well-known and, in retrospect, programmatic message, he affirms "a new ecological awareness... beginning to emerge which, rather than being downplayed, ought to be encouraged to develop into concrete programs and initiatives" (John Paul II 1989, #1). Perhaps it is not too strong to suggest that in this statement, he extended the legitimacy of concern for the environment to broader cultural forms of environmentalism itself, rejecting extremes of de-centering anthropologies but nonetheless stressing the vocation of the human, distinct from other creatures, is not an absolute dominion. Emphasizing the need for "a morally coherent worldview," John Paul II links "respect for life, and above all for the dignity of the human person" to "the fact that there is an integrity to creation," an objective order to the universe which not only must be respected, but which constitutes a common heritage which must be safeguarded according to the reality of global interdependence. Again he emphasizes the need for a "new solidarity" (1989, #4) in addressing global inequalities and for pursuing a "courageous reform of structures as well as new ways of relating among peoples and States" (#11).

In this affirmation of environmental concern, John Paul II bridges the distance between religious language of "conversion" and exhortations toward large-scale social reform in authentic development, by specifying the kinds of intermediate institutions which actually effect change: "An education in ecological responsibility is urgent: responsibility for oneself, for others and for the earth.... A true education in responsibility entails a genuine conversion in ways of thought and behavior. Churches and religious bodies, non-governmental and governmental organizations, indeed all members of society have a precise role to play in such education. The first educator, however, is the family, where the child learns to respect his neighbor and to love nature"

(#13). He concludes by mentioning themes which will, retrospectively, be seen as hallmarks of his teachings, such as the aesthetic significance of creation, and the emerging opportunities for ecumenical and interreligious cooperation in the protection of the environment. In this regard, *Peace with God the Creator, Peace with All of Creation* signals a turn—in the documentary heritage of this pope, if already an existing orientation personally—toward a more integral and solidary ecological praxis at the level of the formation of the person in society.

The social encyclical, *Centisimus annus*, which John Paul II issued in 1991 in response to changes in the world situation with the fall of Communism, again seeks to read the social, economic, and political issues of the day in terms of the fundamental "anthropological error." Its seminal analysis of both socialism and capitalism has become central in its treatment of "the phenomenon of consumerism" as reflective of a culture's "overall understanding of life" (#36) and the human relation to nature :

> Equally worrying is the ecological question which accompanies the problem of consumerism and which is closely connected to it. In his desire to have and to enjoy rather than to be and to grow, man consumes the resources of the earth and his own life in an excessive and disordered way. At the root of the senseless destruction of the natural environment lies an *anthropological error*, which unfortunately is widespread in our day. Man, who discovers his capacity to transform and in a certain sense create the world through his own work, forgets that this is always based on *God's prior and original gift of the things that are*. Man thinks that he can make arbitrary use of the earth, subjecting it without restraint to his will, as though the earth did not have its own requisites and a prior God-given purpose, which man can indeed develop but must not betray. Instead of carrying out his role as a cooperator with God in the work of creation, man sets himself up in place of God and thus ends up provoking a rebellion on the part of nature, which is more tyrannized than governed by him. In all this, one notes first the poverty or narrowness of man's outlook, motivated as he is by a desire to possess things rather than to relate them to the truth, and lacking that disinterested, unselfish and aesthetic attitude that is *born of wonder in the presence of being and of the beauty which enables one to see in visible things the message of the invisible God who created them"* (John Paul II 1991, #37, italics added).

The contemplative appreciation of the gift of existence, manifest in "lifestyles in which the quest for truth, beauty, goodness and communion with others for the sake of common growth" (#36) is formative, reflects a conception of human flourishing grounded in a deep awareness of the natural world, comprehended in relation to transcendent origin.

Consideration of John Paul's environmental legacy would be incomplete without mentioning the encyclical *Evangelium vitae*. In this long paean to "the Gospel of life" in its confrontation with "the culture of death" (1995 #24), the topic of ecology is again interwoven throughout his analysis. His reading of "the eclipse of the sense of God and of man," which he describes as "the heart of the tragedy being experienced by modern man," (#21) reflects his attentiveness and participation, per our introduction, in the twentieth century debate concerning the roots of the environmental crisis: "[O]nce all reference to God has been removed, it is not surprising that *the meaning of everything else becomes profoundly distorted*. Nature itself, from being 'mater' (mother), is now reduced to being 'matter,' and is subjected to every kind of manipulation.... By living 'as if God did not exist', man not only loses sight of the *mystery of God*, but also of the *mystery of the world* and the *mystery of his own being*" (#22, italics added). Again, under the heading of "signs of hope and invitation to commitment" we are not surprised to find among the "welcome signs" of hope "the growing attention being paid to the quality of life and to ecology" (#27).

Discerning the "truth" of the "gift" of created existence, learning again to read "the book of nature," became the focus of the 1998 encyclical, *Fides et ratio*, where John Paul develops further his insistence that "Faith and reason are like two wings on which the human spirit rises to the contemplation of truth" (Introduction). That Joseph Ratzinger was a major contributor to the writing of *Fides et ratio* is commonly known. Commenting on the encyclical after its appearance, Cardinal Ratzinger said: "It is the peculiarity of Christianity, in the realm of religions, that it claims to tell us the truth about God, the world and man, and lays claim to being *religio vera*, the religion of truth... If I had briefly to state the main intention of the encyclical, I would say it is trying to rehabilitate the question of truth" (2004, 184). However, once again, this "truth" is not an abstraction. It is a matter of love encountered in the Mystery of Christian faith. It is this experiential, or spiritual, orientation toward nature's beauty in John Paul II's own life and loves that we should like to underscore, as we conclude this overview of his teachings. For in his pastoral addresses—especially those directed to young people—the heart of his Christian faith, his joy in encountering "the gospel in kayaks" (Weigel 1999, 102), shines through as the basis of the social teachings he inherited and developed.

In 1984, he addressed youth in Viterbo, Italy, in a pastoral rendering of relational anthropology: "Another aspect of Christ's newness that you can witness in the world is a new relationship with the environment.... It is now necessary to learn to look at nature with new eyes.... Sun, stars, water, air, plants and animals are gifts with which God has made comfortable and beautiful the home prepared for man.... Young people, be witnesses of the risen Christ in the newness of your personal life, in the newness of your relationship with others, in the newness of your relationship with the environment (John Paul II 1984). At a Sunday Angelus in Cadore, John Paul again made explicit his belief in the role of nature in the integral formation of the person, as well as his affirmation of environmental concern: "Symptomatic of our time is the fact that, in the face of what has been called the danger of an 'environmental holocaust', *a great cultural movement* has been started to protect and rediscover the natural environment. Young people especially must be sensitized to this need. The respectful enjoyment of nature should be considered an important part of their educational development. Whoever really wants to find himself must learn to savor nature, whose charm is intimately linked with the silence of contemplation. *The rhythms of creation are so many paths of extraordinary beauty along which the sensitive, believing heart easily catches the echo of the mysterious, loftier beauty that is God himself, the Creator, the source of life and all reality*" (1993, 1; italics added). The "rhythms of creation," the "gift" of existence which is the fundamental "truth" reflected in the very structure of reality, will become the distinctive orientation of Benedict's legacy.

POPE BENEDICT XVI (2005–): THE "GIFT" OF THE "TRUTH"

Pope Benedict XVI, has been called the first "green pope." In the relatively short period of his pontificate (by comparison with nearly three decades of his predecessor's), "the consistency and regularity with which he has emphasized the environmental message" is remarkable.

"'We must awaken consciences,' he has said. 'We have to face up to this great challenge and find the ethical capacity to change the situation of the environment for the good'" (Koeniz-Bricker 2009, 8). In his most formal teachings, but also in addresses to assemblies of youth, to the diplomatic corps, his weekly audiences, and his dialogues with priests, he returns again and again to the environmental theme.

No one should have been surprised to find that John Paul II's successor had accepted the basic contours of the Polish pope's environmental legacy. However, the manner in which Benedict XVI has received and carried forward the dimension of nature as the realm of an integral and solidary humanism, the realm of conversion *ad intra* and encounter *ad extra*, is remarkable and full of promise. Benedict has consistently opened the issue of the environment to the deeper question of the human relation to nature, explicitly linking it with the recovery of the whole breadth of reason, and with this, the rediscovery of "the truly human conscience" (Benedict XVI 2008a). He has called for the cultivation of "an education in listening," that encompasses "the opening of the inner senses" to "the voice of being," to the "inner laws of creation," "the voice of the Creator" (ibid. 2007a). To accomplish this, he has proposed what he terms a "combination of a secular approach and a religious approach" in which the "voice of the earth" and the "great moral experiences of humanity...born from the encounter with the other, with the community," become ways in which we learn again to hear the "voice of being" (ibid.). For Benedict, contemporary environmental awareness represents a fundamental human experience in which the transcendent openness of the person may once again be recognized, honored, and evangelically engaged.

During his vacation in 2008, when asked how to increase the Christian community's sense of responsibility for the environment, he reaffirmed the importance of a "renewal of the doctrine of Creation" in its "indissoluble bond" with redemption. "The Redeemer is the Creator and if we do not proclaim God in his full grandeur...we also diminish the value of the Redemption....This is why, for me, the renewal of the doctrine of Creation and a new understanding of the inseparability of Creation and Redemption are of supreme importance. We must recognize anew: he is the *Creator Spiritus*, the Reason that exists in the beginning, from which all things are born and of which our own reason is but a spark..." (Benedict XVI 2008b). In these same remarks, he again identified environmental awareness as a public context in

which a "sensitivity" to the Christian gospel already exists, and where
Catholics must strive to live their responsibilities in a way that "credibly convey[s] faith as an orientation in our time." For Benedict, it is
the wholeness of Christian faith itself which must be recovered and
lived anew.

As if to offer, *ad intra*, a catechesis in this wholeness, the "indissoluble bond" between creation and redemption, Benedict's progression of
encyclicals began from the richness of Christian particularity. In *Deus
caritas est*, wishing to convey in his first encyclical the central meaning of Christianity, the new pope spoke of love: God is love and the
source of love as an energy to transform the world. Bringing care for
the environment into the heart of Catholic belief and practice, we find:
"Faith, worship and ethos are interwoven into a single reality which
takes shape in our encounter with God's agape.... A Eucharist which
does not pass over into the concrete practice of love is intrinsically
fragmented" (Benedict XVI 2005b, #14). Shortly after *Deus caritas
est*, this Eucharistic orientation was expanded in the apostolic exhortation, *Sacramentum caritatis* , where Benedict reminded the faithful that
"the Eucharistic form of life can help us foster a real change in the way
we approach history and the world":

> [T]o develop a profound Eucharistic spirituality that is also ca-
> pable of significantly affecting the fabric of society, the Christian
> people, in giving thanks to God through the Eucharist, should be
> conscious that they do so in the name of all creation, aspiring to
> the sanctification of the world and working intensely to that end.
> The Eucharist itself powerfully illuminates human history and the
> whole cosmos.... The world is not something indifferent, raw ma-
> terial to be utilized simply as we see fit. Rather, it is part of God's
> good plan, in which all of us are called to be sons and daughters
> in the one Son of God, Jesus Christ (Eph 1:4-12). The justified
> concern about threats to the environment present in so many parts
> of the world is reinforced by Christian hope, which commits us to
> working responsibly for the protection of creation. The relationship
> between the Eucharist and the cosmos helps us to see the unity of
> God's plan and to grasp the profound relationship between creation
> and the "new creation" inaugurated in the resurrection of Christ, the
> new Adam. Even now we take part in that new creation by virtue
> of our Baptism (Col 2:12ff). Our Christian life, nourished by the
> Eucharist, gives us a glimpse of that new world—new heavens and
> a new earth (Benedict XVI 2005a, #92).

If indeed the Eucharist is that sacrament in which Christians encounter Christ the Redeemer as "the center of the universe and of history," Benedict's next encyclical, *Spe salvi*, develops that performative aspect of Christian hope to which the Eucharist "challenges" the faith community. Differentiating Christian hope from human utopias, the pope identifies the hope that comes from faith in God as a motivation for building our common planetary future. This is a hope for the world, not just for individual souls. In this second encyclical, Benedict also engages in historical analysis of the environmental crisis reminiscent of the many genealogies of environmental history; he identifies Francis Bacon's celebration of human artifice over nature as a fatal turn, laying the groundwork for the cult of "progress" with all its ambiguities and providing the justification for human domination of the earth (Benedict XVI 2007b, #16-18). Science and politics then become "a critique of the earth itself" (#20) in a materialist view of all life which reduces the human person to the economic realm. The encyclical's central thesis is well stated in the following excerpt: "This great hope can only be God who encompasses the whole of reality and who can bestow upon us what we, by ourselves, cannot attain. The fact that it comes to us as a gift is actually part of hope. God is the foundation of hope: not any god, but the God who has a human face and who has loved us to the end, each one of us and humanity in its entirety. His Kingdom is not an imaginary hereafter, situated in a future that will never arrive; his Kingdom is present *wherever he is loved* and *wherever his love reaches us*" (Benedict XVI 2007b, #31, italics added).

The God "who encompasses the whole of reality," God's presence in love, leads us to consideration of *Caritas in veritate* . In this first social encyclical of his pontificate, Benedict writes: "The Christian revelation of the unity of the human race presupposes a *metaphysical interpretation of the 'humanum' in which relationality is an essential element*" (2009a, #55). We find the terms of this relational anthropology presented in a manner that reaffirms John Paul's diagnosis of the "anthropological error" of dominion/mastery, and which seeks to recuperate all relationality: the self in relation to nature, to others, and by extension, to society. These relations are recuperated because each is recognized as originating from the God who is the Mystery of Love. In this relation of origin, each thereby is reaffirmed in its intrinsic integrity, and from this relation of origins with the Creator, each reflects the structure of reality as itself relational. Every "other," whether nature's non-human creatures or another human, *every Other* embodies the Mystery of loving reciprocity, and thus becomes our encounter with the Mystery of our common existence. In its presentation of the gift-nature of existence and its corollary of the human vocation to self-gift, we find Benedict's exposition of how the most particular (theological) aspect of Christianity becomes also the most universal, as metaphysics, and also how this putatively most abstract, in fact becomes deeply experiential and profound in its applications to social issues.

In order to accomplish this comprehensiveness and comprehensibility, the particular and universal offered to "all persons of goodwill" in a social encyclical, Benedict must harvest his own participation in and interpretation of the twentieth century renewals in theology, especially developments in the theology of grace, to which we adverted in our essay's introduction. In an insightful analysis aptly entitled "Grace Reloaded: *Caritas in Veritate's* Theological Anthropology," Philipp Gabriel Renczes, S.J. states: "In light of both the dramatic events connected to the 2008-2009 world-wide economic crisis and the long-term challenges posed by globalization and climate change, much can be said in favor of this desire to go 'back to the basics' concerning the conditions of human co-existence, a concern that clearly reflects Ratzinger's priority agenda for Christian theology and life in the modern world" (Renczes 2010, 274). Drew Christiansen, S.J. suggests the encyclical's treatment of natural law should be read in light of the 2009 International Theological Commission's document, "The Search for a Universal Ethics: A New Look at the Natural Law," a "comprehensive and balanced statement" especially important "for a time when, in response to globalization, many, particularly in the business and environmental communities, are exploring the possibilities of a common ethic" (2010, 13). Christiansen cites the ITC document which states: "We should reach the point of saying, behind our religious convictions and the diversity of our cultural presuppositions, what are the fundamental values for our common humanity, in a manner to work together to promote comprehension, reciprocal recognition and peaceful cooperation between all the members of the human family (No. 116)" (ibid. 14).

To reach those "fundamental values," Benedict articulates the heart of Christian experience presented in *Deus caritas est*—trinitarian love in its own metaphysical *fundamentum*: wonder before the mystery of existence. "Charity in truth places man before the astonishing experience of gift…. The human being is made for gift, which expresses and makes present his transcendent dimension…Because it is a gift received by everyone, charity in truth is a force that builds community" (Benedict XVI 2005b, #34). As Renczes puts it, "Grace precisely contains the idea of a continuous, dynamic structure that sustains human existence in its relation to ultimate meaning. Attesting to grace suggests that life makes sense as 'being-received' from a preceding, greater love and as handed over to others" (Renczes 2010, 280). The 'God-who-precedes,' made known in Christ, "remains shrouded in mystery" (281), and we are reminded that an apophatic dimension remains in the midst of Christianity's particularity that is a mandate for comprehensibility in the universal vocation to self-gift, to communion. Thus, Benedict has offered "the elaboration of *a social doctrine* out of humanity's finality to create that results from the human person's capacity to give as a consequence of having him/herself gratuitously received (283, italics added)…. Love as grace is…ontologically received creativity (284)." This "revalorized theology of grace" should be expected to "imprint its features onto concrete teachings" in a manner commensurate with "the fundamental ways that human beings engage reality" (284). Both "nature" as "cosmological" (the environment) and "anthropological" (the human person) reflect this gratuitous relation of origins in which the Creator's self-gift is a "grammar" calling forth reciprocal gift of self in a relation of "covenant between human beings and the environment, which should mirror the creative love of God, from whom we come and towards whom we are journeying" (Benedict XVI 2009a, #50). Technology, human creativity, "must serve to reinforce the covenant between human beings and the environment" (#69).

Yet how does this renewal in the theology of grace, its language of "mystery" and the "grammar" of "gift," translate into social action, especially in the controverted area of the environment and climate change? Christiansen points out that for Benedict "the key norm for social action [in continuity with his recent predecessors] is the 'vocation' to integral human development" (2010, 11), the whole person, in relation to the whole human family, in relation to the whole of reality via the solidarity that this constitutive relationality betokens. It is the experience of love and gift, in Christian life and the life of persons of goodwill, that is a dynamism of charity's efficacy in the transformation of society. Christiansen asserts the "transformative social policy proposals" carried in this encyclical derive from the theological reading of charity's role in the social order via the principle of the common good:

"Benedict makes explicit what is sometimes less obvious in Catholic social teaching, namely, that the common good is primarily effected through institutions…(CV 7). The common good is 'the institutional path—we might also call it the political path—of charity' (7). For anyone still tempted to think that Benedict does not favor a structural approach to social justice, the encyclical's treatment of the common good is strong evidence to the contrary. Just as charity 'directly encounters the neighbor,' so also it expresses itself through institutions and their reform in a way that is essentially 'political'. With Paul VI, Benedict also believes in the urgency of institutional reform. 'It is Christ's charity that urges us on' (CV 20)" (Christiansen 2010, 12-13).

That urgency was again underscored in Benedict's January 2010 World Day of Peace Message, *If You Want to Cultivate Peace, Protect Creation* (2009c). Here Benedict restates the principle elements of his approach to the environment: the covenant between humans and the environment, the rejection of absolute dominion, unity in shared responsibility for the environment as "God's gift to all people, especially the poor and future generations," and the contemplative aspect of the human relation to nature which "helps us understand our vocation and worth as human beings" (#2). Acknowledging the legacy of John Paul II whom Benedict describes as having "emphasized our relationship, as God's creatures, with the universe all around us" (#3), he affirms his predecessor's appeal for a "new solidarity" which "is all the more pressing today, in the face of signs of a growing crisis which it would be irresponsible not to take seriously" (#4). Confronted with the problems associated with "such realities as climate change" and other urgent examples of global environmental degradation, he asks "Can we remain indifferent?" Calling again for "a profound cultural renewal" in which humanity "rediscover[s] those values which can serve as the solid base for building a brighter future for all" (#5), he reinforces the reciprocal relation between human ecology and the well-being of the natural environment, hallmarks of John Paul II's legacy now re-expressed in *Caritas in veritate*'s metaphysical key.

Benedict describes globalization as "the principle new feature" of the contemporary situation, "the explosion of worldwide interdependence" that presents both a crisis and an opportunity and requires the "guidance of charity" (2009a, #33). The "altogether new and creative challenge" that the vocation to charity in truth presents "is about broadening the scope of reason and making it capable of knowing and directing these powerful new forces, animating them within the perspective of that 'civilization of love' *whose seed God has planted in every people, in every culture"* (#33, italics added). However, broadening the scope of reason presents a challenge, first and foremost, to conversion *ad intra*. When the Catholic contribution to confronting human-forced climate change and environmental degradation is considered in light of the cultural dynamics of our "contemporary situation," we encounter not only a legacy of promise from these two popes, but also an aspect of poignance.

CONCLUSION:
PROMISE & POIGNANCE, *AD INTRA* & *AD EXTRA*

In an essay entitled "The Future of Catholic Social Thought," John A. Coleman, S.J., observes: "A vital tradition is brought forward only as it really addresses the new questions or signs of the times. A tradition of social thought is challenged to translate this address, beyond large generalities couched in anthropological assumptions, into policy and middle-axioms to guide policy studies, advocacy, and social choices" (2005, 535). By their continued reference to the environmental crisis, John Paul II and Benedict XVI have made the church part of the world-wide conversation which now is beginning to take place with new urgency. However, that urgency becomes occasion for reading both promise and poignance in the legacies of these two popes. It is a problem of reception of teachings, conversion as much *ad intra* as witness *ad extra*.

Christiansen rightly calls Benedict's challenge in *Caritas in veritate* "the most radical teaching on economic life any modern pope has given" (2010, 4), and queries "whether its vision of society as gift and communion can penetrate economics and commerce" (19)? Indeed, can it transform politics, civic life, and international forms of governance necessary to confront a crisis of the scope and complexity of human-forced climate change? If it is to do so, it will have to be "carried" into these spheres in the first place by Christians who seek to live credibly within not just the particularity, but also the integral and solidary humanism the popes' teachings convey. The magisterium's insistence upon safeguarding "an authentic 'human ecology' and "the inviolability of human life at every stage" (Benedict 2009c, #12) can only gain in credibility when Catholic witness reflects the "indivisible relationship between God, human beings and the whole of creation" (2009c, #14) that also encounters in generosity all persons of goodwill and the integrity of their concern for the gravity and meaning of the human destruction of the natural world. Does Christian credibility really advance when too many within the Church, in today's politicized rhetoric, continue to insist upon caricatures of environmental*ism* reduced to extremes of ecocentric or biocentric expressions? Such caricatures overlook not only the Church's "balanced" approach (Benedict 2009c, #13), but moreover, its humanistic concern for the millions worldwide whose moral sensibility recognizes the natural world as locus of common transcendent experience, where humans encounter the Mystery of existence and engage universal questions of meaning. Can we find a unity of voice, *ad intra*, that celebrates this enduring human reverence for the gift of existence, the preciousness of life, the beauty of nature which both popes recognize, as our shared encounter and point of contact from which Christians enter into their vocation of solidarity and service to the world? Are the sociological and institutional carriers of Catholic social teaching themselves so polarized as to be yet capable of delivering a moderating, mediating, edifying word to a globalized world whose dynamics of division often seem more amplified than its voices for solidarity?

Perhaps Christians will have to re-learn that apophatic dimension which Benedict XVI's theology of grace embraces, a renewed humility before the Mystery of both creation and those *other* persons of goodwill with whom we are called to join in confronting environmental degradation. This gathering around transcendent meaning, value intuited in the beauty of love and the love of beauty, makes nature one of the spaces, one of the poles in the primary pairing of the dignity of the human person and of creation, where this graced encounter must happen, where "new forms of communion, fraternity, gratuitousness" must emerge. Will climate change prove to be the crisis which opens a new opportunity to discover this on a global scale? In a recent address, Cardinal William Levada, Prefect of the Congregation for the Doctrine of the Faith for which he is Ratzinger's successor, expressed the urgency of a new apologetics for the church in the twenty-first century. A renewed fundamental theology where faith and reason, credibility and truth come together must be thought through again in light of "new situations, new generations, new cultures" (Levada 2010). In continuity with both popes, one of the new situations he identifies is environmental concern, and he further asserts that "the shape" of "a new apologetics for the new millennium should focus on the beauty of God's creation" (Levada 2010). However, he notes the original aspect of apologetics which required that it proceed "with courtesy and respect" was forgotten by Christians, and thus even the term is rejected today. Cardinal Levada cites the reflections of Avery Dulles, S.J. who wrote on the history of apologetics: "Finally apologists came to recognize that every Christian harbors within himself a secret infidel. At this point apologetics became, to some extent, a dialogue between the believer and the unbeliever in the heart of the Christian himself" (Dulles 1971, xvi cited in Levada).

There is a perpetual struggle between Christian particularity and universalism, between comprehensiveness and comprehensibility, in the journey of the Church through history. In moments of cultural crisis, this perpetual struggle is revealed within the hearts of Christians themselves. There is profound promise in the teachings of Benedict XVI and John Paul II which exhort us all toward ecological conversion and renewed solidarity. However, there is a certain Christian courage required for the generosity of universal encounter, integral Christian humanism, even as faith's experience of wonder before the Mystery of Being will necessarily take on concrete Christian expression. The world awaits what each religion can say toward a universal ethic, but also what each can say from its own particularity. These two popes have sought to serve this perennial, difficult task, and nowhere will their legacies be more tested than on the question of human-forced climate change.

When acknowledging both the promise and the poignance of their legacies, perhaps we do well to recall the words of their mutual friend and collaborator, the Swiss theologian Hans Urs von Balthasar, who exemplifies efforts in the renewal of metaphysics for which Benedict calls. In this task of renewal he warns that "no concept of human.... wholeness can be constructed from the function of a mastery and domination of the world...," for the world "will only be filled with love by making ultimate decisions for love, and never through the ameliorizations of technology" (Balthasar 1991, 653-54). "Christians," he tells us, "can only adequately answer God's universal engagement with the world in the love of Jesus Christ for them by lending their own love, in the *concretissimum* of the encounter with their brother, that universal breadth of Being which...the metaphysical act possesses and is"

(654-55). Perhaps, in generous, courageous Christian charity enacted in universal solidarity, the message of conversion, both personal and structural, can find efficacy in forestalling the consequences of human-forced climate change. Perhaps Catholics who inherit the teaching legacies of John Paul II and Benedict XVI can join with other persons of goodwill and attest that, as Cardinal Levada has said, "The relativization of truth is not the necessary precondition of real dialogue; the desire to know the other in the fullness of his or her humanity is" (2010).

Can social Catholicism present a witness to the world which defies cultural dynamics of polarization and, with generous courage, fosters a new springtime commensurate with the global moment? Can we find it within ourselves as Christians to courageously initiate, as prerequisite for confronting the ecological crisis, a different sort of climate change?

SOURCES

Balthasar, Hans Urs von. 1991. *The Realm of Metaphysics in the Modern Age.* Vol. 5 of *The Glory of the Lord: A Theological Aesthetics.* San Francisco: Ignatius Press.

Benedict XVI. 2005a. *Sacramentum caritatis.* Post-Synodal Apostolic Exhortation on the Eucharist as the Source and Summit of the Church's Life and Mission, February 22. Accessed from http://www.vatican.va/holy_father/benedict_xvi/apost_exhortations/documents/hf_ben-xvi_exh_20070222_sacramentum-caritatis_en.html on November 20, 2010.

————. 2005b. *Deus Caritas est (God is Love).* December 25. Accessed from http://www.vatican.va/holy_father/benedict_xvi/encyclicals/documents/hf_ben-xvi_enc_20051225_deus-caritas-est_en.html on November 20, 2010.

————. 2007a. Meeting with Clergy of the Diocese of Belluno-Feltre and Treviso, July 24. Accessed from http://www.vatican.va/holy_father/benedict_xvi/speeches/2007/july/documents/hf_ben-xvi_spe_20070724_clero-cadore_en.html on November 20, 2010.

————. 2007b. *Spe Salvi.* November 30. Accessed from http://www.vatican.va/holy_father/benedict_xvi/encyclicals/documents/hf_ben-xvi_enc_20071130_spe-salvi_en.html on November 20, 2010.

————. 2008a. Meeting with the Parish Priests and the Clergy of the Diocese of Rome, February 7. Accessed from http://www.vatican.va/holy_father/benedict_xvi/speeches/2008/february/documents/hf_ben-xvi_spe_20080207_clergy-rome_en.html on November 20, 2010.

————. 2008b. Meeting with Clergy of the Diocese of Bolzano-Bressanone, August 6. Accessed from http://www.vatican.va/holy_father/benedict_xvi/speeches/2008/august/documents/hf_ben-xvi_spe_20080806_clero-bressanone_en.html on November 20, 2010.

————. 2009a. *Caritas in Veritate*. June 29. Accessed from http://www.vatican.va/holy_father/benedict_xvi/encyclicals/documents/hf_ben-xvi_enc_20090629_caritas-in-veritate_en.html, November 20, 2010.

————. 2009b. Videostatement of His Holiness Benedict XVI to the UN 2009 Summit on Climate Change. September 23. Accessed from http://www.vatican.va/holy_father/benedict_xvi/messages/pont-messages/2009/documents/hf_ben-xvi_mes_20090924_summit-climat-change_en.html on November 20, 2010.

————. 2009c. *If You Want to Cultivate Peace, Protect Creation*. 2010 World Day of Peace Message, December 8. Accessed from http://www.vatican.va/holy_father/benedict_xvi/messages/peace/documents/hf_ben-xvi_mes_20091208_xliii-world-day-peace_en.html on November 20, 2010.

Christiansen, Drew, and Walter Grazer, eds. 1990. *"And God Saw That It Was Good": Catholic Theology and the Environment*. Washington, DC: United States Catholic Conference.

Christiansen, Drew. 2010. "Metaphysics and Society: A Commentary on *Caritas in veritate*." *Theological Studies* 71 (March): 3-28.

Coleman, John A. 2005. "The Future of Catholic Social Thought." In *Modern Catholic Social Teaching: Commentaries & Interpretations*, ed. Kenneth R. Himes, OFM., 522-44. Washington, DC: Georgetown University Press.

Dulles, Avery. 1971. *A History of Apologetics*. New York: Corpus.

Himes, Michael J., and Kenneth R. Himes, OFM. 1993. *Fullness of Faith: The Public Significance of Theology*. New York: Paulist Press.

John Paul II. 1979. *Redemptor hominis*. March 4. Accessed from http://www.vatican.va/holy_father/john_paul_ii/encyclicals/documents/hf_jp-ii_enc_04031979_redemptor-hominis_en.html on November 20, 2010.

————. 1984. Address to Youth, Viterbo Italy. Accessed from http://www.vatican.va/holy_father/john_paul_ii/speeches/1984/may/documents/hf_jp-ii_spe_19840527_giovani_it.html on November 20, 2010.

————. 1987. *Sollicitudo rei socialis*. Encyclical on the Twentieth Anniversary of "Populorum Progression." December 30. Accessed from http://www.vatican.va/holy_father/john_paul_ii/encyclicals/documents/hf_jp-ii_enc_30121987_sollicitudo-rei-socialis_en.html on November 20, 2010.

————. 1989. *Peace with God the Creator, Peace with All of Creation.* Message on the 1990 World Day of Peace. December 8. Accessed from http://www.vatican.va/holy_father/john_paul_ii/messages/peace/documents/hf_jp-ii_mes_19891208_xxiii-world-day-for-peace_en.html on November 20, 2010.

————.1991. *Centesimus Annus.* Encyclical on the 100th Anniversary of *Rerum Novarum*m, May 1. Accessed from http://www.vatican.va/holy_father/john_paul_ii/encyclicals/documents/hf_jp-ii_enc_01051991_centesimus-annus_en.html on November 20, 2010.

————. 1993. "Prayer, Work and Nature Must Be Harmonized." *L'Osservatore Romano* 28, July 14, 1-2.

————. 1994. *Crossing the Threshold of Hope.* New York: Alfred A. Knopf.

————. 1995. *Evangelium vitae.* Encyclical on the Value and Inviolability of Human Life, March 25. Accessed from http://www.vatican.va/holy_father/john_paul_ii/encyclicals/documents/hf_jp-ii_enc_25031995_evangelium-vitae_en.html on November 20, 2010.

Judt, Tony. 2009. "What is Living and What is Dead in Social Democracy?" *The New York Review of Books*, December 17, 2009. Accessed from http://www.nybooks.com/articles/archives/2009/dec/17/what-is-living-and-what-is-dead-in-social-democrac/ on November 20, 2010.

Koeniz-Bricker, Woodeene. 2009. *Ten Commandments for the Environment. Pope Benedict XVI Speaks Out for Creation and Justice.* Notre Dame: Ave Maria Press.

Lasher, Connie. 2004. "The Contemplative Glance of Faith": Hans Urs von Balthasar's Contribution to a Catholic Theology of Ecological Identity. PhD. diss., Boston College. ProQuest (UMI 3135966).

Levada, William Cardinal. 2010. "The Urgency of a New Apologetics for the Twenty-First Century." Accessed from http://www.zenit.org/article-29086?l=english on April 29.

Miller, Michael J., ed. 1996. *The Encyclicals of John Paul II.* Huntington, IN: Our Sunday Visitor Publishing Division.

Murphy, Charles M. 1989. *At Home on Earth: Foundations for a Catholic Ethic of the Environment.* New York: Crossroad.

Pontifical Council for Justice and Peace (PCJC). 2005. *Compendium of the Social Doctrine of the Church.* Washington, DC: USCCB.

Ratzinger, Joseph. 1969. "The Dignity of the Human Person." In *Commentary on the Documents of Vatican II*, ed. Herbert Vorgrimler, vol. 5: 115-63. New York: Herder and Herder.

————. 2004. *Truth and Tolerance: Christian Belief and World Religions*. San Francisco: Ignatius Press.

Renczes, Phillip Gabriel. 2010. "Grace Reloaded: *Caritas in Veritate's* Theological Anthropology." *Theological Studies* 71: 273-290.

Schindler, David L. 2000. "Reorienting the Church on the Eve of the New Millennium." In *The Legacy of Pope John Paul II: His Contribution to Catholic Thought*, ed. Geoffrey Gneuhs, 94-127. New York: Crossroad.

Weigel, George. 1999. *Witness to Hope: The Biography of John Paul II*. New York: HarperCollins.

Worster, Donald. 1985. *Nature's Economy: A History of Ecological Ideas*. Cambridge: Cambridge University Press.

SOLIDARITY, SUBSIDIARITY, &
PREFERENCE FOR THE POOR

EXTENDING CATHOLIC SOCIAL TEACHING IN
RESPONSE TO THE CLIMATE CRISIS

Jame Schaefer

Grounding Catholic social thought is the belief that each human person has an intrinsic dignity and a desire for relationships with others that should lead to cooperating for their mutual good, making decisions individually and collectively to achieve it, and showing preference for the poor, vulnerable, and suffering in decision-making and actions. Rooted in the sacred scriptures, reflected upon at least implicitly for centuries by theologians, and taught by bishops of the Roman Catholic Church (the magisterium), principles to guide our decision-making and actions began to be considered collectively toward the end of the 19ᵗʰ century as "Catholic social teaching" through which the bishops direct the faithful in living lovingly in relation to one another because of their relationship with God. Most prominent among the popes who issued explicit directives in response to societal problems during their times was Leo XIII. In 1891, he underscored in *Rerum Novarum* the dignity, rights, and responsibilities of humans laboring in the newly burgeoning industrial economy. Subsequent popes have conveyed their social teachings primarily through encyclicals in which they address new and continuing issues, and other bishops have reflected upon these papal documents when issuing pastoral statements to their constituents. Basic to these teachings is the necessity to cherish the life and inherent dignity of the human person from conception to bodily death. As the Catholic bishops in the United States insist: "This central Catholic principle requires that we measure every policy, every institution, and every action by whether it protects human life and enhances human dignity, especially for the poor and vulnerable" (USCCB 1998).

One of the latest principles to emerge from these teachings in response to escalating environmental problems is the necessity for the faithful to show respect for God by protecting and caring for God's creation. Pope John Paul II issued the first dedicated statement on this principle in his message celebrating the 1990 World Day of Peace entitled *Peace with God the Creator, Peace with All of Creation*. In this message, he lamented the ongoing destruction of the natural environment, declared its adverse affects on human life as a moral problem for which people at all levels of endeavor are responsible to address, and directed the faithful to "respect and watch over" God's creation "in light of that greater and higher fraternity that exists within the human family" (1989, #16). He integrated this teaching in numerous encyclicals and statements, and many bishops from around the world issued pastoral statements on this teaching to the faithful in areas in which they serve (Whittington 1994). Among these statements is one that directly addresses the climate crisis—*Global Climate Change: A Plea for Dialogue, Prudence, and the Common Good* (USCCB 2001)— and is included in this collection of essays. As indicated in the essay by Msgr. Charles Murphy and Connie Lasher, Pope Benedict XVI carried into his encyclicals, messages, and statements the principle to care for God's creation, capping them to the present with his message on the 2010 World Day of Peace, *If You Want to Cultivate Peace, Protect Creation* (2009c). These magisterial documents show their authors' grasp of basic scientific facts about the adverse effects that environmental degradation and destruction have had on human life in the past and present. They also recognize scientific projections that point to a more bleak picture for future generations whose well-being cannot be separated from the well-being of other species, ecological systems, and the biosphere of Earth. From this informed perspective, the bishops urge thinking and acting more responsibly toward other constituents of our planet.

Though caring for God's creation addresses environmental concerns in the interests of humanity now and into the future, other principles of Catholic social teaching can be extended to effectively address anthropogenic causes of climate change. Particularly significant are the subjects of this essay: (1) the *solidarity* of all people because of their shared human dignity; (2) *subsidiarity* as a process for addressing concerns at incremental levels of governance beginning with the individual; and, (3) *preference for poor and vulnerable people* when making and executing decisions. I begin by exploring these three principles sequentially, providing a brief overview of each in magisterial documents and pointing to the bishops' applications of these principles to environmental concerns. The fourth part focuses on extending each principle in an attempt to more effectively address the climate crisis. Assumed throughout is the basic Catholic understanding that the human person possesses a special dignity in relation to God that should be exemplified in responsible and loving relationships with other persons. Also assumed are basic scientific findings on the effects of human-forced climate change discussed in the introduction and reinforced in other essays of this anthology.

SOLIDARITY—SOCIETAL COLLABORATION TO ACHIEVE THE COMMON GOOD

From the patristic period onward, the bishops conveyed their understanding that the faithful should work together to achieve their common good. The identification and explanation of the concept of solidarity by the bishops appeared explicitly in Pope Pius XII's *Summi Pontificatus* as a law rooted in Christian charity that binds humans to collaborate in achieving a mutual good because of their common origin, their rational nature, and the redeeming action of Jesus on the cross (1939, #15, 35, 72). Bringing about and maintaining international peace in human solidarity was a major mutual good to which Pius XII referred at the outbreak of World War II, throughout this war, and during its aftermath (see Doran 1996, 83-84).

Subsequent popes and other bishops continued to use, apply, and enrich the meaning of solidarity when addressing issues pertinent to their times. For example, In *Mater et Magistra*, Pope John XXIII identified solidarity as a guiding principle for wealthy nations to address hunger, misery, and poverty in other nations, for materially impoverished nations to work together in addressing their mutual problems, and for establishing workers' unions through which relations between workers and employers can be addressed to their mutual advantage (1961, #23, 157, 190). The bishops of the Second Vatican Council used the concept of solidarity several times in *Gaudium et Spes* to emphasize the need for the faithful to strive for loving and respectful relationships among themselves as "brothers" and among people throughout the world to achieve the common good of their "universal brotherhood" (1965, #32, 37, 38, 42). Advancing solidarity as a duty especially to the poor, the bishops of Latin America who met in Medellín, Colombia in 1968 attributed the widespread suffering and poverty in their area to the lack of solidarity that was epitomized by the failure to criticize and correct the ongoing injustice, oppression, and "intolerable situation" within which poor people are enmeshed (CELAM 1970, 217).

Pope Paul VI was also concerned about impoverished people and nations. In *Populorum Progressio,* he pointed to the inequities among materially rich and poor nations that thwart achieving peace in the world, expressed the need for all people to have opportunities to develop themselves, and described solidarity as a "duty" that the wealthy are obligated to embrace: "This duty concerns first and foremost the wealthier nations. Their obligations stem from the human and supernatural brotherhood of man, and present a three-fold obligation: (1) mutual solidarity—the aid that the richer nations must give to developing nations; (2) social justice—the rectification of trade relations between strong and weak nations; (3) universal charity—the effort to build a more humane world community, where all can give and receive, and where the progress of some is not bought at the expense of others" (1969, #44). Wealthy nations acting in solidarity with poor nations could achieve "spiritual growth" that, together with economic growth, will "contribute immeasurably to the preservation of world peace" (#73). Identifying solidarity as a call from God, the pope urged the faithful to be alert to God's calling and to respond through their relations with other persons (#42).

Having reflected on the concept of solidarity before becoming the 265th head of the Roman Catholic Church and the onset of the Polish Solidarity Movement, Pope John Paul II reinforced and advanced his predecessors' teachings in *Redemptor Hominis, Laborem Exercens, Sollicitudo Rei Socialis, Centesimus Annus,* and several World Day of Peace messages (Bilgrien 1999). He stressed solidarity primarily as an attitude and as a moral virtue. As an attitude, solidarity assumes an individual and a group's recognition of the dignity and rights of human persons that disposes the individual to work cooperatively with others toward the common good of all persons (1987, #38-39). As a moral virtue, the pope insisted, solidarity should be practiced by individuals and groups with "a firm and persevering determination" to achieve "the good of all and of each individual, because we are all really responsible for all" (#38).

Pope Benedict XVI has continued to address solidarity with emphasis on God's love as a stimulus to the faithful to demonstrate their relatedness to all people as one family in search of its material and spiritual common good. Especially concerned about impoverished people and nations struggling to develop in a globalized economy, he applauded in his encyclical *Deus Caritas est* a growing sense of solidarity among peoples throughout the world that has been fostered by governmental agencies and humanitarian organizations through subsidies, tax relief, and making resources available to people who need them. He expressed his special gratitude to volunteer organizations for their loving service and lauded the participation of young people in these efforts that he interpreted as "a formation in solidarity and in readiness to offer others not simply material aid but their very selves" (2005c, #30). From his perspective, achieving solidarity requires recognizing the interrelatedness and interdependence of humans to one another physically, socially, and economically, working together lovingly and respectfully to develop economically in ways that avoid subordinating the aid-receivers to the aid-givers, and remaining cognizant of the effects that decisions made today will have on future generations (#43).

According to at least two other pontiffs, the human family goes beyond current generations to include future people. The future of humanity concerned Paul VI deeply, especially in light of the deepening rift between rich and poor people and projections of even greater poverty among them in the future. He recognized that each of us has obligations to others not only in the present: "We are the heirs of earlier generations, and we reap benefits from the efforts of our contemporaries; we are under obligation to all men. Therefore we cannot disregard the welfare of those who will come after us to increase the human family. The reality of human solidarity brings us not only benefits but also obligations" (1969, #17). Fretting over the "excessive" and "disordered" consumption of Earth's "resources" by present generations, John Paul II cautioned the faithful about their "capacity to transform" the world through technology while forgetting that the world is "God's prior and original gift" of which they cannot make "arbitrary use." When using God's "gift" of the world today, he continued, the faithful must be conscious of its duties and obligations to future generations (1991a, #37). Benedict XVI underscored Earth as "a precious gift of the Creator" to humans (2009b) and "our common home" from which "future generations have the right to reap its benefits" responsibly as should current generations (2007b). In his 2010 World Day of Peace message, he lamented the misuse of the goods of Earth and encouraged "a greater sense of intergenerational solidarity" so future people are not "saddled with the cost of our use of common environmental resources" (2009c, #8). Studies questioning the availability of food for future generations prompted him in his message on World Food Day to urge governments to provide "adequate funding" that will facilitate "the reactivation of [agricultural] production cycles, despite the deterioration of climatic and environmental conditions" (2010b).

Teachings on solidarity have been issued collectively by other bishops. In *The Catechism of the Catholic Church*, solidarity is identified as "a law" linked to charity (1994, #361), "a principle" also articulated in terms of "friendship" or "social charity" (#1939), "a virtue" for spreading spiritual goods and developing temporal goods (#1942, 1948, and 2407), and "a duty" that rich nations have to the poor out of charity and justice (#2439). The Pontifical Council for Justice and Peace, which is comprised predominantly of bishops and cardinals, considers solidarity both as a social principle that stresses the interdependence among individuals and peoples who must avoid perpetrating injustice and as "*an authentic moral virtue*" that is manifested by "*a firm and persevering determination* to commit oneself to the *common good*" (2004, 85). All references to solidarity and its precursors in these documents assume the traditional theological understanding that humans are intrinsically social by nature and intended by God to live in cooperative relationships with one another to achieve their mutual good.

Thus, solidarity has a rich heritage in Catholic social teaching by the bishops of the Church in the 20th to early 21st centuries. While the basic idea that the faithful are called to recognize and act in collaboration with all people for their common good regardless of race, ethnicity, religion, or any other categories of differences among peoples, the concept of solidarity has been nuanced in magisterial teachings when applied to the particular circumstances the bishops address in the contexts of their times. Solidarity has been categorized variously as a law that binds people together in friendship, a duty to one another with preference for the poor to achieve the common good, a social principle for recognizing the oneness of all in the journey of life, an attitude that disposes the individual and group to seek the common good, and a moral virtue to be developed in the individual and demonstrated throughout one's lifetime. Yet no one category seems adequate. Drawing from magisterial teachings discussed above, a multi-faceted definition of solidarity is warranted: *The social bond grounded in Christian love that exists within and among persons through awareness of their shared human nature, their interdependence in an increasingly global society, their special obligation to impoverished people and nations, and the moral commitment to strive virtuously to make and execute informed decisions for the common good of all people in the present and the future through dialogue, collaboration, aid to the poor, and service to one another individually and collectively.* Foundational to magisterial teachings about solidarity is the understanding that humans have a unique dignity among creatures as having been made in the "image and likeness" of God (Gen 1:26) and graced with the ability to relate freely, responsibly, and lovingly with one another and to God for their common good and the good of humanity.

In nations all over the world, the term solidarity has been used to call people together for common causes. Among the most well known of these efforts is the Independent and Self-Governing Trade Union Solidarność formed in 1980 at the shipyards in Gdańsk, Poland. Other groups that are Catholic-based rely upon the term as exemplified by members of the Society of Jesus who exchange social justice and ecology news, stimulate contacts, and promote networking throughout the world (Jesuit Social Apostolate 2010). The Catholic faithful have been called together in biological regions to address environmental concerns (e.g., Society of Jesus Oregon Province 2006 and Catholic Bishops of the Columbia River Watershed 2005), and they are called together in solidarity to address the climate crisis (e.g., Catholic Coalition on Climate Change 2010 and Caritas International 2010).

The sustainability of the global climate is a major common cause to which people can think about themselves as called by God to respond in solidarity with one another as individuals, groups, and nations. Recognition of human interrelatedness and interdependence is key to responding to this call to seek their mutual good. So also is the obligation of the present generation to future members of the human family so they can inherit a climate that will sustain them. As the United

States Catholic bishops taught in *Global Climate Change*: "Our obligations to the one human family stretch across space and time. They tie us to the poor in our midst and across the globe, as well as to future generations" (USCCB 2001).

SUBSIDIARITY—AN ORGANIZING PRINCIPLE FOR MAKING DECISIONS

The principle of subsidiarity is closely linked to magisterial pronouncements about solidarity. Surfacing in the bishops' teachings as the organizing means through which persons can achieve and exercise solidarity for their common good, the principle of subsidiarity was explicitly taught by Pope Pius XI in *Quadragesimo Anno*. In this encyclical, he reinforced and applied for his age of rapid industrial expansion, the entrenchment of large workers unions, class conflicts, and the spread of communism Pope Leo XIII's prior teaching in *Rerum Novarum* about the roles of laborers, owners of industry, and government. Pius XI was troubled by the diminished role of individuals and small labor associations that had been able to accomplish goals by their own initiative and the increased role of large associations with "unbridled ambition for power" propelled by "greed for [economic] gain" (1931, #109) that assumed decision-making on matters that could be handled at more local levels. This situation was gravely wrong, the pope taught, because it violates the harmonious functioning of an orderly society:

> As history abundantly proves, it is true that on account of changed conditions many things which were done by small associations in former times cannot be done now save by large associations. Still, that most weighty principle, which cannot be set aside or changed, remains fixed and unshaken in social philosophy: Just as it is gravely wrong to take from individuals what they can accomplish by their own initiative and industry and give it to the community, so also it is an injustice and at the same time a grave evil and disturbance of right order to assign to a greater and higher association what lesser and subordinate organizations can do (#79).

To help society function in an orderly manner, the pope continued, drawing upon reflections by Thomas Aquinas in *Summa contra Gentiles* 3.71, secular government must limit its purview to matters it can handle that individuals and smaller associations cannot:

The supreme authority of the State ought, therefore, to let subordinate groups handle matters and concerns of lesser importance, which would otherwise dissipate its efforts greatly. Thereby the State will more freely, powerfully, and effectively do all those things that belong to it alone because it alone can do them: directing, watching, urging, restraining, as occasion requires and necessity demands. Therefore, those in power should be sure that the more perfectly a graduated order is kept among the various associations, in observance of the principle of 'subsidiary function,' the stronger social authority and effectiveness will be the happier and more prosperous the condition of the State (#80).

Foundational to his understanding of these roles is a primary commitment to the dignity of human persons who should be free to form associations and to collaborate responsibly with others in making and carrying out decisions that are helpful for their self development as creatures who are striving for their common good. The associations formed do not supplant or subordinate the individual. They are intended to perform tasks that the individual alone cannot. When associations are formed by individuals to perform these tasks, the purview and responsibilities of the individual persist while the individual participates in the associations formed for the common good.

Magisterial teaching on the principle of subsidiarity continued beyond Pius XI's seminal efforts through encyclicals issued by subsequent popes and pastoral statements released by other bishops. For example, in *Mater et Magistra*, Pope John XXIII considered the "principle of subsidiary" as the "guiding principle" for the "work of directing, stimulating, co-ordinating, supplying and integrating" (1961, #53) efforts to find "appropriate solutions to the many social problems" of his time (#50). These problems included the lack of opportunity for workers to participate in management and to share in profits (#75, 77, 91), growing intervention by government in the personal lives of individuals and families (#60), and increasing government ownership of property that might reduce private ownership "beyond measure" or completely destroy it (#117). Like Pius XI, John XXIII was clear about the person's role in initiating associations to handle economic and political affairs and participating responsibly in these associations: "[I]n the economic order first place must be given to the personal initiative of private citizens working either as individuals or in association with each other in various ways for the furtherance of common interests" (#51). He also outlined the tasks that are appropriate for action by higher associations that lower associations cannot accomplish, and he underscored the dictum that these actions should not deprive the individual of his or her freedom to act. These actions by higher associations must "augment" human freedom "while effectively guaranteeing the protection" of the person's "essential personal rights. Among these is a man's right and duty to be primarily responsible for his own upkeep and that of his family" (#55).

During the Second Vatican Council, the bishops endorsed the principle of subsidiarity when discussing economic development. It must remain under human determination, the bishops directed, and not left to a judgment of a few people or nations that are economically and/or politically powerful: "It is necessary...that at every level the largest possible number of people and, when it is a question of international relations, all nations have an active share in directing that development. There is need as well of the coordination and fitting and harmonious combination of the spontaneous efforts of individuals and of free groups with the undertakings of public authorities" (Second Vatican Council 1965, #65). Citizens have the "right and duty...to contribute to the true progress of their own community according to their ability," and to do so freely. Exercising this right and dispensing this duty falls within the "basic rights of individual persons and groups" that should not be obstructed (ibid.). However, developing nations, economically advanced nations, and the international community have vital roles to play in the development process when following the principle of subsidiarity. Developing nations should "express and secure the total human fulfilment of their citizens" as the object for progress. Advanced nations should help the developing nations in discharging their responsibilities by respecting their need to support themselves from the income they receive on the sale of their "homemade products." And, "suitable organizations should be set up to foster and regulate international business affairs, particularly with the underdeveloped countries, and to compensate for losses resulting from an excessive inequality of power among the various nations" so the developing nations can "advantageously pursue their own economic advancement" (#86).

Pope John Paul II applied the principle of subsidiarity to the "[m]alfunctions and defects" of the welfare state in an encyclical that he issued to celebrate the 100th anniversary of Leo XIII's *Rerum Novarum*. In *Centesimus Annus*, John Paul attributed welfare state problems to "an inadequate understanding of the tasks proper to the State" and the failure to respect the principle of subsidiarity. A community of "a higher order should not interfere in the internal life of a community of a lower order, depriving the latter of its functions," he wrote. Instead, the "higher order" community should support the lower order community when needed and help "coordinate its activity with the activities of the rest of society, always with a view to the common good (1991a, #48). In *Sollicitudo Rei Socialis*, he emphasized the various forms of exploitation and suppression of the individual's right to develop economically, socially, and politically, especially in the developing countries. The denial or limitation of these rights "diminishes, or in practice absolutely destroys...the creative subjectivity of the citizen" and results in "passivity, dependence and submission to the bureaucratic apparatus" (1988, #15).

Composed predominantly of bishops and cardinals appointed by the pope, the Pontifical Council for Justice and Peace summarized magisterial teachings on the principle of subsidiarity in *Compendium of the Social Doctrine of the Church* (PCJP 2004, 81-83). Key to the principle of subsidiarity is promoting the dignity of the human person, and this is impossible "without showing concern for the family, groups, associations, local territorial realities" and other organizations formed to address issues. As the PCJP states, these collectives constitute an "aggregate of economic, social, cultural, sports-oriented, recreational, professional and political expressions to which people spontaneously give life and which make it possible for them to achieve effective social growth (#185, 81). All associations formed to address concerns that cannot be handled by individuals or groups at a more local level should adopt "*attitudes of help ('subsidium')*" whereby they assist the local associations through support, promotion, and development without supplanting their "initiative, freedom and responsibility" (#186, 81). Because every person, family, and association makes a unique contribution to the community, the principle of subsidiarity protects them from abuses by associations with greater power so they do not destroy "the spirit of freedom and initiative" of associations they are supposed to help for their common good (#187, 82). When a higher-level authority takes over a function that a lower-level authority or association cannot initiate (e.g., stimulate the economy and redress a serious injustice), this intervention is exceptional and "must not continue any longer than is absolutely necessary" so "the primacy of the person" is maintained as expressed in society (#188, 82-83). Implicit in the principle of subsidiarity is the duty of citizens to participate freely and responsibly in the cultural, economic, political, and social life of the community to which they belong for the common good (#189, 83).

In *Caritas in Veritate*, Benedict XVI stressed the importance of embracing the principle of subsidiarity to construct "a new order of economic productivity" (2009a, #41). The order he envisions is oriented toward socially responsible human self-development guided by "a dispersed political authority" that operates effectively on different levels of governance and activity (#41), organized to accomplish particular tasks for the common good, and subsidizes others when needed without infringing on their freedom. Key to maintaining these different levels of activity and governance is the inalienable freedom of human persons to actuate themselves, to relate to others for their common good, and to demonstrate charity to others by offering to assist them when they are unable to accomplish tasks on their own (#57). Linking his understanding of solidarity to the principle of subsidiarity, the pope stressed the need for assistance programs at all levels to involve recipients of aid from initiation to completion of the programs (#58) with the aim of seeking the common good (#64) locally to globally. The goods of Earth are a common good of all people, he cautioned, and all nations should "choose the path of dialogue," cooperate responsibly with one another, and "act in harmony" to "reassess the high levels of consumption" by technologically advanced countries, the hunger for energy by emerging nations, and "the search for alternative sources of energy and for greater energy efficiency" (2007b).

In summation, subsidiarity surfaces in magisterial teachings as *a societal organizing principle through which individual persons freely associate with one another at increasing levels of governance to accomplish a common good that cannot otherwise be achieved.* The principle of subsidiarity assumes the following: (1) Respect for the inherent dignity of the human person is a societal priority; (2) the person is naturally social and only able to become fully himself/herself in solidarity with others; (3) an association formed by persons exist to provide help (*subsidium*) to individuals so they can assume responsibility for their self-fulfillment and relationships with others; (4) an association formed by other associations to address goals that cannot otherwise be accomplished should help those associations so they can assume

responsibility for their efforts; (5) an association formed by other associations does not replace or diminish the work of the forming association; and, (6) intervention of an association into associations or
individuals that formed it is only appropriate when helping them help
themselves (see Komanchak 1988, 301-2). Thus, from the individual
to the highest level of association formed, each has its own purpose,
purview, task and authority to address issues that the individual or
lower association is incapable of addressing to achieve the good of all
(see Doran 1996, 212).

Like the concept and practice of solidarity, the principle of subsidiarity has significance for addressing environment concerns. Humans
are not solely individuals. We are social creatures who form and participate in associations to seek our common good at increasing levels
of governance—families, neighborhood associations, municipal, town,
county, state and federal governments, and regional to international
organizations. We depend upon these various levels to achieve goals
that one less encompassing association cannot. We can address many
environmental issues in our homes and businesses, but environmental
problems usually transcend political boundaries and require the cooperation of other people, associations, and governing bodies to address
effectively. The principle of subsidiarity can guide people in making
and executing decisions at appropriate levels while not absolving individuals, families, and local associations of their responsibilities for
decisions they can make and actions they can take to mitigate the effects of human-forced changes in the global climate. While a plethora
of examples can attest to collective action at several levels to address
problems, the climate crisis presents a challenge where action is needed concurrently at all levels.

OPTION FOR THE POOR

Admonitions to attend to the poor and vulnerable permeate the prophetic books of the Hebrew Bible and the teachings and life of Jesus
the Christ depicted in books of the New Testament. Drawing from
scriptural texts (e.g., Exodus 22:20-26, Leviticus 19:9-10, Job 34:20-
28, Proverbs 31:8-9, Sirach 4:1-10, Isaiah 25:4-5, Isaiah 58:5-7, Matthew 25:34-40, Luke 4:16-21 and 6:20-23, and 1 John 3:17-18), the
bishops of the Roman Catholic Church have consistently taught that
a basic test of society from a Christian perspective is how its most
vulnerable people are faring, and they instruct the faithful to put the

needs of the poor and vulnerable first when making and carrying out decisions individually and in association with others at all levels of governance. The account of the Last Judgment in Matthew 25:31-46 serves as a poignant reminder for our having to account to God for how we responded to the poor and vulnerable in our midst and throughout the world (USCCB 2010b). Drawing upon these passages either explicitly or implicitly, the bishops have stressed the need to show preference for the poor, suffering, and vulnerable in our deliberations and actions at all levels of our lives.

Showing Preference for Impoverished People

The imperative that we show preference for impoverished people resounds in so many papal encyclicals and bishops' pastoral statements (USCCB 2010d) that only a few can be mentioned here. In *Rerum Novarum*, Pope Leo XIII stressed the special consideration that should be given for the poor: "[W]hen there is question of defending the rights of individuals, the poor and badly off have a claim to especial consideration. The richer class have many ways of shielding themselves, and stand less in need of help from the State; whereas the mass of the poor have no resources of their own to fall back upon, and must chiefly depend upon the assistance of the State" (1891, #37). He was particularly conscious of the conditions in which laborers were working and living as Europe and North America transitioned from an agriculture to an industrial economy.

In *Mater et Magistra*, Pope John XXIII urged economically developed countries whose people have an abundance of wealth to help the developing countries whose people are poor and hungry:

> Perhaps the most pressing question of our day concerns the relationship between economically advanced commonwealths and those that are in process of development. The former enjoy the conveniences of life; the latter experience dire poverty. Yet, today men are so intimately associated in all parts of the world that they feel, as it were, as if they are members of one and the same household. Therefore, the nations that enjoy a sufficiency and abundance of everything may not overlook the plight of other nations whose citizens experience such domestic problems that they are all but overcome by poverty and hunger, and are not able to enjoy basic human rights (1961, #157).

Suffering people throughout the world prompted the Second Vatican Council bishops to consider the best ways in which the Church should respond. They professed that "God intended the earth with everything contained in it for the use of all human beings and peoples" (1965, #69). When using these goods of Earth, the faithful should consider the goods they legitimately possess not only as their own, but also as common insofar as they should be able to benefit not only the individual person but also others. "On the other hand," the bishops taught, "the right of having a share of earthly goods sufficient for oneself and one's family belongs to everyone. The Fathers and Doctors of the Church held this opinion, teaching that men are obliged to come to the relief of the poor and to do so not merely out of their superfluous goods [citing Ss. Basil, Augustine, Gregory the Great, Bonaventure, and Albert the Great]. If one is in extreme necessity, he has the right to procure for himself what he needs out of the riches of others" (#69). The bishops called upon individuals and governments to aid the many poor in the world so they can help themselves: "Since there are so many people prostrate with hunger in the world, this sacred council urges all, both individuals and governments, to remember the aphorism of the Fathers, 'Feed the man dying of hunger, because if you have not fed him, you have killed him' [citing Gratiam in *Decretum* 21] and really to share and employ their earthly goods, according to the ability of each, especially by supporting individuals or peoples with the aid by which they may be able to help and develop themselves" (#69). As discussed above, helping the poor and vulnerable help themselves is integral to the principle of subsidiarity when propelled by a realistic and spiritually-inspired sense of solidarity.

Continuing to express concern for the poor and vulnerable, Pope Paul VI quoted 1 John 3:17, "He who has the goods of this world and sees his brother in need and closes his heart to him, how does the love of God abide in him?" in *Populorum Progressio* to teach: "Everyone knows that the Fathers of the Church laid down the duty of the rich toward the poor in no uncertain terms. As St. Ambrose put it: 'You are not making a gift of what is yours to the poor man, but you are giving him back what is his. You have been appropriating things that are meant to be for the common use of everyone. The earth belongs to everyone, not to the rich' (Ambrose 1933, PL 14.747, c. 12, n. 53). These words indicate that the right to private property is not absolute and unconditional" (1967, #23). On the 80th anniversary of

Rerum Novarum, Paul VI issued *Octogesima Adveniens* in which he told the more fortunate to give to the less fortunate: "In teaching us charity, the Gospel instructs us in the preferential respect due to the poor and the special situation they have in society: the more fortunate should renounce some of their rights so as to place their goods more generously at the service of others" (1971, #23). He urged "a renewed education in solidarity" in order to develop a "deeper feeling of respect for and service to others" that should lead the faithful to seek the common good of all (#23).

The bishops of the United States expressed their deep concern for the poor in one of the world's most wealthy nations when issuing a pastoral statement on the economy after an extensive period of study and consultation throughout the country in the 1980s. In *Economic Justice for All*, they identified giving preference for the poor as "the single most urgent economic claim on the conscience of the nation" and a moral obligation from a justice perspective (NCCB 1986, #86):

> All members of society have a special obligation to the poor and vulnerable. From the Scriptures and church teaching, we learn that the justice of a society is tested by the treatment of the poor. The justice that was the sign of God's covenant with Israel was measured by how the poor and unprotected—the widow, the orphan, and the stranger—were treated. The kingdom that Jesus proclaimed in his word and ministry excludes no one. Throughout Israel's history and in early Christianity, the poor are agents of God's transforming power. "The Spirit of the Lord is upon me, therefore he has anointed me. He has sent me to bring glad tidings to the poor" (Luke 4:18). This was Jesus' first public utterance. Jesus takes the side of those most in need. In the Last Judgment, so dramatically described in St. Matthew's Gospel, we are told that we will be judged according to how we respond to the hungry, the thirsty, the naked, the stranger. As followers of Christ, we are challenged to make a fundamental "option for the poor"—to speak for the voiceless, to defend the defenseless, to assess life styles, policies, and social institutions in terms of their impact on the poor. This "option for the poor" does not mean pitting one group against another, but rather, strengthening the whole community by assisting those who are the most vulnerable. As Christians, we are called to respond to the needs of *all* our brothers and sisters, but those with the greatest needs require the greatest response (NCCB 1986, #16).

Yet the purpose of opting for the poor goes beyond giving them goods to consume, the bishops explained: "The primary purpose of this special commitment to the poor is to enable them to become active participants in the life of society. It is to enable all persons to share in and contribute to the common good.... The 'option for the poor,' therefore, is not an adversarial slogan that pits one group or class against another. Rather it states that the deprivation and powerlessness of the poor wounds the whole community. The extent of their suffering is a measure of how far we are from being a true community of persons. These wounds will be healed only by greater solidarity with the poor and among the poor themselves" (NCCB 1986, #88). The needs of the poor take priority over "the desires of the rich," the bishops proclaimed, just as "the rights of workers take priority over the maximization of profits; the preservation of the environment over uncontrolled industrial expansion; the production to meet social needs over production for military purposes" (#94).

Throughout his papacy, Pope John Paul II lamented the plight of the poor amidst the growing wealth of a few and championed the preferential option for the poor as a major Catholic social teaching. Examples of his concern can be found in *Sollicitudo Rei Socialis:* "A consistent theme of Catholic social teaching is the option or love of preference for the poor. Today, this preference has to be expressed in worldwide dimensions, embracing the immense number of the hungry, the needy, the homeless, those without medical care and those without hope" (1987, #42). He appealed to rich nations and people to recognize their moral obligations to impoverished people: "Therefore political leaders, and citizens of rich countries considered as individuals, especially if they are Christians, have the moral obligation, according to the degree of each one's responsibility, to take into consideration, in personal decisions and decisions of government, this relationship of universality, this interdependence which exists between their conduct and the poverty and underdevelopment which exists between their conduct and the poverty of so many millions of people" (#9). In *Centesimus Annus*, he taught that love for others must first be "love for the poor, in whom the Church sees Christ himself," and this love "is made concrete in the promotion of justice" (1991a, #58). Many archbishops and bishops have been promoting environmental justice through various programs within their dioceses (e.g., Catholic Diocese of Columbus 2010). Pope Benedict XVI followed his predecessors' concerns for the poor, emphasizing in *Caritas in Veritate* the love we must have for the poor and vulnerable amidst the rich and powerful: "While the poor of the world continue knocking on the doors of the rich, the world of affluence runs the risk of no longer hearing those knocks, on account of a conscience that can no longer distinguish what is human. God reveals man to himself; reason and faith work hand in hand to demonstrate to us what is good, provided we want to see it; the natural law, in which creative Reason shines forth, reveals our greatness, but also our wretchedness insofar as we fail to recognize the call to moral truth" (2009a, #75).

Recognizing the Connection between Environmental Abuse & the Human Poor

Pope John Paul II linked the plight of the poor and the degradation of the environment in his message celebrating the 1990 World Day of Peace: [T]he earth is ultimately *a common heritage, the fruits of which are for the benefit of all....* It is manifestly unjust that a privileged few should continue to accumulate excess goods, squandering available resources, while masses of people are living in conditions of misery at the very lowest level of subsistence. Today, the dramatic threat of ecological breakdown is teaching us the extent to which greed and selfishness—both individual and collective—are contrary to the order of creation, an order which is characterized by mutual interdependence" (1989, #8). Each person has "a grave responsibility to preserve this order for the well-being of future generations," he continued, repeating again in this message that "*the ecological crisis is a moral issue*" (#15).

Reflecting on John Paul II 1990 message, the Catholic bishops of the United States issued a pastoral statement entitled *Renewing the Earth* in which they found that "the ecological problem is intimately connected to justice for the poor." They agreed with the pope that the goods of Earth should be "a common patrimony," and they shared his concern that these goods run the risk of being monopolized by a few who degrade and sometimes destroy them, "thereby creating a loss for all humanity" (USCCB 1991, 3F quoting John Paul II 1991b). "Poor people are even more vulnerable," the bishops continued, and they "offer a special test of our solidarity" when addressing environmental problems.

> The painful adjustments we have to undertake in our own economies
> for the sake of the environment must not diminish our sensitivity

to the needs of the poor at home and abroad. The option for the poor embedded in the Gospel and the Church's teaching makes us aware that the poor suffer most directly from environmental decline and have the least access to relief from their suffering. Indigenous peoples die with their forests and grasslands. In Bhopal and Chernobyl, it was the urban poor and working people who suffered the most immediate and intense contamination. Nature will truly enjoy its second spring only when humanity has compassion for its own weakest members (USCCB 1991).

The bishops also addressed the connection between the poor and environmental degradation in *Global Climate Change* as noted above: "[T]he common good requires solidarity with the poor who are often without the resources to face many problems, including the potential impacts of climate change. Our obligations to the one human family stretch across space and time. They tie us to the poor in our midst and across the globe, as well as to future generations. The commandment to love our neighbor invites us to consider the poor and marginalized of other nations as true brothers and sisters who share with us the one table of life intended by God for the enjoyment of all" (USCCB 2001). Following the intentions of this document, the Catholic bishops of the United States are currently engaged in promoting "the needs of the poor and vulnerable at the center of climate legislation" using the argument that "poor people should not bear an undue burden of the impacts of climate change or the global adjustments needed to address it" (USCCB 2010a). Many other bishops around the world individually and collectively within political boundaries and biological regions have issued their reflections of John Paul II's 1990 message, most of which appeal to the faithful to be cognizant of their actions that can adversely affect the poor and vulnerable in their midst and especially in economically underdeveloped countries (Whittington 2004).

Pope Benedict XVI has advanced thinking about the poor in relation to ecological problems in his World Day of Peace messages, encyclicals, homilies, and various statements issued from the Vatican. Commemorating the 20th anniversary of his predecessor's 1990 World Day of Peace message, Benedict XVI dedicated his message on the 2010 World Day of Peace to addressing the numerous threats to peace and authentic human development posed by misuse of Earth and its natural goods provided by God (2009c, #1). The use of these goods is "a shared responsibility for all humanity, especially the poor and future generations" (#2, see #8). In *Caritas in Veritate*, he linked the development of people and economies to the natural environment:

> Today the subject of development is also closely related to the duties arising from *our relationship to the natural environment*. The environment is God's gift to everyone, and in our use of it we have a responsibility towards the poor, towards future generations and towards humanity as a whole. When nature, including the human being, is viewed as the result of mere chance or evolutionary determinism, our sense of responsibility wanes. In nature, the believer recognizes the wonderful result of God's creative activity, which we may use responsibly to satisfy our legitimate needs, material or otherwise, while respecting the intrinsic balance of creation. If this vision is lost, we end up either considering nature an untouchable taboo or, on the contrary, abusing it. Neither attitude is consonant with the Christian vision of nature as the fruit of God's creation" (2009a, #48).

Recognizing the connection between energy and the natural environment, he shared his concern about obstacles to the economic development of poor countries set by nations, powerful groups, and companies that hoard non-renewable energy sources found within those countries. They lack the means to access these sources and to finance research into alternatives, the pope lamented, so they are exploited and conflicts erupt between and among them. Thus, the international community has "an urgent duty" to develop institutional means through which the exploitation of non-renewable resources can be regulated and poor countries must be involved in the process so all can plan together for the future (#49). Finally, recognizing that the climate crisis is projected to affect the production of food and its availability to the poor in areas already afflicted with food shortages, he urged the international community to be united against hunger, to overcome obstacles of self-interest, and "to make room for a fruitful *gratuitousness*, manifested in international cooperation as an expression of genuine fraternity" (2010b).

Thus, "preferential option for the poor" looms large in Catholic social teaching and is often connected with teachings on the principles of solidarity and subsidiarity. The faithful are enjoined to show utmost concern individually and collectively for the poor and vulnerable. Their needs in life must be met, and they must be helped to meet them. Doing so requires those who have more than they need to give to those who do not and to aim this assistance toward helping the impoverished people help themselves so their dignity as humans is not compromised. Aid to the impoverished is required at incremental levels of governance, following the principle of subsidiarity, when individual and collective help is inadequate. Because the availability of goods to meet the needs of future generations may be compromised by current overuse and abuse, generations to come must be included among the poor and factored into decisions made today.

EXTENDING CATHOLIC SOCIAL TEACHING IN RESPONSE TO THE CLIMATE CRISIS

Solidarity, subsidiarity, and preferential option for the poor hold considerable promise for dealing with environmental concerns generally and with the climate crisis specifically. Catholic social teaching about the solidarity of humans can motivate the faithful to care about and for other species, their habitats, the air, the land, and waters for the good of all persons now and in the future. Teachings about subsidiarity provide a hierarchical organizing method for addressing environmental concerns when drawing incrementally upon individuals and associations at various societal levels to protect, ameliorate, and mitigate adverse effects on human persons. Teachings about showing preference for impoverished and vulnerable persons can be effective when making decisions and taking action that protects, aids, and empowers persons who are most severely affected by environmental degradation currently and projected to be adversely affected in the future. Thus, Catholic social teachings in these three categories appear sufficient when focusing on the good of human persons, and much can be accomplished when functioning from this anthropocentric perspective.

However, are these teachings sufficient when they are exclusively centered on the human common good? Are they too centered on valuing the human intrinsically while only valuing other species and biological systems instrumentally for how they can be used to achieve the human common good? Are these teachings sufficiently relevant to the climate crisis when considering the long-term effects on humans, other species, ecological systems, and the biosphere? Are they sufficiently realistic and intellectually honest when recognizing that *Homo sapiens* evolved from and with other species over millions of years on a planet that had its beginning with other planets and solar systems approximately 14 billion years ago and when acknowledging that humans are radically dependent upon other species, ecological systems, and the biosphere to sustain our lives and efforts to flourish? Are they sufficiently helpful for dealing with a complicated and seemingly intractable global problem caused by many human-induced sources—the climate crisis?

If the answers to these questions are not resoundingly positive as the climate crisis looms, a theological thought experiment is warranted in an attempt to determine if Catholic social teaching about solidarity, subsidiarity, and preferential option for the poor can be more relevant, more realistic, more intellectually honest, and more helpful. Alternatives are to expand solidarity to include other species and ecological systems, to reconfigure the principle of subsidiarity so decision-making and acting is based on biological regions and the biosphere, and to consider endangered species and degraded ecological systems among the poor and impoverished.

From the Solidarity of Persons to Earth Solidarity

The emergence of *Homo sapiens* is well documented by data, reports, and discussions in the scientific literature, incorporated in the newly burgeoning discourse on the relationship between theology and the natural sciences, and popularized by the media. From an initial beginning of the universe, its expansion, the subsequent death of stars that yielded elements essential to life, the formation of billions of galaxies of which at least one had a solar system with a planet within which simple forms of life emerged, and an evolutionary process within Earth that yielded increasingly complex species, ours emerged from earlier Hominoidea with the capacity to talk about and reflect on our place in existence (Schaefer 2009, 165-70). Our physical connection with earlier forms of life is indisputable. As law historian and diplomat Arvid Pardo recalled: "[T]he dark oceans were the womb of life: from the protecting oceans, life emerged. We still bear in our bodies—in our blood, in the salty bitterness of our tears—the marks of this remote past" (1967). Our interconnection with other species in the ecological systems of which we are constituents is also beyond doubt. And, our dependence on other species, the air, land, and water for sustaining our lives is radical. With them, we constitute Earth. We live and function in a biosphere of ecological systems, marginal areas, and myriad plant and animal species. Together, we constitute the "Earth community," as moral theologian Larry Rasmussen insists poignantly (1996).

However, the climate crisis suggests that we are not living in solidarity with the other species and abiota that constitute the Earth community. As Pope Benedict XVI noted, "[w]e have usurped" God's creation, we "want to dominate it," and we "want unlimited possession of the world" (2005a). In an address to members of the Pontifical Academy of Sciences and the Pontifical Academy of Social Sciences who were commencing a study of the human person, he underscored the fact that humans are "part of nature." Yet we are distinct. As "free subjects," he continued, "who have moral and spiritual values," humans "transcend nature" as creatures who have "a superior dignity and a shared mission toward the whole of creation" (2005b).

How can this shared mission be understood in light of the real and projected effects that humans are forcing on the global climate? How can the "superior dignity" of humans be understood when we are intricately interconnected with other species and abiota, all of whom will be adversely affected now and into the future one way of another? When we are radically dependent on them for the basics they supply that are necessary for our lives and well-being: air to breathe, water to drink, land on which to live, food to eat, shelter in which to live, and myriad other basics? How can we demonstrate our role as part of nature while transcending nature?

We can strive to live in solidarity with all constituents of the Earth community. Living in solidarity with them means choosing to make decisions now for the common good of all species, abiota, ecosystems they constitute, and the biosphere into the future. In light of the ongoing disruption of the global climate, choosing to make decisions for the good of all will require a change in attitude. The change that is required will move beyond the sense of solidarity of all humans that popes John Paul II and Benedict XVI have emphasized. Earth solidarity requires a more expansive focus that includes other species, abiota, ecological systems, and the biosphere. Earth solidarity requires a conversion from an anthropocentric attitude that they are merely instruments intended for human use to a planetary attitude that prompts us to intrinsically value other species, the air, the land, waters, ecological systems, and the biosphere within which all function as contributors to and benefactors of a life-sustaining climate.

In addition to a change in attitude from anthropocentric to planetary, recognition of human interconnections with and radical dependence upon other constituents of the Earth community should provide the impetus for demonstrating Earth solidarity as a moral virtue. Developed and practiced consistently by the faithful individually and collectively, solidarity will guide the faithful to think *more intelligently* informed by climate science knowing that the life-supporting climate of Earth is under siege; to make decisions *more prudently* about the possible ways of mitigating human-forced climate change and to make the best possible decisions expeditiously but cautiously in the interest of present and future inhabitants of Earth while remaining vigilant to modifying decisions as new knowledge is discovered; to act *more justly* toward them so they are able to obtain what they need from one another and the abiotic environment to survive, flourish, and contribute to the planetary common good—a life-sustaining climate; to use them *more moderately* knowing that we cannot use them up without adversely affecting the functioning of the global climate; to be *humble* when knowing the cosmological-biological history of our species' emergence from and with other species and abiota and our radical dependence on them for our health and well-being; and to act *boldly and courageously* intelligent, prudent, temperate, and just by facing the facts that we are disrupting the global climate and by persisting in making and implementing decisions geared toward mitigating the real and anticipated effects and adapting to conditions that cannot be mitigated due to our abuses and overuses.

Developing an attitude of solidarity with all other biota and abiota of Earth does not diminish the dignity of the human person. Earth solidarity elevates the dignity of the human and the dignity of human associations. Human dignity is elevated when activating two of the characteristics that distinguish *Homo sapiens*—the capacity to make informed decisions and the will to execute them—and applying these characteristics to address the climate crisis. Though much can be accomplished when making informed decisions in the interests of humans now and in the future, making them in the interests of all constituents of Earth would demonstrate an attitude that recognizes the emergence of our species from and with other species over cosmological and biological time, the human interconnection with them today in light of the mutually experienced effects of changes humans are forcing on the global climate, and the radical dependence humans have on other species and abiota for the necessities of life and pleasure in living.

What motivates Earth solidarity? Nothing less than love for God and for the Earth community of many varied and interconnected creatures that God loves and calls us to love in solidarity with them (Schaefer 2009, 255-63).

From Politically-Defined to Bioregionally-Defined Subsidiarity

Because subsidiarity has been understood as a societal organizing principle through which individual persons freely associate with one another at increasing levels of governance to facilitate the accomplishment of a common good that cannot otherwise be achieved, decision-making is assumed to occur within human-drawn boundaries at increasingly higher levels (family, neighborhood, municipality, county, state, national and international). However, these political boundaries do not seem appropriate for making decisions about either mitigating changes humans are forcing on the global climate or adapting to them (see O'Brien 2008). The past, current, and projected adverse effects on oceanic coasts, lake basins, river valleys, wetlands, and other ecological systems often cross political boundaries, and each needs to be addressed in its totality as emphasized in scientific and economic reports (e.g., International Bank 2010).[1] Associations for decision-making pertaining to the climate crisis are better centered around around biological regions that encompass marginal areas around an ecosystem. From a bioregional perspective, decisions can be made about the most prudent actions to mitigate the adverse effects of human-forced climate change or, if necessary, adapting to changes that cannot be mitigated.

How can the principle of subsidiarity be followed when attempting to mitigate human-forced climate change in a biological region and the greater biosphere? I will attempt to sketch bioregional subsidiarity based on my experience as an environmental group organizer and appointee to several policy positions at various levels of governance.

Because the entire biological region must be addressed, representatives of increasing levels of contact with the region from the most local (e.g., the habitats of species in or on the banks of a river) to more encompassing (e.g., the entire river and banks) to the all-encompassing (e.g., the river basin with additional representation from areas marginal to

1   See also USEPA 2010; IPCC 2007; and Kling 2003.

the ecosystem that may be affected or may affect the region) will identify the actions that need to be taken at their levels where they have the
most immediate contact with and control over their actions in relation
to the region. The interests of future generations of humans and of
other species, habitats, and the ecological systems they constitute will
be advocated by representatives who volunteer or are appointed for
that purpose. Decisions will be made and actions taken accordingly at
those levels to achieve their goals. Persons in association with one another at each level of activity will remain cognizant of its goal, achievements, and shortcomings in relation to the region, prudently alert to
the need for modifying actions as new scientific data are compiled and
interpreted, and responsible for sharing their findings, decisions, and
actions with other levels that are working in solidarity toward mitigating the effects of the climate crisis on the region. If a more local or less
encompassing level cannot achieve a goal, help will be provided by the
next more encompassing level. All levels will continue to reassess the
extent to which their specific goals are being met, share their assess-
ments with other levels culminating in the most encompassing level of
the ecological region, and modify their goals to assure that the current
circumstances, possible consequences, and scientific projections about
present and future effects are considered.

Bioregional subsidiarity will require increasingly encompassing levels of the region to help those that are less encompassing to achieve
their mitigation and adaptation goals. Based on climate science data
and interpretations, this process will be ongoing for many decades if
not centuries in an attempt to recover from the adverse effects human
activities are causing today.

Efforts could be made in several bioregional areas of North America. One in particular has considerable potential—the Great Lakes
region. Encompassing the mid-central parts of the United States and
Canada to their shared eastern coast, this region is the subject of study
and recommendations by the International Joint Commission which
has been charged by the two governments to address their shared
boundary waters. The IJC identified as a "key priority" the mitigation
of adverse effects of climate change (Great Lakes Water Quality Board
2003). However, much has yet to be accomplished to produce a plan
of action at various physical levels of the bioregion.

Unfortunately for the biological regions that exist within the United
States, federal legislation has yet to be enacted to address the climate

crisis. Though hopes were high for passing a comprehensive climate bill during the Obama administration when the Democratic Party dominated both the House of Representatives and the Senate, one had not passed as the Congressional year closed in 2010, and there is little likelihood that climate legislation will be passed during the 2011-13 period with a Republican-based House and a slim majority of Democrats in the Senate.

At this point, hope may be justified in small bioregions. Underway are two efforts with which I am familiar that signify hope for addressing the climate crisis. One effort is among materially poor people in the mountain-enveloped river valley of El Cercado, Dominican Republic. Faith communities organized in "farming associations" consisting of groups of "families" of approximately twenty-five persons are in the process of implementing a plan to mitigate the effects they are experiencing from changes forced on the climate. They are growing organic crops, replanting mango and coffee trees on hillsides made barren of the native mahogany and other trees by multi-national lumber companies and the Trujillo regime, planning to build up the soil sufficiently so native trees can be planted some day, piping clean spring water to their homes so they can adapt to drought, and constructing solar ovens to minimize the carbon output from traditional ways of cooking. Their efforts are motivated by a Christian faith-based discernment process through which they identified problems caused by human-forced climate changes, considered possibilities for addressing them, reflected on these possibilities informed by basic Gospel values and goals, decided on projects to implement, are implementing them, and will be evaluating the outcome in light of basic Gospel values and goals. Another effort is occurring in the mountainous Petén Department of Guatemala where potable water projects drawing on rivers and springs are at various stages of implementation along with sustainable farming, constructing solar bricks, and using them in solar ovens. Hope for future action may also be gleaned from a plan developed by undergraduate students in the Capstone Seminar for the Interdisciplinary Minor in Environmental Ethics at Marquette University who chose to focus on mitigating human-forced climate change through agricultural practices in southeastern Wisconsin (Brunette et al. 2008).

Preference for All Poor & Vulnerable Members of the Earth Community

As already explained, the well-expressed Catholic social teaching of opting to give preference for the poor, suffering, and vulnerable means putting poor, suffering, and vulnerable humans first when making and executing decisions. Including the future poor, suffering, and vulnerable humans who will be born into the adversities of changes that human activities are forcing on the climate today is a reasonable extension of this principle. Doing so is a matter of justice because the next generation and generations after will be affected but had no part in forcing these changes, yet some will be more poor, some will suffer more, and all will be vulnerable to these adverse effects to some extent. Thus, present and future poor, suffering, and vulnerable people should be given preference when decisions are made to mitigate the climate crisis and to implement these decisions.

Yet where do the present and future poor, suffering, and vulnerable animal and plant species, ecological systems, and the biosphere fit into this principle? Should their impoverishment, suffering, and vulnerability to the effects of human-forced climate change be ignored? Should accelerated rates of species endangerment and extinction caused by these changes be ignored? When considering the fact that humans are utterly dependent upon other species and abiota that constitute the ecosystems within which we live, separating their well-being from ours is impossible. However, we could continue to view them strictly for their usefulness to us and not intrinsically for their value in themselves and their contributions to the functioning of the ecosystems of which we also are constituents. If we continue to value them instrumentally and not intrinsically, are we not showing our ignorance of our radical dependence upon them, a dependence that should elicit our gratitude to them and to God for making their existence possible? If we continue to value them instrumentally and not intrinsically, are we not also showing our ignorance of the cosmological to biological history out of which humans emerged from and with other species? If we continue to value them instrumentally and not intrinsically, are we not also continuing an anthropocentric mindset that has been so damaging in the past, is damaging in the present, and will be damaging in the future?

If the answer to these questions is "yes," an extension of the principle to prefer poor, suffering, and vulnerable humans is warranted to include other species, ecological systems, and the biosphere when making decisions about mitigating the climate crisis and executing these decisions. Extending this principle not only shows that we value other species and systems intrinsically. Extending this principle indicates that we value other species and systems instrumentally for the contributions they make to one another as well as to human-well being. Including them in this principle demonstrates our inseparability from them in this life, our utter dependence on them to continue our lives, and our indisputable interconnections with them in the web of life.

Including other poor, suffering, and vulnerable species, ecological systems, and the biosphere in this long-underscored Catholic social teaching principle does not denigrate the dignity of the human person. Including them *elevates* the dignity of persons by capitalizing upon what distinguishes our species—especially the ability to make informed decisions and the freedom with which to execute them in solidarity with others for our planetary common good—a life-sustaining climate. Including these "new poor," as theologian Sallie McFague characterizes them altogether as "nature" (1997, 170) brings other species and systems directly into our consciousness and into our deliberations as we struggle to address the disruption of the global climate that our actions are forcing on the Earth community.

CONCLUSION

Catholic social teachings about solidarity, subsidiarity, and preferential option for the poor focus on human solidarity to achieve the human common good, decision-making at incremental levels of human associations that empower humans to achieve their common good, and opting to prefer the human poor, vulnerable, and suffering when making and implementing decisions. Applying these teachings to the climate crisis yields considerable promise for mitigating the adverse effects of changes humans are forcing on the global climate.

However, as the climate crisis looms in the present, as dire predictions about near and far future effects are issued by climate scientists, and as pleas and proposals to mitigate the adverse effects are ignored, some changes are needed to spark the attention of the faithful to take action at their personal and local levels and to demand action at higher levels of governance. Among the possibilities is extending these teachings so they are more realistic, more intellectually honest, more relevant, and more helpful.

By extending solidarity to include other species, ecosystems, and the biosphere, *Earth solidarity* may be experienced as a planetary bond grounded in Christian love that persons consistently show toward others with whom they share Earth by striving virtuously to make and execute informed decisions for their common good—a life sustaining climate. By extending the principle of subsidiarity so decisions are made at various physical levels encompassed by biological regions, *bioregional subsidiarity* may be experienced as an organizing principle through which individual persons at various levels of the ecological region collaborate with individuals who represent future generations of people, species, abiota, and ecological systems in making and executing informed decisions to mitigate the adverse effects of changes humans are forcing on the global climate today. By extending the teaching of preferential option for poor to include *the future poor, suffering, and vulnerable people, other species, and ecological systems*, their interests will be given preference when making and executing decisions aimed at mitigating the climate crisis. Acting on these extended principles should facilitate addressing the climate crisis effectively.

SOURCES

Benedict XVI, Pope. 2005a. Homily of his Holiness Benedict XVI at the Opening Mass of the 11th Ordinary General Assembly of the Synod of Bishops, October 2. Accessed from http://www.vatican.va/holy_father/benedict_xvi/homilies/2005/documents/hf_ben-xvi_hom_20051002_opening-synod-bishops_en.html on September 28, 2010.

————. 2005b. Address of His Holiness Benedict XVI to the Members of the Pontifical Academy of Sciences and the Pontifical Academy of Social Sciences, November 21. Accessed from http://www.vatican.va/holy_father/benedict_xvi/speeches/2005/november/documents/hf_ben_xvi_spe_20051121_academies_en.html on October 1, 2010.

————. 2005c. *Deus Caritas est*. Encyclical Letter on Christian Love, December 25. Accessed from http://www.vatican.va/holy_father/benedict_xvi/encyclicals/documents/hf_ben-xvi_enc_20051225_deus-caritas-est_en.html on September 10, 2010.

————. 2007a. *Spe Salvi*. Encyclical Letter on Christian Hope, November 30. Accessed from http://www.vatican.va/holy_father/benedict_xvi/encyclicals/documents/hf_ben-xvi_enc_20071130_spe-salvi_en.html on September 10, 2010.

———. 2007b. *The Human Family, a Community of Peace.* 2008 World Day of Peace Message, December 8. Accessed from http://www.vatican.va/holy_father/benedict_xvi/messages/peace/documents/hf_ben-xvi_mes_20071208_xli-world-day-peace_en.html on September 10, 2010.

———. 2009a. *Caritas in Veritate.* Encyclical Letter on Integral Human Development in Charity and Truth, June 29. Accessed from http://www.vatican.va/holy_father/benedict_xvi/encyclicals/documents/hf_ben-xvi_enc_20090629_caritas-in-veritate_en.html on September 10, 2010.

———. 2009b. Videostatement of His Holiness Benedict XVI to the UN 2009 Summit on Climate Change, September 22. Accessed from http://www.vatican.va/holy_father/benedict_xvi/messages/pont-messages/2009/documents/hf_ben-xvi_mes_20090924_summit-climat-change_en.html on November 8, 2010.

———. 2009c. *If You Want to Cultivate Peace, Protect Creation.* Message for the celebration of World Day of Peace, 1 January 2010, December 8. Accessed from http://www.vatican.va/holy_father/benedict_xvi/messages/peace/documents/hf_ben-xvi_mes_20091208_xliii-world-day-peace_en.html on September 5, 2010.

———. 2010a Address of His Holiness Benedict XVI at Meeting with Clerical and Lay Representatives of Other Religions, Twickenham, England, September 17. Accessed from http://www.vatican.va/holy_father/benedict_xvi/speeches/2010/september/documents/hf_ben-xvi_spe_20100917_altre-religioni_en.html on October 1.

———. 2010b. Message of His Holiness Benedict XVI to Mr. Jacques Diouf, Director of FAO, on the Occasion of World Food Day 2010, October 15. Vatican City. Accessed from http://www.vatican.va/holy_father/benedict_xvi/messages/food/documents/hf_ben-xvi_mes_20101015_world-food-day-2010_en.html on October 30.

Bilgrien, Marie Vianney, SSND. 1999. *Solidarity: A Principle, an Attitude, a Duty? Or the Virtue for an Interdependent World?* New York: Peter Lang.

Brunette, Ben, Becky Goossen, Nick McDaniels, Chalie Nevárez, and Beth Wilson. 2008. "Mitigating the Climate Crisis through Agriculture in Southeastern Wisconsin." Capstone ARSC 110 Capstone Seminar Report, Interdisciplinary Minor in Environmental Ethics, Marquette University. Accessible from http://www.inee.mu.edu/capstone_2008/Capstone2008.htm.

Caritas International. 2010. "Climate Justice." Accessed from http://www.caritas.org/activities/climate_change/index.html on September 25.

Catholic Bishops of the Columbia River Watershed. 2005. *The Columbia River Watershed: Caring for Creation and the Common Good*. A Pastoral Statement. Accessed from http://www.thewscc.org/columbia-river on October 2, 2010.

Catholic Church. 1994. *The Catechism of the Catholic Church*. Washington DC: United States Catholic Conference.

Catholic Coalition on Climate Change. 2010. "Who's Under Your Carbon Footprint?" Accessed from http://catholicclimatecovenant.org/ on September 25.

CELAM. 1970. *The Church in the Present Day Transformation of Latin America in the Light of the Council*. Vol. 2. Bogata, Colombia.

Coste, René. 1990. "Solidarité." In *Dictionnaire de Spiritualité: Ascétique et Mystique, Doctrine et Histoire*, vol. 14, edited by Marcel Viller, F. Cavallera, J. De Guibert, 999-1006. Paris: G. Beauchesne & Sons.

Diocese of Columbus. 2010. "Environmental Justice." Office for Social Concerns. Accessed from http://www.colsdioc.org/Offices/SocialConcernsOfficefor/Programs/EnvironmentalJustice.aspx on October 1.

Diocese of San Jose. 2010. "Environmental Justice." Accessed from http://www.dsj.org/category/social-justice-categories/environmental-justice on October 1.

Doran, Kevin P. 1996. *Solidarity: A Synthesis of Personalism and Communalism in the Thought of Karol Wojtyla/Pope John Paul II*. New York: Peter Lang.

Dorr, Donald. 1983. *Option for the Poor: A Hundred Years of Vatican Social Teaching*. Dublin: Gill and Macmillan: Orbis Books.

European Union. 2010. "Subsidiarity." Europa: Gateway to the European Union. Accessed from http://europa.eu/scadplus/glossary/subsidiarity_en.htm on September 10.

Great Lakes Water Quality Board. 2003. *Climate Change and Water Quality in the Great Lakes Basin*. International Joint Commission. Ottawa, Canada. Accessed from http://www.great-lakes.net/lists/glin-announce/2003-12/msg00005.html on November 8, 2010.

Henning, Brian G. 2009. "From Despot to Steward: The Greening of Catholic Social Teaching." In The Heart of Catholic Social Teaching: Its Origins and Contemporary Significance, edited by David Matzko McCarthy, 183-93, 209-10. Grand Rapids: Brazos Press.

Independent and Self-Governing Trade Union Solidarno__. 2010. "History in Dates." NSZZ Solidarno__. Accessed from http://www.solidarnosc. org.pl/en/about-us.html on September 14.

Irish Catholic Bishops Conference. 2009. "The Cry of the Earth: A Pastoral Reflection on Climate Change." November 10. Accessed from http:// www.catholicbishops.ie/images/stories/features/Cry_of_the_Earth/ env_pastoral_09_long_final.pdf on October 2, 2010.

Intergovernmental Panel on Climate Change (IPCC). 2007. *Climate Change 2007: Synthesis Report, Summary for Policy Makers, Contribution of Working Groups I, II and III to the Fourth Assessment Report*. Edited by Rajendra K. Pachauri and Andy Reisinger. Geneva: Intergovernmental Panel on Climate Change. Also accessible from http://www.ipcc.ch/publications_and_data/publications_ipcc_fourth_assessment_report_synthesis_report.htm.

International Bank for Reconstruction and Development. 2010. *Climate Risks and Adaptation in Asian Coastal Megacities: A Synthesis Report*. Washington DC: The World Bank. Also accessible from http:// siteresources.worldbank.org/EASTASIAPACIFICEXT/Resources/226300-1287600424406/coastal_megacities_fullreport.pdf.

International Joint Commission. 2004. "The Impact of Climate Change on Ground Water and Surface Water Quality." 12th Biennial Report on Great Lakes Water Quality. Ottawa, Canada. Accessed from http://www.ijc. org/php/publications/html/12br/english/report/physical/climate.html on November 8, 2010.

Jesuit Social Apostolate. 2010. "Headlines 2010/08: News from the Jesuit Social Apostolate." August 31. sjshl@sjcuria.org.

John XXIII, Pope. 1961. *Mater et Magistra*. Encyclical on Christianity and Social Progress, May 15. Accessed from http://www.vatican.va/holy_father/john_xxiii/encyclicals/documents/hf_j-xxiii_enc_15051961_mater_en.html on September 5, 2010.

John Paul II, Pope. 1987. *Sollicitudo Rei Socialis*. Encyclical on the Twentieth Anniversary of *Populorum Progressio*, December 30. Accessed from http:// www.vatican.va/holy_father/john_paul_ii/encyclicals/documents/hf_jp-ii_enc_30121987_sollicitudo-rei-socialis_en.html on September 19, 2010.

———. 1989. *Peace with God the Creator, Peace with All of Creation*. Message for the 1990 World Day of Peace, December 8. Accessed from http:// www.vatican.va/holy_father/john_paul_ii/messages/peace/documents/ hf_jp-ii_mes_19891208_xxiii-world-day-for-peace_en.html on September 5, 2010.

————. 1991a. *Centesimus Annus.* Encyclical on the 100th Anniversary of *Rerum Novarum,* May 1. Accessed from http://www.vatican.va/holy_father/john_paul_ii/encyclicals/documents/hf_jp-ii_enc_01051991_centesimus-annus_en.html on September 19, 2010.

————. 1991b. *Discorso Di Giovanni Paolo Ii Ai Partecipanti Al Premio Internazionale per L'ambiente San Francesco "Cantico Delle Creature,* October 25. Accessible from http://www.vatican.va/holy_father/john_paul_ii/speeches/1991/october/documents/hf_jp-ii_spe_19911025_teologia-francescana_it.html.

Kling, George W. et al. 2003. *Confronting Climate Change in the Great Lakes Region: Impacts on Our Communities and Ecosystems.* Cambridge MA and Washington DC: Union of Concerned Scientists and The Ecological Society of America. Also available from http://ucsusa.org/assets/documents/global_warming/greatlakes_final.pdf.

Komonchak, Joseph A. 1988. "Subsidiarity in the Church: The State of the Question." *The Jurist* 48: 298-349.

Mahoney, Jack, S.J. 1988. "Subsidiarity in the Church." *The Month* 159: 968-74.

Matz, Brian J. 2008. *Patristic Sources and Catholic Social Teaching: A Forgotten Dimension: A Textual, Historical, and Rhetorical Analysis of Patristic Source Citations in the Church's Social Documents.* Dudley MA: Peeters.

McFague, Sallie. 1993. *The Body of God: An Ecological Theology.* Minneapolis: Fortress Press.

National Conference of Catholic Bishops (NCCB). 1995. "A Decade after Economic Justice for All: Continuing Principles, Changing Context, New Challenges." A Pastoral Message of the National Conference of Catholic Bishops on the Tenth Anniversary of the Economic Pastoral, November. Accessed from http://www.osjspm.org/majordoc_us_bishops_statements_economic_justice_for_all.aspx on September 19, 2010.

O'Brien, Kevin. 2008. "Thinking Globally and Thinking Locally: Ecology, Subsidiarity, and Multiscaler Environmentalism." *Journal for the Study of Religion, Nature, and Culture* 2.2: 218-36.

Office for Social Justice. 2010. "Catholic Social Teaching." Archdiocese of St. Paul and Minneapolis. Accessed from http://www.osjspm.org/option_for_the_poor.aspx on September 4, 2010.

Pachauri, R. K. 2007. "Acceptance Speech for the Nobel Peace Prize Awarded to the Intergovernmental Panel on Climate Change (IPCC)," Oslo, Norway, December 10. Accessed from http://www.ipcc.ch/graphics/

speeches/nobel-peace-prize-oslo-10-december-2007.pdf on September 20, 2010.

Pachauri, Rajendra. 2008. Speech at the Opening Session of the World Economic Forum, Davos, Switzerland, January 23. Accessed from http://www.ipcc.ch/graphics/speeches/pachauri-davos-january-2008.pdf on September 20, 2010.

Paul VI, Pope. 1967. *Populorum Progressio.* Encyclical on the Development of Peoples, March 26. Accessed from http://www.vatican.va/holy_father/paul_vi/encyclicals/documents/hf_p-vi_enc_26031967_populorum_en.html on September 5, 2010.

————. 1971. *Octogesima Adveniens.* 80th Anniversary of the Encyclical Rerum Novarum, May 14. Accessed from http://www.vatican.va/holy_father/paul_vi/apost_letters/documents/hf_p-vi_apl_19710514_octogesima-adveniens_en.html on September 19, 2010.

Pius XI, Pope. 1931. *Quadragesimo Anno.* Encyclical on Reconstruction of the Social Order, May 15. Accessed from http://www.vatican.va/holy_father/pius_xi/encyclicals/documents/hf_p-xi_enc_19310515_quadragesimo-anno_en.html on September 17, 2010.

Pius XII, Pope. 1939. *Summi Pontificatus.* Encyclical on the Unity of Human Society, October 20. Accessed from http://www.vatican.va/holy_father/pius_xii/encyclicals/documents/hf_p-xii_enc_20101939_summi-pontificatus_en.html on September 5, 2010.

Pontifical Council for Justice and Peace. 2004. *Compendium of the Social Doctrine of the Church.* Washington DC: United States Conference of Catholic Bishops.

Rasmussen, Larry. 1996. *Earth Community, Earth Ethics.* Maryknoll: Orbis Books.

Schaefer, Jame. 2010. "Environmental Degradation, Social Sin, and the Common Good." In *God, Creation, and Climate Change: A Catholic Response to the Environmental Crisis,* edited by Richard W. Miller, 69-94. Maryknoll. Orbis Books.

————. 2009. *Theological Foundations for Environmental Ethics: Reconstructing Patristic and Medieval Concepts.* Washington, DC: Georgetown University Press.

————. 2005. "Valuing Earth Intrinsically and Instrumentally: A Theological Framework for Environmental Ethics." *Theological Studies* 66.4: 783-814.

Schilling, Theodor. 1995. *Subsidiarity as Rule and Principle, Or Taking Subsidiarity Seriously.* New York: Jean Monnet Center. Accessed from http://

centers.law.nyu.edu/jeanmonnet/papers/95/9510ind.html on September 19, 2010.

Second Vatican Council. 1965. *Gaudium et Spes: Pastoral Constitutiion on the Church in the Modern World*. Promulgated by His Holiness, Pope Paul VI, December 7. Accessed from http://www.vatican.va/archive/hist_councils/ii_vatican_council/documents/vat-ii_cons_19651207_gaudium-et-spes_en.html on September 5, 2010.

Silecchia, Lucia A. 2008. "The 'Preferential Option for the Poor': An Opportunity and a Challenge for Environmental Decision-Making." *University of St. Thomas Law Journal* 5.1: 87-143.

Sobrino, Jon, and Juan Hernández Pico. 1985. *Theology of Christian Solidarity*. Orbis Press: Maryknoll.

Society of Jesus Oregon Province. 2006. Regional Sustainable Development: A Plan of Action. Accessed from http://www.nwjesuits.org/WhoWeAre/Colombia/PDF%20Files/Regional%20Sustainable%20Development—Plan%20of%20Action.pdf on October 1, 2010.

United States Catholic Bishops (USCB). 1986. *Economic Justice for All*. A Pastoral Letter on Catholic Social Teaching and the U.S. Economy. Accessed from http://www.osjspm.org/economic_justice_for_all.aspx on September 24, 2010.

United States Conference of Catholic Bishops (USCCB). 1991. *Renewing the Earth: an Invitation to Reflection and Action on Environment in Light of Catholic Social Teaching*. A Pastoral Statement of the United States Catholic Conference of Bishops, November 14. Accessed from http://www.usccb.org/sdwp/ejp/bishopsstatement.shtml#3 on September 25, 2010.

———. 1998. *Sharing Catholic Social Teaching: Challenges and Directions*. No. 5-281. Washington DC: United States Conference of Catholic Bishops.

———. 2001. *Global Climate Change: A Plea for Dialogue, Prudence, and the Common Good*, June 15. Accessed from http://www.usccb.org/sdwp/international/globalclimate.shtml on September 20, 2010.

———. 2010a. "Catholics Confront Global Poverty: Global Climate Change and our Catholic Response." Accessed from http://www.usccb.org/sdwp/globalpoverty/pdfs/Climate_Change.pdf on September 10.

———. 2010b. "Our Catholic Faith in Action: Transforming the World: Option for the Poor." Accessed from http://www.usccb.org/campus/teaching-option-poor.shtml on September 24.

———. 2010c. "Papal and Vatican Documents on Catholic Social Teaching." Accessed from http://www.usccb.org/sdwp/vaticanencyclicals.shtml on September 5, 2010.

————. 2010d. "Catholic Teaching and Principles: Option for the Poor." Justice, Peace and Human Development. Accessed from http://www.usccb.org/sdwp/catholicteachingprinciples.shtml#3 on September 29.

————. 2010e. "Climate Change Justice and Health Initiative: Letters and Action Alerts." Accessed from http://www.usccb.org/sdwp/ejp/climate/lettersalerts.shtml on October 2.

United States Environmental Protection Agency. 2010. "Climate Change-Science." Washington DC: USEPA. Accessed from http://epa.gov/climatechange/science/index.html on November 19.

West Coast Energy Limited. 2005. *Vulnerability Assessment of the North East Atlantic Shelf Marine Ecoregion to Climate Change.* Trevor Baker, Project Manager. Washington DC: World Wildlife Fund. Also available at http://www.wwf.org.uk/filelibrary/pdf/climatechangeandseas01.pdf.

Whittington, Heather A. 2004. "The Catholic Church on Ecological Degradation." Interdisciplinary Minor in Environmental Ethics, Marquette University. Accessible from http://www.inee.mu.edu/CatholicChurchonEnvironmentalDegradation.htm.

World Socialist Web Site. 2002. "Workers Struggles: Europe & Africa." International Committee of the Fourth International (ICFI), October 18. Accessed from http://www.wsws.org/articles/2002/oct2002/labo-o18.shtml on September 15, 2010.

World Synod of Catholic Bishops. 1971. *Justice in the World.* Accessed from http://www.osjspm.org/majordoc_justicia_in_mundo_offical_test.aspx on September 8, 2010.

Zenit.org. 2010. "European Parliament Debates 'Caritas in Veritate.'" Vatican News Service, September 16. Accessed from http://www.zenit.org/article-30369?l=english on September 16.

18

GLOBAL CLIMATE CHANGE

A PLEA FOR DIALOGUE, PRUDENCE, & THE COMMON GOOD[1]

United States Conference of Catholic Bishops

June 15, 2001

INTRODUCTION

As people of faith, we are convinced that "the earth is the Lord's and all it holds" (Ps 24:1). Our Creator has given us the gift of creation: the air we breathe, the water that sustains life, the fruits of the land that nourish us, and the entire web of life without which human life cannot flourish. All of this God created and found "very good." We believe our response to global climate change should be a sign of our respect for God's creation.

The continuing debate about how the United States is responding to questions and challenges surrounding global climate change is a test and an opportunity for our nation and the entire Catholic community. As bishops, we are not scientists or public policymakers. We enter this debate not to embrace a particular treaty, nor to urge particular technical solutions, but to call for a different kind of national discussion. Much of the debate on global climate change seems polarized and partisan. Science is too often used as a weapon, not as a source of wisdom. Various interests use the airwaves and political process to minimize or exaggerate the challenges we face. The search for the

---

common good and the voices of poor people and poor countries sometimes are neglected.

At its core, global climate change is not about economic theory or political platforms, nor about partisan advantage or interest group pressures. It is about the future of God's creation and the one human family. It is about protecting both "the human environment" and the natural environment (John Paul II 1991, #38). It is about our human stewardship of God's creation and our responsibility to those who come after us. With these reflections, we seek to offer a word of caution and a plea for genuine dialogue as the United States and other nations face decisions about how best to respond to the challenges of global climate change.

The dialogue and our response to the challenge of climate change must be rooted in the virtue of prudence. While some uncertainty remains, most experts agree that something significant is happening to the atmosphere. Human behavior and activity are, according to the most recent findings of the international scientific bodies charged with assessing climate change, contributing to a warming of the earth's climate. Although debate continues about the extent and impact of this warming, it could be quite serious ....[2] Consequently, it seems prudent not only to continue to research and monitor this phenomenon, but to take steps now to mitigate possible negative effects in the future.

As Catholic bishops, we seek to offer a distinctively religious and moral perspective to what is necessarily a complicated scientific, economic, and political discussion. Ethical questions lie at the heart of the challenges facing us. John Paul II insists, "We face a fundamental question which can be described as both ethical and ecological. How can accelerated development be prevented from turning against man? How can one prevent disasters that destroy the environment and threaten all forms of life, and how can the negative consequences that have already occurred be remedied?" (John Paul II 1996).

Because of the blessings God has bestowed on our nation and the power it possesses, the United States bears a special responsibility in its stewardship of God's creation to shape responses that serve the entire human family. As pastors, teachers, and citizens, we bishops

---

2   Editor's Note: In its original online format, this statement referred readers to an informative side bar entitled "The Science of Global Climate Change" which is available at http://www.usccb.org/sdwp/international/globalclimate.shtml#introduction.

seek to contribute to our national dialogue by examining the ethical implications of climate change. We offer some themes from Catholic social teaching that could help to shape this dialogue, and we suggest some directions for the debate and public policy decisions that face us. We do so with great respect for the work of the scientists, diplomats, business and union representatives, developers of new technologies, environmental leaders, and policymakers who have been struggling with the difficult questions of climate change for many years.

While our own growing awareness of this problem has come in part from scientific research and the public debate about the human contribution to climate change, we are also responding to the appeals of the Church in other parts of the world. Along with Pope John Paul II, church leaders in developing countries—who fear that affluent nations will mute their voices and ignore their needs—have expressed their concerns about how this global challenge will affect their people and their environment. We also hear the call of Catholic youth and other young people to protect the environment.

Therefore, we especially want to focus on the needs of the poor, the weak, and the vulnerable in a debate often dominated by more powerful interests. Inaction and inadequate or misguided responses to climate change will likely place even greater burdens on already desperately poor peoples. Action to mitigate global climate change must be built upon a foundation of social and economic justice that does not put the poor at greater risk or place disproportionate and unfair burdens on developing nations.

SCIENTIFIC KNOWLEDGE &
THE VIRTUE OF PRUDENCE

As Catholic bishops, we make no independent judgment on the plausibility of "global warming." Rather, we accept the consensus findings of so many scientists and the conclusions of the Intergovernmental Panel on Climate Change (IPCC) as a basis for continued research and prudent action.... Scientists engaged in this research consistently acknowledge the difficulties of accurate measurement and forecasting. Models of measurement evolve and vary in reliability. Researchers and advocates on all sides of the issue often have stakes in policy outcomes, as do advocates of various courses of public policy. News reports can oversimplify findings or focus on controversy rather than areas of consensus. Accordingly, interpretation of scientific data and conclusions in public discussion can be difficult and contentious matters. Responsible scientific research is always careful to recognize uncertainty and is modest in its claims. Yet over the past few decades, the evidence of global climate change and the emerging scientific consensus about the human impact on this process have led many governments to reach the conclusion that they need to invest time, money, and political will to address the problem through collective international action.

The virtue of prudence is paramount in addressing climate change. This virtue is not only a necessary one for individuals in leading morally good lives, but is also vital to the moral health of the larger community. Prudence is intelligence applied to our actions. It allows us to discern what constitutes the common good in a given situation. Prudence requires a deliberate and reflective process that aids in the shaping of the community's conscience. Prudence not only helps us identify the principles at stake in a given issue, but also moves us to adopt courses of action to protect the common good. Prudence is not, as popularly thought, simply a cautious and safe approach to decisions. Rather, it is a thoughtful, deliberate, and reasoned basis for taking or avoiding action to achieve a moral good.

In facing climate change, what we already know requires a response; it cannot be easily dismissed. Significant levels of scientific consensus—even in a situation with less than full certainty, where the consequences of not acting are serious—justifies, indeed can obligate, our taking action intended to avert potential dangers. In other words, if enough evidence indicates that the present course of action could jeopardize humankind's well-being, prudence dictates taking mitigating or preventative action.

This responsibility weighs more heavily upon those with the power to act because the threats are often greatest for those who lack similar power, namely, vulnerable poor populations, as well as future generations. According to reports of the IPCC, significant delays in addressing climate change may compound the problem and make future remedies more difficult, painful, and costly. On the other hand, the impact of prudent actions today can potentially improve the situation over time, avoiding more sweeping action in the future.

CLIMATE CHANGE & CATHOLIC SOCIAL TEACHING

God has endowed humanity with reason and ingenuity that distinguish us from other creatures. Ingenuity and creativity have enabled us to make remarkable advances and can help us address the problem of global climate change; however, we have not always used these endowments wisely. Past actions have produced both good works and harmful ones, as well as unforeseen or unintended consequences. Now we face two central moral questions:

1) How are we to fulfill God's call to be stewards of creation in an age when we may have the capacity to alter that creation significantly, and perhaps irrevocably?

2) How can we as a "family of nations" exercise stewardship in a way that respects and protects the integrity of God's creation and provides for the common good, as well as for economic and social progress based on justice?

Catholic social teaching provides several themes and values that can help answer these questions.

The Universal Common Good

Global climate is by its very nature a part of the planetary commons. The earth's atmosphere encompasses all people, creatures, and habitats. The melting of ice sheets and glaciers, the destruction of rain forests, and the pollution of water in one place can have environmental impacts elsewhere. As Pope John Paul II has said, "*We cannot interfere in one area of the ecosystem without paying due attention both to the consequences of such interference in other areas and to the well being of future generations*" (1990, #6). Responses to global climate change should reflect our interdependence and common responsibility for the future of our planet. Individual nations must measure their own self-interest against the greater common good and contribute equitably to global solutions.

Stewardship of God's Creation & the Right to Economic Initiative & Private Property

Freedom and the capacity for moral decision making are central to what it means to be human. Stewardship—defined in this case as the ability to exercise moral responsibility to care for the environment—requires freedom to act. Significant aspects of this stewardship include the right to private initiative, the ownership of property, and the exercise of responsible freedom in the economic sector. Stewardship requires a careful protection of the environment and calls us to use our intelligence "to discover the earth's productive potential and the many different ways in which human needs can be satisfied" (John Paul II 1991, #32).

We believe economic freedom, initiative, and creativity are essential to help our nation find effective ways to address climate change. The United States' history of economic, technological innovation, and entrepreneurship invites us to move beyond status quo responses to this challenge. In addition, the right to private property is matched by the responsibility to use what we own to serve the common good. Our Catholic tradition speaks of a "social mortgage" on property and, in this context, calls us to be good stewards of the earth (John Paul II 1988, #42). It also calls us to use the gifts we have been given to protect human life and dignity, and to exercise our care for God's creation.

True stewardship requires changes in human actions—both in moral behavior and technical advancement. Our religious tradition has always urged restraint and moderation in the use of material goods, so we must not allow our desire to possess more material things to overtake our concern for the basic needs of people and the environment. Pope John Paul II has linked protecting the environment to "authentic human ecology," which can overcome "structures of sin" and which promotes both human dignity and respect for creation (1991, #38). Technological innovation and entrepreneurship can help make possible options that can lead us to a more environmentally benign energy path. Changes in lifestyle based on traditional moral virtues can ease the way to a sustainable and equitable world economy in which sacrifice will no longer be an unpopular concept. For many of us, a life less focused on material gain may remind us that we are more than what we have. Rejecting the false promises of excessive or conspicuous consumption can even allow more time for family, friends, and civic responsibilities. A renewed sense of sacrifice and restraint could make an essential contribution to addressing global climate change.

Protecting the Environment for Future Generations

The common good calls us to extend our concern to future generations. Climate change poses the question "What does our generation owe to generations yet unborn?" As Pope John Paul II has written, "there is an order in the universe which must be respected, and...the human person, endowed with the capability of choosing freely, has a grave responsibility to preserve this order for the well-being of future generations" (1997, 12).

Passing along the problem of global climate change to future generations as a result of our delay, indecision, or self-interest would be easy. But we simply cannot leave this problem for the children of tomorrow. As stewards of their heritage, we have an obligation to respect their dignity and to pass on their natural inheritance, so that their lives are protected and, if possible, made better than our own.

Population & Authentic Development

Population and climate change should be addressed from the broader perspective of a concern for protecting human life, caring for the environment, and respecting cultural norms and the religious faith and moral values of peoples. Population is not simply about statistics. Behind every demographic number is a precious and irreplaceable human life whose human dignity must be respected.

The global climate change debate cannot become just another opportunity for some groups—usually affluent advocates from the developed nations—to blame the problem on population growth in poor countries. Historically, the industrialized countries have emitted more greenhouse gases that warm the climate than have the developing countries. Affluent nations such as our own have to acknowledge the impact of voracious consumerism instead of simply calling for population and emissions controls from people in poorer nations.

A more responsible approach to population issues is the promotion of "authentic development," which represents a balanced view of human progress and includes respect for nature and social well-being (John Paul II 1988, chap. 4).[3] Development policies that seek to reduce poverty with an emphasis on improved education and social conditions for women are far more effective than usual population reduction programs and far more respectful of women's dignity (Second Vatican Council 1996, #50-51).

We should promote a respect for nature that encourages policies fostering natural family planning and the education of women and men rather than coercive measures of population control or government

---

3   This chapter of the encyclical *On Social Concern* gives a more complete definition of the concept of authentic development.

incentives for birth control that violate local cultural and religious norms.

Caring for the Poor and Issues of Equity

Working for the common good requires us to promote the flourishing of all human life and all of God's creation. In a special way, the common good requires solidarity with the poor who are often without the resources to face many problems, including the potential impacts of climate change. Our obligations to the one human family stretch across space and time. They tie us to the poor in our midst and across the globe, as well as to future generations. The commandment to love our neighbor invites us to consider the poor and marginalized of other nations as true brothers and sisters who share with us the one table of life intended by God for the enjoyment of all.

All nations share the responsibility to address the problem of global climate change. But historically the industrial economies have been responsible for the highest emissions of greenhouse gases that scientists suggest are causing the warming trend. Also, significant wealth, technological sophistication, and entrepreneurial creativity give these nations a greater capacity to find useful responses to this problem. To avoid greater impact, energy resource adjustments must be made both in the policies of richer countries and in the development paths of poorer ones.

Most people will agree that while the current use of fossil fuels has fostered and continues to foster substantial economic growth, development, and benefits for many, there is a legitimate concern that as developing countries improve their economies and emit more greenhouse gases, they will need technological help to mitigate further atmospheric environmental harm. Many of the poor in these countries live in degrading and desperate situations that often lead them to adopt environmentally harmful agricultural and industrial practices. In many cases, the heavy debt burdens, lack of trade opportunities, and economic inequities in the global market add to the environmental strains of the poorer countries. Developing countries have a right to economic development that can help lift people out of dire poverty. Wealthier industrialized nations have the resources, know-how, and entrepreneurship to produce more efficient cars and cleaner industries. These countries need to share these emerging technologies with the less-developed countries and assume more of the financial responsibility that would enable poorer countries to afford them. This would help developing countries adopt energy-efficient technologies more rapidly while still sustaining healthy economic growth and development.[4] Industries from the developed countries operating in developing nations should exercise a leadership role in preserving the environment.

No strategy to confront global climate change will succeed without the leadership and participation of the United States and other industrial nations. But any successful strategy must also reflect the genuine participation and concerns of those most affected and least able to bear the burdens. Developing and poorer nations must have a genuine place at the negotiating table. Genuine participation for those most affected is a moral and political necessity for advancing the common good.

The Public Policy Debate & Future Directions

Catholic social teaching calls for bold and generous action on behalf of the common good. "Interdependence," as Pope John Paul II has written, "must be transformed into *solidarity* …. Surmounting every type of *imperialism* and determination to preserve their *own hegemony*, the stronger and richer nations must have a sense of moral *responsibility* for the other nations, so that a *real international system* may be established which will rest on the foundation of the *equality* of all peoples and on the necessary respect for their legitimate differences" (USCCB 1993, #39).

The common good is built up or diminished by the quality of public debate. With its scientific, technological, economic, political, diplomatic, and religious dimensions, the challenge of global climate change may be a basic test of our democratic processes and political institutions. We respect the inquiry and dialogue which has been carried forward by a wide variety of scientists, diplomats, policy makers, and advocates, not only in the United States but around the world. These efforts should not be demeaned or distorted by disinformation or exaggeration. Serious dialogue should not be jeopardized by public relations tactics that fan fears or pit nations against one another. Leaders in every sector should seek to build a scientifically based consensus

---

4   See also treatment of this topic in *Stewardship: A Disciple's Response* (Washington, DC: United States Conference of Catholic Bishops, 1993), 27.

for the common good; avoid merely representing their own particular interests, industries, or movements; and act responsibly to protect future generations and the weak.

In the past decade, a continuing process of international diplomacy has led to agreements on principles and increasingly on procedures. In 1992, more than 160 nations, including the United States, ratified the first international treaty on global climate change at Earth Summit in Rio de Janeiro, Brazil, which was known as the United Nations Framework Convention on Climate Change (UNFCCC). In 1997, parties to the UNFCCC including the United States negotiated the Kyoto Protocol, which established mandatory emission reduction targets, market-based procedures for meeting those targets, and timetables for industrialized nations.

Without endorsing the specifics of these agreements and processes, we Catholic bishops acknowledge the development of these international negotiations and hope they and other future efforts can lead to just and effective progress. However, serious deliberations must continue to bring about prudent and effective actions to ensure equity among nations.

As an act of solidarity and in the interest of the common good, the United States should lead the developed nations in contributing to the sustainable economic development of poorer nations and to help build their capacity to ease climate change. Since our country's involvement is key to any resolution of these concerns, we call on our people and government to recognize the seriousness of the global warming threat and to develop effective policies that will diminish the possible consequences of global climate change. We encourage citizens to become informed participants in this important public debate. The measures we take today may not greatly moderate climate change in the near future, but they could make a significant difference for our descendants.

We also hope that the United States will continue to undertake reasonable and effective initiatives for energy conservation and the development of alternate renewable and clean-energy resources. New technologies and innovations can help meet this challenge. While more needs to be done to reduce air pollution, through the use of improved technologies and environmental entrepreneurship, the United States has made significant environmental gains over the last several decades. Our hope is that these technologies along with other resources can be shared with developing countries.

Within the United States, public policy should assist industrial sectors and workers especially impacted by climate change policies, and it should offer incentives to corporations to reduce greenhouse gas emissions and assistance to workers affected by these policies.

We encourage all parties to adopt an attitude of candor, conciliation, and prudence in response to serious, complex, and uncertain challenges. We hope the continuing dialogue within and among the diverse disciplines of science, economics, politics, and diplomacy will be guided by fundamental moral values: the universal common good, respect for God's creation, an option for the poor, and a sense of intergenerational obligation. Since religious values can enrich public discussion, this challenge offers opportunities for interfaith and ecumenical conversation and cooperation.

Finally, we wish to emphasize the need for personal conversion and responsibility. In our pastoral reflection *Renewing the Earth*, we wrote the following:

> Grateful for the gift of creation...we invite Catholics and men and women of good will in every walk of life to consider with us the moral issues raised by the environmental crisis.... These are matters of powerful urgency and major consequence. They constitute an exceptional call to conversion. As individuals, as institutions, as a people, we need a change of heart to preserve and protect the planet for our children and for generations yet unborn (USCCB 1992, 3).[5]

Each of us should carefully consider our choices and lifestyles. We live in a culture that prizes the consumption of material goods. While the poor often have too little, many of us can be easily caught up in a frenzy of wanting more and more—a bigger home, a larger car, etc. Even though energy resources literally fuel our economy and provide a good quality of life, we need to ask about ways we can conserve energy, prevent pollution, and live more simply.

CONCLUSION

Our national debate over solutions to global climate change needs to move beyond the uses and abuses of science, sixty-second ads, and exaggerated claims. Because this issue touches so many people, as well as the planet itself, all parties need to strive for a civil and constructive debate about U.S. decisions and leadership in this area.

---

5   See also the treatment of this theme in USCCB 1993, #46.

As people of religious faith, we bishops believe that the atmosphere that supports life on earth is a God-given gift, one we must respect and protect. It unites us as one human family. If we harm the atmosphere, we dishonor our Creator and the gift of creation. The values of our faith call us to humility, sacrifice, and a respect for life and the natural gifts God has provided. Pope John Paul II reminds us in his statement *The Ecological Crisis: A Common Responsibility* that "respect for life and for the dignity of the human person extends also to the rest of creation, which is called to join man in praising God" (John Paul II 1990, #16). In that spirit of praise and thanksgiving to God for the wonders of creation, we Catholic bishops call for a civil dialogue and prudent and constructive action to protect God's precious gift of the earth's atmosphere with a sense of genuine solidarity and justice for all God's children.

SOURCES

John Paul II, Pope. 1988. *On Social Concern (Sollicitudo Rei Socialis.* Washington, DC: United States Conference of Catholic Bishops.

————. 1990. *The Ecological Crisis: A Common Responsibility.* Washington, DC: United States Conference of Catholic Bishops.

————. 1991. *On the Hundredth Anniversary of Rerum Novarum (Centesimus Annus).* Washington, DC: United States Conference of Catholic Bishops.

————. 1996. "International Solidarity Needed to Safeguard Environment." Address by the Holy Father to the European Bureau for the Environment. *L'Osservatore Romano,* June 26.

————. 1997. "The Exploitation of the Environment Threatens the Entire Human Race." Address to the Vatican Symposium on the Environment (1990). In *Ecology and Faith: The Writings of Pope John Paul II,* ed. Sr. Ancilla Dent, OSB. Berkhamsted, England: Arthur James.

Second Vatican Council. 1996. *Pastoral Constitution on the Church in the Modern World (Gaudium et Spes),* nos. 50-51. In *Vatican Council II: The Conciliar and Post Conciliar Documents,* ed. Austin Flannery, new rev. ed., 1st vol. Northport, NY: Costello Publishing.

United States Conference of Catholic Bishops. 1992. *Renewing the Earth: An Invitation to Reflection and Action on Environment in Light of Catholic Social Teaching.* Washington, DC: USCCB.

————. 1993. Stewardship: A Disciple's Response. Washington, DC: United States Conference of Catholic Bishops.

COMMENTS ON THE USCCB'S

GLOBAL CLIMATE CHANGE

BY

Jane Elyse Russell, O.S.F.

In June, 2001, the United States Catholic Conference (now the United States Conference of Catholic Bishops, USCCB)[1] approved a document entitled *Global Climate Change: A Plea for Dialogue, Prudence, and the Common Good.* The authors offered "no independent judgment on the plausibility of 'global warming'" (5), but they made a strong plea for people to *act* with prudent foresight upon the general consensus of scientists as articulated in the conclusions of the Intergovernmental Panel on Climate Change. They built a case for action based on standard principles of Catholic social teaching. The familiarity of the principles cited (e.g., the universal common good and protecting the environment for future generations) makes for less-than-exciting reading, but the document presents a rock-solid case for thinking about and acting to address human-forced climate change without delay.

In this introduction, I review the background of the document, summarize its main points, and sample its "effective history" to determine its impact on American Catholic consciousness and public policy activity.

---

1 The United States Conference of Catholic Bishops describes its mission "to support the ministry of bishops with an emphasis on evangelization, by which the bishops exercise in a communal and collegial manner certain pastoral functions entrusted to them by the Lord Jesus of sanctifying, teaching, and governing" (www.usccb.org). This mission includes collaborative action as a conference "on vital issues confronting the Church and society." Clearly, their actions indicate that they recognize environmental concerns as a vital issue.

BACKGROUND OF THE STATEMENT

Although, as Thomas Massaro notes, "it is surprising how seldom ecological concerns are actually mentioned in the encyclicals" (2008, 159), Catholic social teaching has been moving in recent decades toward affirmation of respect for the environment and concern about environmental degradation. Pope John Paul II's message on the 1990 World Day of Peace, *Peace with God the Creator, Peace with All of Creation*, marked a milestone as the first papal statement dedicated to environmental concern. Its conclusion is often quoted: "Respect for life and for the dignity of the human person extends also to the rest of creation, which is called to join man in praising God" (1989, # 16). With Ecumenical Patriarch Bartholomew I, the pope issued a *Common Declaration on Environmental Ethics* in which they called for environmental "repentance" in thinking and "conversion" in living less consumptively (2002).

In the United States, partly as a follow-up to John Paul II's 1990 message, members of the U. S. Catholic Conference committees on International Policy and on Domestic Policy met in June 1990 to address concerns "about the religious and moral dimensions of the environmental crisis" that had been raised by several U.S. bishops and by bishops' conferences in developing countries (USCC 1991). They began consultations to develop a statement that eventually was issued under the title of *Renewing the Earth: An Invitation to Reflection and Action on Environment in Light of Catholic Social Teaching*. Adopted subsequently by the full assembly of the U. S. bishops in November 1991, this statement helped inspire the writing of other regional bishops' statements on particular regional environmental concerns (e.g. *At Home in the Web of Life* issued by the Bishops of Appalachia in 1995 and the 2001 letter on *The Columbia River Watershed* issued by the bishops of the Northwestern United States and British Columbia, Canada).

Despite the trend in statements about the environment, Catholics in the U.S. may be surprised that their bishops have ventured to pronounce on climate change in particular, considering the ongoing disagreements among political leaders over the extent of and appropriate responses to climate change. A detailed history of the bishops' decision to write about the climate crisis is not yet available, but the 2001 statement mentions several factors that contributed to the bishops' "growing awareness of this problem": (1) publications of scientific research and public debate on the topic, (2) concern expressed by Pope John Paul II and bishops from developing countries, and (3) "the call of Catholic youth and other young people to protect the environment" (3). The initial paragraph of the document notes that it originated, like the 1992 pastoral letter, from the Domestic and International Policy Committees, and included collaboration with the members of the bishops' Committee on Doctrine and the Committee on Science and Human Values (USCC 2001, reverse of title page). The degree of episcopal firepower invested in this document may indicate the importance and/or sensitivity of the issue it addresses.

CONTENT OF *GLOBAL CLIMATE CHANGE*

The bishops acknowledge that they bring neither scientific expertise on climate change nor responsibility for public policy, but they want to convey a sense of the moral weight of the situation. "As Catholic bishops, we seek to offer a distinctively religious and moral perspective to…a complicated scientific and political discussion" (USCC 2001, 2). The first action for which they call is "a different kind of national discussion" (1), one that is less strident, partisan, and self-interested.

They stress repeatedly that the virtue of prudence counsels swift action, "even in a situation with less than full certainty" (6), to forestall or minimize future harm. Summarizing some of "The Science of Climate Change" from assessment reports of the Intergovernmental Panel on Climate Change, the bishops conclude that it would be "prudent not only to continue to research and monitor this phenomenon, but to take steps now to mitigate possible negative effects in the future" (2; see 5-6 and 14-15).

To support their call for prudent action, the bishops invoke several principles from the long tradition of Catholic Social Teaching:

1. "The Universal Common Good" is an overriding issue, since "global climate is by its very nature a part of the planetary commons" (7).

2. Under the nuanced heading of "Stewardship of God's Creation and the Right to Economic Initiative and Private Property," the bishops attempt a characteristic Catholic balancing act. On the one hand, economic freedom and private property are good things, and private initiatives may bring about just the technological innovations "that can lead us to a more environmentally benign energy path" (9). On the other hand, Catholic tradition sees "a 'social mortgage' on private property,"

to use John Paul II's phrase. This means that all are responsible "to use what we own to serve the common good" (8). Our human actions must change, "both in moral behavior and technical advancement," if we are to fulfill our God-given responsibility to steward earth's gifts as a heritage for the future. Balancing the two aspects is not easy, but this section ends with an emphasis on sacrificial changes in lifestyle that are necessary, especially in materially developed countries.

3. "Protecting the Environment for Future Generations" constitutes another way of calling for prudent action that is mindful of the temporal dimensions of the common good.

4. Speaking of "Population and Authentic Development," the bishops dispute one common response to the climate-change threat: blaming it on "population growth in poor countries" (10). They remind readers that Americans' own "voracious consumerism" has historically caused more greenhouse gas emissions than anything happening in poorer countries. They suggest that "improved education and social conditions for women" will, in any case, do more to curb population growth than coercion or offering incentives for contraception (10).

5. "Caring for the Poor and Issues of Equity" brings to the fore one major theme that other voices may neglect: the need to give "developing and poor nations" a prominent place in any climate-change negotiations and to assist them with technologies needed to lessen environmental damage. "To avoid greater impact [on the global climate], energy resource adjustments must be made both in the policies of richer countries and in the development paths of poorer ones" (12). If wealthier countries have the resources and know-how "to produce more efficient cars and cleaner industries," for example, they should share these with less-developed countries, along with the resources to afford them (12).

The bishops round out their statement with a strong call for "constructive civil debate" on this urgent public issue (17), putting aside both disinformation and fear-mongering exaggeration. They review favorably the history of international discussions and agreements on climate change up to the Kyoto Protocol of 1997. "Without endorsing the specifics of these agreements and processes" (14), they express a hope that future negotiations can lead to significant agreements, always including the developing countries. They offer a few public-policy hints for the United States (e.g., "offer incentives to corporations to reduce greenhouse gas emissions") (15) but do not make specific recommendations.

After additional calls for personal conversion and lifestyle changes (15), they end their statement as they began it, with a resounding affirmation of the earth and its atmosphere as a "God-given gift." Earth is a gift to be protected for the present and future of "all God's children" (17).

IMPACT

Global Climate Change: A Plea for Dialogue, Prudence and the Common Good is a good statement, especially compared with the outright denial of human-forced climate change one hears from some political leaders today. Contemplating the bishops' 'plea' almost ten years later, one wonders how much impact it has had on U.S. Catholic thought. We hear about and observe groups of Catholics who are active in alternative-energy and climate-change issues, though the extent to which the bishops' statement has influenced their activities remains elusive. How has it influenced parish initiatives or preaching? How often have the bishops themselves referred back to it or used it, for example to challenge the Bush Administration's rejection of the Kyoto Protocols?

When questioned about the influence the 2001 statement has had, Cecilia Calvo, Project Coordinator of the Environmental Justice Program for the USCCB, commented: "It is difficult to measure the impact of the bishops' statement" (Calvo 2010). Nevertheless, she proceeded to list a number of positive outcomes, especially on the educational front. Between her communication and the USCCB website, one may note the following activities more or less directly related to the climate-change statement:

The webpage for the Bishops' Environmental Justice Program initiated late in 1993 has a section on climate (www.usccb.org/sdwp/ejp/climate) with solid information on pertinent issues and diocesan educational and action projects that were facilitated by grants from the USCC in 2006. One of the funded projects occurred in my diocese of Charlotte, though I did not realize that the workshop I attended had been funded by the U.S. bishops.

According to Ms. Calvo, "[t]he Bishops' statement is shared with many national Catholic organizations through the Catholic Covenant for Climate Change and the Catholic Climate Covenant campaign" that was initiated in 2006 with support of the USCCB. Calvo described the covenant as "an education and action initiative that has been shared with every Catholic parish and school in our country" in order to "increase awareness among Catholics of a Catholic approach to climate change" (2010).

The bishops' statement is also shared by way of "another educational joint Catholic Relief Services/USCCB campaign called Catholics Confront Global Poverty." This campaign aims "to look at various issues including climate change that affect people in poverty" (Calvo 2010).

The USCCB has shared its statement with interfaith partners in the National Religious Partnership for the Environment. Working with the Coalition on the environment and Jewish Life, the National Council of Churches of Christ, and the Evangelical Environmental Network, the bishops aim "to educate Members of Congress about the moral implications of climate change" (Calvo 2010).

The Environmental Justice web page includes links to many examples of USCCB letters and action alerts to educate members of Congress from 2005 through 2010. One example is the statement of John L. Carr, Secretary, USCCB Department of Social Development and World Peace, entitled "Religious and Moral Dimensions of Global Climate Change: Before the Senate Committee on Environment and Public Works, 110th Congress (June 2, 2007)."

In summary, while climate change advocacy is not the highest policy priority anyone would associate with the Catholic Church in the United States, the record shows that its bishops' conference has indeed been stirring this issue on the back burner (no pun intended) for the last ten years. Since neither Republican nor Democratic administrations has succeeded in enacting effective policies to reduce the threat of human-forced changes to the global climate, we may hope that the bishops and other communities of faith will bring this concern to the forefront before it is too late. Now more than ever, we need the change of heart and the prudent actions that will "preserve and protect the planet for our children and for generations yet unborn" (USCC 1991, 3; see USCCB 15.)

SOURCES

Calvo, Cecilia. 2010. USCCB Department of Justice, Peace and Human Development. E-mail to the author, October 6.

John Paul II, Pope. 1989. *Peace with God the Creator, Peace with All of Creation.* Message on the 1990 World Day of Peace, December 8. Accessed from http://www.vatican.va/holy_father/john_paul_ii/messages/peace/documents/hf_jp-ii_mes_19891208_xxiii-world-day-for-peace_en.html on October 17, 2010.

———. with Ecumenical Patriarch His Holiness Bartholomew I. 2002. *Common Declaration on Environmental Ethics,* June 10. Accessed from http://www.vatican.va/holy_father/john_paul_ii/speeches/2002/june/documents/hf_jp-ii_spe_20020610_venice-declaration_en.html on October 17, 2010.

Massaro, Thomas. 2008. *Living Justice: Catholic Social Teaching in Action.* Classroom edition. Lanham, MD: Rowman & Littlefield.

United States Catholic Conference/ National Conference of Catholic Bishops (USCC). 1992. *Renewing the Earth.* Washington, DC: United States Catholic Conference, Inc.

United States Catholic Conference of Catholic Bishops[2] (USCCB). 2001. *Global Climate Change: A Plea for Dialogue, Prudence, and the Common Good.* Washington, DC: United States Conference of Catholic Bishops.

2 The U.S. Catholic bishops adopted this name in July 2001.

19

IF YOU WANT TO CULTIVATE PEACE,
PROTECT CREATION

MESSAGE OF HIS HOLINESS POPE BENEDICT XVI FOR
THE CELEBRATION OF THE WORLD DAY OF PEACE[1]
I JANUARY 2010

---

1   *If You Want to Cultivate Peace, Protect Creation* © 2009, Libreria Editrice Vaticana. Downloaded from http://www.vatican.va/holy_father/ benedict_ xvi/messages/peace/documents/hf_ben-xvi_mes _20091208_ xliii-world-day-peace_en.html & authorized for inclusion in this anthology by Professor D. Giuseppe Costa, S.D.B., Director of Libreria Editrice Vaticana, March 2011.

1. At the beginning of this New Year, I wish to offer heartfelt greetings of peace to all Christian communities, international leaders, and people of good will throughout the world. For this XLIII World Day of Peace I have chosen the theme: *If You Want to Cultivate Peace, Protect Creation*. Respect for creation is of immense consequence, not least because "creation is the beginning and the foundation of all God's works,",[1] and its preservation has now become essential for the pacific coexistence of mankind. Man's inhumanity to man has given rise to numerous threats to peace and to authentic and integral human development—wars, international and regional conflicts, acts of terrorism, and violations of human rights. Yet no less troubling are the threats arising from the neglect–if not downright misuse–of the earth and the natural goods that God has given us. For this reason, it is imperative that mankind renew and strengthen "that covenant between human beings and the environment, which should mirror the creative love of God, from whom we come and towards whom we are journeying".[2]

2. In my Encyclical *Caritas in Veritate*, I noted that integral human development is closely linked to the obligations which flow from *man's relationship with the natural environment*. The environment must be seen as God's gift to all people, and the use we make of it entails a shared responsibility for all humanity, especially the poor and future generations. I also observed that whenever nature, and human beings in particular, are seen merely as products of chance or an evolutionary determinism, our overall sense of responsibility wanes.[3] On the other hand, seeing creation as God's gift to humanity helps us understand our vocation and worth as human beings. With the Psalmist, we can exclaim with wonder: "When I look at your heavens, the work of your hands, the moon and the stars which you have established; what is man that you are mindful of him, and the son of man that you care for him?" (Ps 8:4-5). Contemplating the beauty of creation inspires us to recognize the love of the Creator, that Love which "moves the sun and the other stars".[4]

3. Twenty years ago, Pope John Paul II devoted his Message for the World Day of Peace to the theme: *Peace with God the Creator, Peace with All of Creation*. He emphasized our relationship, as God's creatures, with the universe all around us. "In our day", he wrote, "there is a growing awareness that world peace is threatened ... also by a lack of *due respect for nature*". He added that "*ecological awareness*, rather than

being downplayed, needs to be helped to develop and mature, and find fitting expression in concrete programmes and initiatives".[5] Previous Popes had spoken of the relationship between human beings and the environment. In 1971, for example, on the eightieth anniversary of Leo XIII's Encyclical *Rerum Novarum*, Paul VI pointed out that "by an ill-considered exploitation of nature (man) risks destroying it and becoming in his turn the victim of this degradation". He added that "not only is the material environment becoming a permanent menace—pollution and refuse, new illnesses and absolute destructive capacity—but the human framework is no longer under man's control, thus creating an environment for tomorrow which may well be intolerable. This is a wide-ranging social problem which concerns the entire human family".[6]

4. Without entering into the merit of specific technical solutions, the Church is nonetheless concerned, as an "expert in humanity", to call attention to the relationship between the Creator, human beings and the created order. In 1990 John Paul II had spoken of an "ecological crisis" and, in highlighting its primarily ethical character, pointed to the "urgent moral need for a new solidarity".[7] His appeal is all the more pressing today, in the face of signs of a growing crisis which it would be irresponsible not to take seriously. Can we remain indifferent before the problems associated with such realities as climate change, desertification, the deterioration and loss of productivity in vast agricultural areas, the pollution of rivers and aquifers, the loss of biodiversity, the increase of natural catastrophes and the deforestation of equatorial and tropical regions? Can we disregard the growing phenomenon of "environmental refugees", people who are forced by the degradation of their natural habitat to forsake it—and often their possessions as well—in order to face the dangers and uncertainties of forced displacement? Can we remain impassive in the face of actual and potential conflicts involving access to natural resources? All these are issues with a profound impact on the exercise of human rights, such as the right to life, food, health and development.

5. It should be evident that the ecological crisis cannot be viewed in isolation from other related questions, since it is closely linked to the notion of development itself and our understanding of man in his relationship to others and to the rest of creation. Prudence would thus dictate a *profound, long-term review of our model of development*, one which would take into consideration the meaning of the economy

and its goals with an eye to correcting its malfunctions and misapplications. The ecological health of the planet calls for this, but it is also demanded by the cultural and moral crisis of humanity whose symptoms have for some time been evident in every part of the world. [8] Humanity needs a *profound cultural renewal*; it needs to *rediscover those values which can serve as the solid basis* for building a brighter future for all. Our present crises—be they economic, food-related, environmental or social—are ultimately also moral crises, and all of them are interrelated. They require us to rethink the path which we are travelling together. Specifically, they call for a lifestyle marked by sobriety and solidarity, with new rules and forms of engagement, one which focuses confidently and courageously on strategies that actually work, while decisively rejecting those that have failed. Only in this way can the current crisis become an opportunity for discernment and new strategic planning.

6. Is it not true that what we call "nature" in a cosmic sense has its origin in "a plan of love and truth"? The world "is not the product of any necessity whatsoever, nor of blind fate or chance... The world proceeds from the free will of God; he wanted to make his creatures share in his being, in his intelligence, and in his goodness".[9] The *Book of Genesis*, in its very first pages, points to the wise design of the cosmos: it comes forth from God's mind and finds its culmination in man and woman, made in the image and likeness of the Creator to "fill the earth" and to "have dominion over" it as "stewards" of God himself (cf. Gen 1:28). The harmony between the Creator, mankind and the created world, as described by Sacred Scripture, was disrupted by the sin of Adam and Eve, by man and woman, who wanted to take the place of God and refused to acknowledge that they were his creatures. As a result, the work of "exercising dominion" over the earth, "tilling it and keeping it", was also disrupted, and conflict arose within and between mankind and the rest of creation (cf. Gen 3:17-19). Human beings let themselves be mastered by selfishness; they misunderstood the meaning of God's command and exploited creation out of a desire to exercise absolute domination over it. But the true meaning of God's original command, as the *Book of Genesis* clearly shows, was not a simple conferral of authority, but rather a summons to responsibility. The wisdom of the ancients had recognized that nature is not at our disposal as "a heap of scattered refuse".[10] Biblical Revelation made us see that nature is a gift of the Creator, who gave it an inbuilt order and enabled man

to draw from it the principles needed to "till it and keep it" (cf. Gen. 2:15).[11] Everything that exists belongs to God, who has entrusted it to man, albeit not for his arbitrary use. Once man, instead of acting as God's co-worker, sets himself up in place of God, he ends up provoking a rebellion on the part of nature, "which is more tyrannized than governed by him".[12] Man thus has a duty to exercise responsible stewardship over creation, to care for it and to cultivate it.[13]

7. Sad to say, it is all too evident that large numbers of people in different countries and areas of our planet are experiencing increased hardship because of the negligence or refusal of many others to exercise responsible stewardship over the environment. The Second Vatican Ecumenical Council reminded us that "God has destined the earth and everything it contains for all peoples and nations".[14] The goods of creation belong to humanity as a whole. Yet the current pace of environmental exploitation is seriously endangering the supply of certain natural resources not only for the present generation, but above all for generations yet to come.[15] It is not hard to see that environmental degradation is often due to the lack of far-sighted official policies or to the pursuit of myopic economic interests, which then, tragically, become a serious threat to creation. To combat this phenomenon, economic activity needs to consider the fact that "every economic decision has a moral consequence" [16] and thus show increased respect for the environment. When making use of natural resources, we should be concerned for their protection and consider the cost entailed—environmentally and socially—as an essential part of the overall expenses incurred. The international community and national governments are responsible for sending the right signals in order to combat effectively the misuse of the environment. To protect the environment, and to safeguard natural resources and the climate, there is a need to act in accordance with clearly-defined rules, also from the juridical and economic standpoint, while at the same time taking into due account the solidarity we owe to those living in the poorer areas of our world and to future generations.

8. *A greater sense of intergenerational solidarity* is urgently needed. Future generations cannot be saddled with the cost of our use of common environmental resources. "We have inherited from past generations, and we have benefited from the work of our contemporaries; for this reason we have obligations towards all, and we cannot refuse to interest ourselves in those who will come after us, to enlarge the

human family. Universal solidarity represents a benefit as well as a duty. *This is a responsibility that present generations have towards those of the future,* a responsibility that also concerns individual States and the international community".[17] Natural resources should be used in such a way that immediate benefits do not have a negative impact on living creatures, human and not, present and future; that the protection of private property does not conflict with the universal destination of goods;[18] that human activity does not compromise the fruitfulness of the earth, for the benefit of people now and in the future. In addition to a fairer sense of intergenerational solidarity there is also an urgent moral need for a renewed sense of *intragenerational solidarity,* especially in relationships between developing countries and highly industrialized countries: "the international community has an urgent duty to find institutional means of regulating the exploitation of non-renewable resources, involving poor countries in the process, in order to plan together for the future".[19] *The ecological crisis shows the urgency of a solidarity which embraces time and space.* It is important to acknowledge that among the causes of the present ecological crisis is the historical responsibility of the industrialized countries. Yet the less developed countries, and emerging countries in particular, are not exempt from their own responsibilities with regard to creation, for the duty of gradually adopting effective environmental measures and policies is incumbent upon all. This would be accomplished more easily if self-interest played a lesser role in the granting of aid and the sharing of knowledge and cleaner technologies.

9. To be sure, among the basic problems which the international community has to address is that of energy resources and the development of joint and sustainable strategies to satisfy the energy needs of the present and future generations. This means that technologically advanced societies must be prepared to encourage more sober lifestyles, while reducing their energy consumption and improving its efficiency. At the same time there is a need to encourage research into, and utilization of, forms of energy with lower impact on the environment and "a world-wide redistribution of energy resources, so that countries lacking those resources can have access to them".[20] The ecological crisis offers an historic opportunity to develop a common plan of action aimed at orienting the model of global development towards greater respect for creation and for an integral human development inspired by the values proper to charity in truth. I would advocate the

adoption of a model of development based on the centrality of the human person, on the promotion and sharing of the common good, on responsibility, on a realization of our need for a changed life-style, and on prudence, the virtue which tells us what needs to be done today in view of what might happen tomorrow.[21]

10. A sustainable comprehensive management of the environment and the resources of the planet demands that human intelligence be directed to technological and scientific research and its practical applications. The "new solidarity" for which John Paul II called in his Message for the 1990 World Day of Peace [22] and the "global solidarity" for which I myself appealed in my Message for the 2009 World Day of Peace [23] "re essential attitudes in shaping our efforts to protect creation through a better internationally-coordinated management of the earth's resources, particularly today, when there is an increasingly clear link between combatting environmental degradation and promoting an integral human development. These two realities are inseparable, since "the integral development of individuals necessarily entails a joint effort for the development of humanity as a whole". [24] At present there are a number of scientific developments and innovative approaches which promise to provide satisfactory and balanced solutions to the problem of our relationship to the environment. Encouragement needs to be given, for example, to research into effective ways of exploiting the immense potential of solar energy. Similar attention also needs to be paid to the world-wide problem of water and to the global water cycle system, which is of prime importance for life on earth and whose stability could be seriously jeopardized by climate change. Suitable strategies for rural development centred on small farmers and their families should be explored, as well as the implementation of appropriate policies for the management of forests, for waste disposal and for strengthening the linkage between combatting climate change and overcoming poverty. Ambitious national policies are required, together with a necessary international commitment which will offer important benefits especially in the medium and long term. There is a need, in effect, to move beyond a purely consumerist mentality in order to promote forms of agricultural and industrial production capable of respecting creation and satisfying the primary needs of all. The ecological problem must be dealt with not only because of the chilling prospects of environmental degradation on the horizon; the real motivation must be the quest for authentic

world-wide solidarity inspired by the values of charity, justice and the common good. For that matter, as I have stated elsewhere, "technology is never merely technology. It reveals man and his aspirations towards development; it expresses the inner tension that impels him gradually to overcome material limitations. *Technology in this sense is a response to God's command to till and keep the land* (cf. Gen 2:15) that he has entrusted to humanity, and it must serve to reinforce the covenant between human beings and the environment, a covenant that should mirror God's creative love".[25]

11. It is becoming more and more evident that the issue of environmental degradation challenges us to examine our life-style and the prevailing models of consumption and production, which are often unsustainable from a social, environmental and even economic point of view. We can no longer do without a real change of outlook which will result in *new life-styles*, "in which the quest for truth, beauty, goodness and communion with others for the sake of common growth are the factors which determine consumer choices, savings and investments".[26] Education for peace must increasingly begin with far-reaching decisions on the part of individuals, families, communities and states. We are all responsible for the protection and care of the environment. This responsibility knows no boundaries. In accordance with the *principle of subsidiarity* it is important for everyone to be committed at his or her proper level, working to overcome the prevalence of particular interests. A special role in raising awareness and in formation belongs to the different groups present in civil society and to the non-governmental organizations which work with determination and generosity for the spread of ecological responsibility, responsibility which should be ever more deeply anchored in respect for "human ecology". The media also have a responsibility in this regard to offer positive and inspiring models. In a word, concern for the environment calls for a broad global vision of the world; a responsible common effort to move beyond approaches based on selfish nationalistic interests towards a vision constantly open to the needs of all peoples. We cannot remain indifferent to what is happening around us, for the deterioration of any one part of the planet affects us all. Relationships between individuals, social groups and states, like those between human beings and the environment, must be marked by respect and "charity in truth". In this broader context one can only encourage the efforts of the international community to ensure progressive disarmament and

a world free of nuclear weapons, whose presence alone threatens the
life of the planet and the ongoing integral development of the present
generation and of generations yet to come.

12. *The Church has a responsibility towards creation*, and she considers it her duty to exercise that responsibility in public life, in order
to protect earth, water and air as gifts of God the Creator meant for
everyone, and above all to save mankind from the danger of self-destruction. The degradation of nature is closely linked to the cultural
models shaping human coexistence: consequently, "when 'human ecology' is respected within society, environmental ecology also benefits".
[27] Young people cannot be asked to respect the environment if they
are not helped, within families and society as a whole, to respect themselves. The book of nature is one and indivisible; it includes not only
the environment but also individual, family and social ethics.[28] Our
duties towards the environment flow from our duties towards the person, considered both individually and in relation to others.

Hence I readily encourage efforts to promote a greater sense of ecological responsibility which, as I indicated in my Encyclical *Caritas in
Veritate*, would safeguard an authentic "human ecology" and thus forcefully reaffirm the inviolability of human life at every stage and in every
condition, the dignity of the person and the unique mission of the family, where one is trained in love of neighbour and respect for nature.
[29] There is a need to safeguard the human patrimony of society. This
patrimony of values originates in and is part of the natural moral law,
which is the foundation of respect for the human person and creation.

13. Nor must we forget the very significant fact that many people
experience peace and tranquillity, renewal and reinvigoration, when
they come into close contact with the beauty and harmony of nature.
There exists a certain reciprocity: as we care for creation, we realize
that God, through creation, cares for us. On the other hand, a correct
understanding of the relationship between man and the environment
will not end by absolutizing nature or by considering it more important than the human person. If the Church's magisterium expresses
grave misgivings about notions of the environment inspired by ecocentrism and biocentrism, it is because such notions eliminate the difference of identity and worth between the human person and other
living things. In the name of a supposedly egalitarian vision of the "dignity" of all living creatures, such notions end up abolishing the distinctiveness and superior role of human beings. They also open the way

to a new pantheism tinged with neo-paganism, which would see the source of man's salvation in nature alone, understood in purely naturalistic terms. The Church, for her part, is concerned that the question be approached in a balanced way, with respect for the "grammar" which the Creator has inscribed in his handiwork by giving man the role of a steward and administrator with responsibility over creation, a role which man must certainly not abuse, but also one which he may not abdicate. In the same way, the opposite position, which would absolutize technology and human power, results in a grave assault not only on nature, but also on human dignity itself.[30]

14. *If you want to cultivate peace, protect creation.* The quest for peace by people of good will surely would become easier if all acknowledge the indivisible relationship between God, human beings and the whole of creation. In the light of divine Revelation and in fidelity to the Church's Tradition, Christians have their own contribution to make. They contemplate the cosmos and its marvels in light of the creative work of the Father and the redemptive work of Christ, who by his death and resurrection has reconciled with God "all things, whether on earth or in heaven" (Col 1:20). Christ, crucified and risen, has bestowed his Spirit of holiness upon mankind, to guide the course of history in anticipation of that day when, with the glorious return of the Saviour, there will be "new heavens and a new earth" (2 Pet 3:13), in which justice and peace will dwell for ever. Protecting the natural environment in order to build a world of peace is thus a duty incumbent upon each and all. It is an urgent challenge, one to be faced with renewed and concerted commitment; it is also a providential opportunity to hand down to coming generations the prospect of a better future for all. May this be clear to world leaders and to those at every level who are concerned for the future of humanity: the protection of creation and peacemaking are profoundly linked! For this reason, I invite all believers to raise a fervent prayer to God, the all-powerful Creator and the Father of mercies, so that all men and women may take to heart the urgent appeal: *If you want to cultivate peace, protect creation.*

Benedictus PP XVI

From the Vatican, 8 December 2009

NOTES

[1] Catechism of the Catholic Church, 198.

[2] Benedict XVI, Message for the 2008 World Day of Peace, 7.

[3] Cf. No.48.

[4] Dante Alighieri, *The Divine Comedy, Paradiso*, XXXIII, 145.

[5] *Message for the 1990 World Day of Peace*, 1.

[6] Apostolic Letter *Octogesima Adveniens*, 21.

[7] *Message for the 1990 World Day of Peace*, 10.

[8] Cf. Benedict XVI, Encyclical Letter *Caritas in Veritate*, 32.

[9] *Catechism of the Catholic Church*, 295.

[10] Heraclitus of Ephesus (c. 535 – c. 475 B.C.), Fragment 22B124, in H. Diels-W. Kranz, Die Fragmente der Vorsokratiker, Weidmann, Berlin,1952, 6th ed.

[11] Cf. Benedict XVI,Encyclical Letter *Caritas in Veritate*, 48.

[12] John Paul II, Encyclical Letter *Centesimus Annus*, 37.

[13] Cf. Benedict XVI, Encyclical Letter *Caritas in Veritate*, 50.

[14] Pastoral Constitution *Gaudium et Spes*, 69.

[15] Cf. John Paul II, Encyclical Letter *Sollicitudo Rei Socialis*, 34.

[16] Benedict XVI, Encyclical Letter *Caritas in Veritate*, 37.

[17] Pontifical Council for Justice and Peace, *Compendium of the Social Doctrine of the Church*, 467; cf. Paul VI, Encyclical Letter *Populorum Progressio*, 17.

[18] Cf. John Paul II, Encyclical Letter *Centesimus Annus*, 30-31, 43.

[19] Benedict XVI, Encyclical Letter *Caritas in Veritate*, 49.

[20] Ibid.

[21] Cf. Saint Thomas Aquinas, S. Th., II-II, q. 49, 5.

[22] Cf. No. 9.

[23] Cf. No. 8.

[24] Paul VI, Encyclical Letter *Populorum Progressio*, 43.

[25] Encyclical Letter *Caritas in Veritate*, 69.

[26] John Paul II, Encyclical Letter *Centesimus Annus*, 36.

[27] Benedict XVI, Encyclical Letter *Caritas in Veritate*, 51.

[28] Cf. ibid., 15, 51.

[29] Cf. ibid., 28, 51, 61; John Paul II, Encyclical Letter *Centesimus Annus*, 38, 39.

[30] Cf. Benedict XVI, Encyclical Letter *Caritas in Veritate*, 70.

CONTRIBUTORS

Colleen Mary Carpenter, a systematic theologian who specializes in feminism, the arts, and ecology, is Assistant Professor of Theology at St. Catherine University (St. Paul, Minnesota). After receiving her doctoral degree in theology and literature from the University of Chicago, she worked with the School Sisters of Notre Dame at Earthrise Farm, a retreat center and organic farm in western Minnesota. She subsequently accepted a position at St. Kate's where she teaches a variety of undergraduate courses including Christology, sacramentality, theological anthropology, and a new course she developed on Christian ecotheology. Among her publications is an article on Terry Tempist Williams published in *The Way* for which she was awarded the St. Catherine University Denny Prize for Distinction in Writing in 2008. Recent research on Canadian painter Emily Carr has enabled her to merge her interests in ecotheology and the arts.

Anne Clifford, C.S.J. holds the Monsignor James Supple Chair in Catholic Studies at Iowa State University. The foci of her teaching and research are systematic theology, feminist theology, and theology and the natural sciences. She received her master's and doctoral degrees from Catholic Theological Union and The Catholic University of America respectively. Among her contributions to ecological and ecofeminist theology are essays in *Systematic Theology: Roman Catholic Perspectives* (Fortress 2011), *Religion and Science*, vol. 4, *Human Actions at the Intersection of Science and Religion* (Routledge 2010), *Environmental Stewardship, A Critical Primer* (T & T Clark 2006), and *Introducing Feminist Theology* (Orbis 2001). For four years she convened the College Theology Society's section on Theology, the Natural Sciences and Ecology, which she proposed, and served as president of the CTS from 2006-2008.

Denis Edwards is Associate Professor of historical and systematic theology at Flinders University in South Australia, a priest of the Archdiocese of Adelaide who teaches for Catholic Theological College, and a priest in residence in Tranmere Parish. He received a S.D.T.

at Catholic University of America. He serves as a member of the national Lutheran-Roman Catholic Dialogue and the Australian Anglican-Roman Catholic Commission, and he engages in the work of the South Australia Council of Churches. His research has focused on Christology, pneumatology, Trinitarian theology, ecumenical theology, and the dialogue between science and religion. Committed to working with others in the emerging field of ecological theology, he has contributed several monographs including *Breath of Life: A Theology of the Creator Spirit* (Orbis 2004), *Ecology at the Heart of Faith* (Orbis 2006), and *How God Acts: Creation, Redemption and Special Divine Action* (Fortress 2010).

Peter Ellard teaches students to relate religion, science, and the environment at Siena College where he serves as the Associate Vice President for Academic Affairs. After writing his master's thesis on Thomas Berry in dialogue with St. Bonaventure, he proceeded to doctoral studies at Fordham University where he wrote a dissertation on the School of Chartres, published subsequently as *The Sacred Cosmos: Theological, Philosophical and Scientific Conversations in the Twelfth Century School of Chartres* (University of Scranton Press 2007). He has been researching the development of Thomas Berry's thought and recently published a book review of Berry's *The Sacred Universe: Earth Spirituality, and Religion in the Twenty-First Century* in the *International Journal of Environmental Studies*.

Robert Faricy, S.J. is Emeritus Professor of Spirituality at the Pontifical Gregorian University in Rome. Now retired and living at the Jesuit Residence at Marquette University, Father Faricy has written over thirty books, most of which have been translated into several languages, and several hundred articles in the fields of spirituality, theology, and philosophy. His most recent monograph soon to be translated into English is *Gesu e lo Spirito (Jesus and the Spirit)* (Assisi 2006). He frequently ministers in the United States, Europe, and Asia, speaking at conferences and conventions and leading retreats.

Suzanne Franck, C.S.J. is Associate Professor of Religious Studies and Associate Academic Dean at St. Joseph's College (Patchogue, New York) where she teaches courses on religion and ecology, social justice, and Christology. She received her undergraduate degree in

Environmental Science from St. John's University, master's degrees in Education and in Theology from Queens College and Seminary of the Immaculate Conception respectively, and a doctoral degree in Contemporary Systematic Theology from Fordham University. She wrote her dissertation, <u>Humanity and Nature in Light of Recent Bishop's Pastoral Letters and Trends in Ecological Theology</u>, under the direction of Elizabeth Johnson, C.S.J.

William C. French is Associate Professor of Theology at Loyola University Chicago where he teaches ethics courses pertaining to the environment, social justice, and peace and conflict. He focuses his research on religious ethics, ecological ethics and policies, and war and peace issues. He received a master's degree in Religion and Ethics from Harvard University and a doctoral degree in Ethics and Society from the University of Chicago. Among his many publications are articles in the *University of St. Thomas Law Journal, Journal of Ecumenical Studies, Journal of Environmental Ethics,* and *Journal of Religion.*

William P. George is Professor of Theology and Director of the Core Curriculum at Dominican University (River Forest, Illinois). Four years of teaching experience in Zambia, doctoral studies at the University of Chicago Divinity School where he wrote a dissertation on the 1982 Convention on the Law of the Sea, and organizational work in peace studies have fueled an abiding interest in international environmental law in relation to theology. He has published several journal articles and book chapters on international law issues including international regime theory and religious ethics, the bias against religion and theology in international law textbooks, concern for future generations, and the place of international law in Catholic higher education, often building on the theology of Bernard Lonergan and Thomas Aquinas. His undergraduate teaching includes courses in Catholic social thought, economic justice, war and peace in Christianity and Islam, and international law.

Annemarie S. Kidder is Assistant Professor of Biblical Studies and Church History at the Ecumenical Theological Seminary in Detroit. She has degrees from the Academy of the Arts in Berlin, the School of Journalism in Columbia, Missouri, and The Southern Baptist Theological Seminary in Louisville. The author, editor, and translator of over ten books in the areas of biblical studies, historical and Catholic theology, and Christian mysticism, Kidder has most recently published *Making Confession/Hearing Confession: A History of the Cure of Souls* (Liturgical Press 2010), *Etty Hillesum: Essential Writings* (Orbis 2009), and *The Mystical Way in Everyday Life* (Orbis 2010), a collection of sermons, essays, and prayers by Karl Rahner, SJ. She is a Presbyterian pastor, an associate member of the I.H.M. sisters in Monroe, Michigan, and an active member in the Karl Rahner Society.

Richard W. Kropf was ordained a priest in the Diocese of Lansing where he served parishes for several years and studied the thought of Teilhard de Chardin. He earned a Ph.D. at the University of Ottawa and a S.T.D. at Université St-Paul in Ottawa, Canada, and conducted research in France and Israel. After teaching philosophy, religious studies, theology, and psychology in various Michigan colleges and seminaries, he retired to a life of prayer and writing in northern Michigan. He has published several monographs including one coauthored with Joseph Provenzano entitled *Logical Faith: Introducing a Scientific Approach to Religion and Spirituality* (iUniverse 2007, 2009) and another *Breaking Open the Creeds: What Can They Mean for Christians Today?* that is forthcoming from Paulist Press.

Connie Lasher completed this essay while serving as a 2010-11 Bannan Fellow at the Ignatian Center for Jesuit Education and Visiting Professor in the Religious Studies Department at Santa Clara University. Her interdisciplinary research and teaching focus on the relationship between religious humanism and ecological identity. She earned a M.S. in Environmental Studies at Antioch University New England, a M.A. in Theology at Andover Newton Theological School, and a Ph.D. in Systematic Theology at Boston College. She founded and for several years directed the John Paul II Center for Theology & Environmental Studies in Maine. She is a member of the Roman Catholic-United Methodist Church Dialogue on Environmental Stewardship and recently began comparative interreligious collaborations in Japan aimed at exploring Buddhist and Christian models of humanistic education in relation to ecological identity. *Passion for Wholeness: "Ignatian Humanism" and Ecological Identity* is the title of her monograph forthcoming from Herder/Crossroad.

Richard M. Liddy is the University Professor of Catholic Thought and Culture at Seton Hall University, a faculty member in the Department of Religious Studies, and Director of the Center for Catholic Studies that includes the Bernard J. Lonergan Institute. He received his Ph.D. in philosophy from the Gregorian University in Rome where he studied under Bernard Lonergan. Among his publications are two books on Lonergan--*Transforming Light: Intellectual Conversion in the Early Lonergan* (Liturgical Press 1993) and *Startling Strangeness: Reading Lonergan's Insight* (University Press of America 2006). He serves as the editor of *The Lonergan Review* published at Seton Hall University, including Vol. 1, *Generalized Empirical Method: Perspectives from Bernard Lonergan* (Spring 2009) and Vol. 2, *Forging a New Economic Paradigm: Perspectives from Bernard Lonergan* (Spring 2010).

Msgr. Charles M. Murphy directs the diaconate program for the Diocese of Portland, Maine. After receiving a master's degree in education at Harvard University, he earned an S.T.D. from Gregorian University where his dissertation focused on the relationship between action for justice and preaching the gospel in the Synod of 1971 and served for several years as a Rector of the North American College in Rome. He has published articles in several academic journals, including *Theological Studies*. Among his five books pertinent to the focus of this anthology is *At Home on Earth: Foundations for a Catholic Ethic of the Environment* (Crossroad 1989).

Dawn M. Nothwehr, O.S.F. is Professor of Catholic Theological Ethics at Catholic Theological Union in Chicago where she currently focuses her research on global climate change, environmental ethics, the ethics of power, and racial justice. Informing her research and teaching are the concept of mutuality as a formal norm, feminist ethics of power, the relationship of ethics and spirituality, and the Franciscan intellectual tradition. Among her publications after receiving her Ph.D. from Marquette University are *A Franciscan View of the Human Person* (The Franciscan Institute 2005), *Mutuality: A Formal Norm For Christian Social Ethics* (Wipf & Stock 2005), *That They May Be One: Catholic Social Teaching on Racism Tribalism and Xenophobia* (Orbis Books 2008) for which she received a Catholic Press Award in June 2009, and *Ecological Footprints: An Essential Franciscan Guide for Faith and Sustainable Living* forthcoming from Liturgical Press in 2012.

Vincent A. Pizzuto is Associate Professor in the Department of Theology and Religious Studies at the University of San Francisco. He received a S.T.L. master's degree and a S.T.D. in New Testament exegesis from the Katholieke Universiteit Leuven, Belgium. He is the author of *A Cosmic Leap of Faith: An Authorial, Structural, and Theological Investigation of the Cosmic Christology in Col 1:15-20* (Peeters 2006). Articles in which he demonstrates the value of biblical exegesis for addressing current controversial issues in church and society have been published in *Biblical Theology Bulletin*. Ordained to the priesthood in 2006, Pizzuto founded and ministers to a small Christian contemplative community, New Skellig, in northern California.

Jane Elyse Russell, O.S.F. is Associate Professor of Theology at Belmont Abbey College in North Carolina. After receiving her doctoral degree from the University of Notre Dame, she focused her research on ecclesiology with particular attention to the role of small Christian communities about which she wrote in *The Cord*. She has recently concentrated her teaching and lecturing on the interface of theology and science with a focus on the long-term flourishing of the Earth community. One arena of her Earth-friendly praxis has been the care and feeding of a "Green Team" at Belmont Abbey College.

Jame Schaefer teaches, researches, and writes on issues that interface theology and the natural sciences with special attention to religious foundations for environmental ethics. She received her Ph.D. from Marquette University where she currently serves as Associate Professor of Systematic Theology and Ethics, directs the Interdisciplinary Minor in Environmental Ethics, and advises Students for an Environmentally Active Campus. For several years she convened the Theology and Ecology Group of the Catholic Theological Society of America and subsequently its Interest Group on Catholic Theology and Global Warming that generated this anthology. Among her publications is *Theological Foundations for Environmental Ethics: Reconstructing Patristic and Medieval Concepts* (Georgetown University Press 2009). Prior to entering academia, Schaefer held leadership positions in local and regional environmental advocacy groups and served in several energy and environmentally-related policy positions by appointment of county, state, and federal governments.

Daniel P. Scheid is Assistant Professor of Theology at Duquesne University where he teaches students about theological ethics, Christian social ethics, and ecological ethics and focuses his research on ecological and comparative ethics. He received his doctoral degree in Theological Ethics from Boston College after writing his dissertation on the foundation Thomas Aquinas and Ved_nta De_ika provide for interreligious ecological ethics. His publications include articles in the *Encyclopedia of Sustainability,* the *Journal of Vaishnava Studies,* and the *Annual Publication of the College Theology Society.*

Anthony C. Sciglitano, Jr. is Associate Professor of Religion at Seton Hall University where he chairs the Department of Religion and cochairs the University Core Curriculum. In his teaching, he focuses on the works of Hans Urs von Balthasar, theological method, Christianity and secularism, and modern Christian thought. He received his doctorate from Fordham University in Contemporary Systematic Theology and has published articles in *Modern Theology* and *Pro Ecclesia.* His monograph on Hans Urs von Balthasar, entitled *The Cross and the Covenant: An Anti-Marcionite Theology of Religions* is forthcoming in 2011 from Crossroad. He is grateful to Kevin Mongrain for comments that improved this essay.

INDEX